The Role of Ecosystem Services in Sustainable Food Systems

The Role of Ecosystem Services in Sustainable Food Systems

Edited by

Leonard Rusinamhodzi, BSc, MPhil, PhD

Scientist - Systems Agronomist
Sustainable Intensification Program
CIMMYT, Kathmandu
Nepal

Academic Press is an imprint of Elsevier
125 London Wall, London EC2Y 5AS, United Kingdom
525 B Street, Suite 1650, San Diego, CA 92101, United States
50 Hampshire Street, 5th Floor, Cambridge, MA 02139, United States
The Boulevard, Langford Lane, Kidlington, Oxford OX5 1GB, United Kingdom

The Role of Ecosystem Services in Sustainable Food Systems

ISBN: 978-0-12-816436-5

Publisher: Charlotte Cockle
Acquisition Editor: Megan Ball
Editorial Project Manager: Ruby Smith
Production Project Manager: Kiruthika Govindaraju
Cover Designer: Mark Rogers

Contents

Contributors

C. Acuin, PhD
International Rice Research Institute, Metro Manila, Manila, Philippines

O. Angeles, PhD
International Rice Research Institute, Metro Manila, Manila, Philippines

D. Chibamba
Department of Geography and Environmental Studies, The University of Zambia, Lusaka, Zambia

Regis Chikowo, MPhil, PhD
Plant Soil and Microbial Sciences Department, Michigan State University, East Lansing, MI, United States; Crop Science Department, University of Zimbabwe, Harare, Zimbabwe

Vimbayi Chimonyo, MSc, PhD
Plant Soil and Microbial Sciences Department, Michigan State University, East Lansing, MI, United States

Brian Chiputwa, BSc, MSc, PhD
Livelihoods and Gender Specialist, Research Methods Group, World Agroforestry (ICRAF), Nairobi, Kenya

P. Chivenge, PhD
International Rice Research Institute, Metro Manila, Manila, Philippines; Senior Scientist, Sustainable Impact, International Rice Research Institute, Los Baños, Laguna, Philippines

M. Connor, PhD
International Rice Research Institute, Metro Manila, Manila, Philippines

Felix D. Dakora, PhD
Professor, Chemistry Department, Tshwane University of Technology, Pretoria, South Africa

Katrien Descheemaeker
Assistant Professor, Plant Production Systems, Wageningen University, Wageningen, Gelderland, The Netherlands

C.C. Du Preez, PhD
Department of Soil, Crop and Climate Sciences, University of the Free State, Bloemfontein, Free State, South Africa

Anja Gassner
Senior Livelihood Specialist & Head of Research Methods, World Agroforestry (ICRAF), Philippines

Chiwimbo Gwenambira, MSc
Plant Soil and Microbial Sciences Department, Michigan State University, East Lansing, MI, United States

B. Hadi, PhD
International Rice Research Institute, Metro Manila, Manila, Philippines

Sabine Homann-Kee Tui
Senior social scientist, International Crops Research Institute for the Semi-Arid Tropics (ICRISAT), Matopos Research Institute, Bulawayo, Zimbabwe

Hanna J. Ihli
Decision Analyst, Systems Theme, World Agroforestry (ICRAF), Nairobi, Kenya

Peter Jeranyama, BSc Agriculture Honors (Crop Science), MS Crop and Soil Sciences, PhD. Crop and Soil Sciences
Associate Professor, University of Massachusetts Amherst, East Wareham, MA, United States

S. Johnson-Beebout, PhD
International Rice Research Institute, Metro Manila, Manila, Philippines

Jefline Kodzwa
Department of Environmental Sciences and Technology, Chinhoyi University of Technology, Chinhoyi, Zimbabwe

E. Kotzé, PhD
Department of Soil, Crop and Climate Sciences, University of the Free State, Bloemfontein, Free State, South Africa

C.F. Kunda-Wamuwi
Department of Geography and Environmental Studies, The University of Zambia, Lusaka, Zambia

Paramu L. Mafongoya, BSc, MSc, PhD
Professor, School of Agricultural, Earth and Environmental Sciences, University of KwaZulu-Natal, Pietermaritzburg, KwaZulu-Natal, South Africa

Milton T. Makumbe
Henderson Research Station, Department of Research & Specialist Services, Division of Livestock Research, Mazowe, Zimbabwe

Mpelang P. Maredi, BSc
Department of Crop Sciences, Tshwane University of Technology, Pretoria, South Africa

Sipho T. Maseko, PhD
Department of Crop Sciences, Tshwane University of Technology, Pretoria, South Africa

Patricia Masikati
Scientist, World Agroforestry Centre (ICRAF), Elm Road Woodlands, Lusaka, Zambia

Esther Nyaradzo Masvaya
International Crop Research Institute for the Semi-Arid Tropics, Matopos Research Station, Bulawayo, Zimbabwe

Cherian Mathews, PhD
Professor, Chemistry Department, Tshwane University of Technology, Pretoria, South Africa

Irvin Mpofu, BSc Hons, MSc, MBA, PhD
Professor, Animal Production and Technology, Chinhoyi University of Technology, Chinhoyi, Zimbabwe

K.H. Mubanga
Department of Geography and Environmental Studies, The University of Zambia, Lusaka, Zambia

Chipo Plaxedes Mubaya, PhD
Chinhoyi University of Technology, International Collaborations Office, Department of Freshwater and Fishery Science, Chinhoyi, Zimbabwe; International Collaborations, Chinhoyi Uinversity of Technology, Chinhoyi, Zimbabwe

B.M. Mushili
Department of Geography and Environmental Studies, The University of Zambia, Lusaka, Zambia

Mzime Regina Ndebele-Murisa, PhD
Chinhoyi University of Technology, International Collaborations Office, Department of Freshwater and Fishery Science, Chinhoyi, Zimbabwe; START International, Washington, DC, United States

Nilhari Neupane
Independent Agricultural Economist, Kathmandu, Nepal

Justice Nyamangara
Department of Environmental Sciences and Technology, Chinhoyi University of Technology, Chinhoyi, Zimbabwe

P.H. Nyanga
Department of Geography and Environmental Studies, The University of Zambia, Lusaka, Zambia

R. Puskur, PhD
International Rice Research Institute, Metro Manila, Manila, Philippines

Leonard Rusinamhodzi, BSc, M.Phil, PhD
Scientist - Systems Agronomist, Sustainable Intensification Program, CIMMYT, Kathmandu, Nepal

Trinity Senda
International Livestock Research Institute, (ILRI), Nairobi, Kenya

Anil Shrestha
Professor, California State University, Fresno, CA, United States

Gudeta W. Sileshi, BSc, MSc, PhD
Plot 1244 Ibex Meanwood, Lusaka, Zambia; Honorary Research Fellow, School of Agricultural, Earth and Environmental Sciences, University of KwaZulu-Natal, Pietermaritzburg, KwaZulu-Natal, South Africa

Sieg Snapp, MSc, PhD
Plant Soil and Microbial Sciences Department, Michigan State University, East Lansing, MI, United States

Gabriel Soropa
Department of Crop Science and Post-Harvest Technology, Chinhoyi University of Technology, Chinhoyi, Zimbabwe

A. Stuart, PhD
International Rice Research Institute, Metro Manila, Manila, Philippines

B.B. Umar
Department of Geography and Environmental Studies, The University of Zambia, Lusaka, Zambia

Roberto O. Valdivia
Assistant Professor, Senior Researcher, Department of Applied Economics, Oregon State University, Corvallis, OR, United States

C.W. van Huyssteen, PhD
Department of Soil, Crop and Climate Sciences, University of the Free State, Bloemfontein, Free State, South Africa

Andre F. van Rooyen
Principal Scientist, International Crops Research Institute for the Semi-Arid Tropics (ICRISAT), Matopos Research Institute, Bulawayo, Zimbabwe

J.J. van Tol, PhD
Department of Soil, Crop and Climate Sciences, University of the Free State, Bloemfontein, Free State, South Africa

Priscilla Wainaina
Post-doctoral Fellow, Governance Theme, World Agroforestry (ICRAF), Nairobi, Kenya

Preface

The concept of ecosystem services (ESs) in the broad sense has gained popularity in recent years though applications to specific disciplines, and translating it into practice remains a major challenge. What are the benefits that humans freely derive from the natural environment and from properly functioning ecosystems and how do they contribute to production of healthy and nutritious food while improving the environment? This book reveals in simple terms the operational definition, concepts, and applications of ESs with a focus on sustainable food systems. It presents case studies with a focus on both geographical and production system—wide considerations. This book integrates knowledge from several aspects of food systems and identifies the range of ESs associated with those case studies. It simplifies how ESs are defined and perceived using a food system—wide perspective. It is all encompassing and comprehensive reference for researchers, students, scientists, development practitioners, and policymakers, which provides a simplified but well nuanced understanding of what constitutes ESs in the food system. It will stimulate debate around the use of ESs and help in conceptualization and their applications in regions that are still lagging behind. The book is intended to:

- Provide an understanding of the tenets of sustainable food systems
- Reveal knowledge of ESs in relation to sustainable food systems
- Outline tools for identification and quantification for ESs related to sustainable food systems
- Explain strategies to maximize ESs in sustainable food systems
- Identify potential challenges to maximization of ESs in sustainable food systems

The book is organized in four sections and 18 chapters and covers some of the topical issues in sustainable food production systems today. It introduces sustainable food systems and ecosystem services, then presents tools such as systems analysis and sustainability indicators for the reader to understand ESs in a comprehensive way. The middle chapters present different perspectives from case studies of ESs derived from some of the key sustainable food production systems used by farmers. Other topical issues covered by the book include climate change, conservation agriculture (CA), integrated soil fertility management (ISFM), and crop-livestock systems. The upcoming new paradigm of doubled-up legumes systems is presented and including the important topic of economic and social values of ESs. The book concludes with some discussions on challenges of deriving full benefits and how these can be overcome, including knowledge gaps and future research needs around ESs in food systems. It will be an essential read for undergraduate through to postgraduate and professional researchers working in sustainable food systems.

The book is organized as follows:

(1) Introduction
- Chapter 1: Sustainable food systems: scope, relevance, and important milestones—the chapter covers the key features of sustainable food systems, and the important targets of these systems are explained

(2) Definitions and tools to understand food systems and ESs
- Chapter 2 defines ESs in sustainable food systems by providing a simple operational definition, concepts, and applications.
- Chapter 3 outlines indices to identify and quantify ESs in sustainable food systems

(3) Case studies of ESs in sustainable food systems
- Chapter 4 outlines the range of ES from biological nitrogen fixation (BNF) systems harnessing ESs from BNF systems
- Chapter 5 explains the importance role of synthetic fertilizers in enhancing ESs in sustainable food systems
- Chapter 6 describes how to reinforce ESs through CA in sustainable food systems
- Chapter 7 describes ESs from livestock with a focus on different management systems
- Chapter 8 outlines how crop-livestock integration can be used to enhance ESs in sustainable food systems
- Chapter 9 defines doubled-up legume system and outlines the ranges of ESs
- Chapter 10 reveals in detail the ESs in paddy-rice systems

(4) ESs and emerging global issues
- Chapter 11 describes the role of ESs in offsetting the effects of climate change in sustainable food systems. How can climate change impact on ESs and how ESs can be manipulated to offset the effect of climate change on food systems
- Chapter 12 reveals the economic and social values of ESs in sustainable food systems. Outlines some of the nonmonetary (social) benefits of ES, including successes and challenges with valuing ES
- Chapter 13 explains the challenges in maximizing benefits from ESs in sustainable food systems. This is a concluding chapter, integrating lessons from all the chapters, outlining keys insights from current knowledge, and identifying knowledge gaps and future research needs

I am profoundly thankful to all the authors firstly for accepting to be part of the book, their contributions, and their help and cooperation during the manuscript writing and revision process. I am also grateful to Megan Ball, Senior Acquisitions Editor, and Ruby Smith, Editorial Assistant, Food Science Unit, Elsevier Cambridge, USA.

Leonard Rusinamhodzi
Kathmandu, Nepal

CHAPTER

Sustainable food systems: diversity, scope and challenges

1

Peter Jeranyama, BSc Agriculture Honors (Crop Science), MS Crop and Soil Sciences, PhD. Crop and Soil Sciences[1], Anil Shrestha[2], Nilhari Neupane[3]

[1]*Associate Professor, University of Massachusetts Amherst, East Wareham, MA, United States;* [2]*Professor, California State University, Fresno, CA, United States;* [3]*Independent Agricultural Economist, Kathmandu, Nepal*

Introduction

Sustainable agriculture means different things to different people, but the basic goals of sustainable agriculture are environmental health, economic profitability, and social and economic equity (sometimes referred to as the "three legs" of the sustainability stool). Legally, sustainable agriculture is defined in U.S. Code Title 7, Section 3103 to mean an integrated system of plant and animal production practices having a site-specific application that will over the long term (1) satisfy human food and fiber needs, (2) enhance environmental quality and the natural resource base on which the agricultural economy depends, (3) make the most efficient use of nonrenewable resources and on-farm resources and integrate, where appropriate, natural biological cycles and controls, (4) sustain the economic viability of farm operations, and (5) enhance the quality of life for farmers and society as a whole.

In this chapter we focus on the scope, relevance, and important milestones in four distinct sustainable food systems: (1) the Zimbabwean smallholder cropping system, (2) market-oriented vegetable production in periurban of Nepal, (3) the US North Central corn belt, and (4) Massachusetts cranberry production system. These four agricultural systems although very different from each other have a common overarching objective of achieving sustainable yields over an extended period while providing a livelihood to the people involved.

The challenge for agriculture in the coming decades will be to increase productivity to meet the increasing demands for food and fiber while addressing risk and variability, and eco-efficiency will undoubtedly be a major challenge. Yield per unit land area is the simplest and most widely used eco-efficiency measure for field crops. However, there are multiple efficiency measures at play simultaneously, such as water use efficiency (crop yield per unit of water used, e.g., rainfall, stored soil moisture, and/or irrigation), nutrient use efficiency (crop yield per unit nutrient uptake or nutrient supplied), radiation use efficiency (crop biomass produced per unit radiation intercepted), labor efficiency (crop production per unit labor invested), and

The Role of Ecosystem Services in Sustainable Food Systems. https://doi.org/10.1016/B978-0-12-816436-5.00001-9

return on capital (profit as a fraction of capital invested) (Keating et al., 2010). However, these eco-efficiency measures and associated challenges can vary across regions and cropping systems as discussed below.

Case study 1 African smallholder system

The challenge for the African smallholder system, used as an example in this chapter, is to provide better than breakeven yields while making use of minimal external inputs including menial labor. The challenge for the two US production systems is to provide profitable yields "without breaking the bank", sustain the environment, and be socially acceptable.

There is compelling evidence that food production systems are constrained by overwhelmingly low soil fertility in Africa (McCown et al., 1992; Giller et al., 2006; Sanginga and Woomer, 2009), and unless solutions are found to relieve this constraint, eco-efficient use of other natural and human resources will remain low. This is consistent with de Wit (1992) conclusion that resource use efficiency is maximized when the level of all inputs is close to their optima and confirmed by Bindraban et al. (2008) who analyzed temporal trends in crop yields in Africa. In Africa nitrogen (N) is the most limiting factor in crop production, whereas phosphorus (P) and other nutrients are limiting in many situations. Within-farm variability is high, and many farms have small areas of higher nutrient supply, usually associated with household areas and livestock containment yards (Giller et al., 2006).

The growing global population is exerting pressure on the world's agricultural ecosystems to supply adequate food, fiber, and increasingly, fuel. At the same time, the Millennium Ecosystem Assessment (2005) has documented the extent and global scale of the environmental costs associated with intensive agriculture. Degradation of water quality and soil resources are becoming urgent problems, and fossil fuel supplies are finite. Organic agriculture is frequently presented as the only currently available "semi-closed system" that provides a viable alternative to conventional "open system" management that relies on large doses of agrochemical inputs (Pearson, 2007).

Africa represents one of the greatest ironies in modern economic development. In spite of advances in science and technology that have boosted incomes in other regions and enabled other continents to attain economic progress and high standards of living for their citizens, Africa enters the 21st Century as the world's poorest continent. In fact, it is the only continent that has grown poorer in the past 30 years in spite of the rapid developments in science, innovation, and trade. Empirical data on Africa's regression are staggering. In sub-Saharan Africa (SSA) alone, over 50% of the 700 million citizens survive on less than 50 US cents a day. In its seminal study *Can Africa Claim the 21st Century?* the World Bank (2000) states "Despite gains in the second half of the 1990s, Africa enters the 21st century with many of the world's poorest countries. Average per capita income is lower than at the end of the 1960s. Incomes and access to essential services are unequally distributed. The region contains a growing share of the world's absolute poor, who have little power to influence the allocation of resources." According to the *World Development Report 2008,* Africa's rural poverty rate was 82% in 2002. Africa lags behind on most of the development goals, and such prognosis is presented against a backdrop of the great irony of a continent that is endowed with immense natural resources, as well as great cultural, ecological, and economic diversity.

The underperformance of Africa's agricultural sector over the past 3 decades is a major reason why the continent lags other regions. This sector accounts for 30—40% of the continent's gross domestic product, nearly 60% of export income, and employs 75% of the population (Commission for Africa, 2005). As such, agriculture is the primary source of livelihood for many Africans. By far a fundamental characteristic of African agriculture is the fact that smallholder farms account for more than 90% of the continent's agricultural production and are dominated by the poor, a majority of whom are women. Food insecurity remains a major problem, requiring shipments of million metric tons of cereal food aid in some years, in particular, 24 African countries had a food emergency in 2004. The number of hungry and malnourished Africans is expected to increase to 300 million by 2020 (Rosegrant et al., 2001). Basic food security and rural livelihoods have deteriorated, and the continent is entrenched in severe poverty, hunger, and disease. Underlying causes of food insecurity include factors such as declining soil fertility and rising fertilizer prices compounded by limited availability of key resources such as land, cash, knowledge, and labor and by inappropriate solutions being advocated (Giller et al., 2006 (Giller et al., 2009); Sanginga and Woomer, 2009).

Zimbabwe smallholder cropping systems

The subhumid zones of SSA are dominated by maize (*Zea mays* L.)-based mixed crop-livestock systems and cereal-root crop mixed farming system (Dixon et al., 2001). Slash and burn systems have existed in large areas with fallowing for soil fertility regeneration. Such systems often involve four stages, namely clearing, burning, cropping, and abandonment, and are regarded as the basic form of agriculture that illustrates clearly the concept of sustainability in agriculture with respect to the intensity of land use (Nye and Greenland, 1960). However, with increasing population densities these systems have become increasingly unsustainable (Raintree and Warner, 1986). Farmers have pursued different paths to overcome the limitations posed by decreasing soil fertility. Despite extension services advocating different routes to intensification, a historically common approach has been to expand the cultivated area without using nutrient inputs, especially when land is sufficient and market access is limited (cf. Baudron et al., 2012).

Maize dominates Zimbabwe's smallholder cropping systems, although some grain legumes, particularly cowpea (*Vigna unguiculata* L.) and peanuts (*Arachis hypogea* L.) are grown in loose rotations with maize. At current fertilizer costs, most smallholder farmers in subhumid Africa grow maize with little N-P-K fertilizer (Jeranyama et al., 2000). Fertilizer application rates in SSA are low relative to the rest of the world. In 2006 average fertilizer use in Africa was about 8 kg ha^{-1}, a 10th of the global average. In that same year, African Union member states met in Abuja, Nigeria, and adopted the *"Abuja Declaration on Fertilizer for the African Green Revolution,"* pledging to increase fertilizer use to 50 kg $ha^{-1}$by 2015. Over the last few decades, policymakers linking low fertilizer use to low yields have attempted numerous interventions to promote fertilizer use across the continent. Yet, in spite of the Abuja Declaration and its lofty aspirations the average fertilizer application rate in Africa today still languishes between 13 and 20 kg ha^{-1}.

Legume-cereal rotations have long been recognized in southern Africa for restoring soil fertility and increasing crop productivity (MacColl, 1989; Mukurumbira, 1985). Rotations shift the biological balance in the soil, reducing build-up of pests and diseases and sustain productivity of the cropping system (Kumwenda et al., 1996).

Declining soil fertility and crop productivity in the smallholder farms of subhumid Zimbabwe is partly a result of continuous maize production and partly because of inadequate nutrient inputs and management, exacerbated by unreliable rainfall distribution and marginal economics. Increasing human population pressure on limited agricultural land has rendered fallowing to restore soil fertility a nonviable option, while continuous maize has become common on smallholder farms in Zimbabwe (Jeranyama et al., 2007).

When cropped to sole maize, the sandy soils in these smallholder systems in Zimbabwe can supply only about 30 kg N ha^{-1} per cropping season because of critically low levels of soil organic matter (Mapfumo and Mtambanengwe, 1999). Further N mineralization is dependent on annual organic inputs produced in crop residues (mainly groundnut and maize) and retained on the field or cycled through animals (as cattle manure). Continuous cropping of maize at a grain yield above 1 t ha-^{1}yr^{-1} cannot be sustained without frequent and substantial additions of mineral nutrients (Grant,1981; MacColl, 1989), but the high cost and low availability of these inputs result in low use by the smallholders. One alternative to reduce overdependence on mineral fertilizers is to grow maize in rotation with a legume such as groundnuts.

Legume—cereal rotations have long been recognized in southern Africa for restoring soil fertility and increasing crop productivity (MacColl, 1989; Mukurumbira, 1985). Rotations shift the biological balance in the soil, reducing build-up of pests and diseases and sustaining productivity of the cropping system (Kumwenda et al., 1996). In Zimbabwe, substantial research effort was made to increase smallholder soybean (*Glycine max* L. Merr.) production, but the area planted remained small. There was a widely held belief that soybean was an unsuitable crop for the sandy soils predominant in communal farming areas (Mpepereki et al.,1996; Mpepereki et al., 2000). An intensive extension project was established through the University of Zimbabwe from 1996 onwards, assisting farmers to access seed and inputs on demonstrating appropriate agronomic management for soybeans, on local processing of soybeans for food and assisting farmers in marketing their surplus soybean, and to obtain good prices (Rusike et al., 1999). A wide range of cultivars were distributed, including 'improved' varieties that had better yield potential (of 3–4 Mg ha^{-1}) but needed inoculation with rhizobia, as well as "promiscuous" cultivars that had stable yields under less favorable conditions but could form nodules and fix N_2 without inoculation. Yields of maize after soybean were demonstrated to be more than double the yields in continuous cultivation. This extension program led to an increase in sales of soybean from the smallholder sector from 350 Mg in 1996 to 12,500 Mg in 2004.

There are high returns in terms of crop yield to application of small amounts of fertilizer in these areas in spite of the rainfall variability (Twomlow et al., 2008). A major hurdle to higher rates of fertilizer usage is the lack of access to fertilizer by smallholder farmers. At times, poor agronomic practices such as late sowing, poor weed control, and suboptimal plant populations can reduce fertilizer use efficiency. However, significant gains in maize yields in the order of 50%–80% can be achieved from open-pollinated varieties and small inputs of fertilizer by smallholder farmers (Twomlow et al., 2008). Further gains in yield are possible by using improved maize cultivars (Banziger et al., 2000). The African "green revolution" will need to be first a revolution based on improved soil fertility complemented by improved germplasm and improved agronomic practices including better water management. For all these factors to make any meaningful impact, government policies need to evolve to support this emerging sector. Furthermore the research and development community should consider how they will build the two-way information flows from research knowledge and technologies to application in a local context.

While the technical challenges concerning soil fertility in African smallholder agriculture are reasonably well identified (World Bank, 2008), the solutions may lie outside the farm household at the level of input/output markets and the institutions governing them. In developed country agriculture, strategies to enable farmers to better understand and manage risks (in particular climate-related risks) associated with input use have been shown to greatly assist in this transition (Hochman et al., 2009). However, in developing country agriculture, particularly in Africa, this pathway is more complex than simply improved knowledge supporting a farmer's decision to take on higher risks (Meinke et al., 2006). In many situations the input/output markets are poorly developed or distorted to the point that even very modest inputs that could be highly effective have not been possible (Rohrbach et al., 1998).

Case study 2 Vegetable production in periurban region of Nepal

Vegetable area in Nepal increased two-folds from 1,40 1,000 hectares in 1990 to 2,80 1,000 hectares in 2016. During the same time, vegetable production increased from 1.1 to 4.0 million tonnes with corresponding increase in vegetable yields from

8,028 to 13,992 kg ha^{-1} (MoAD, 2016). Currently, over three million land holders are engaged in vegetable production mostly in the hills and the Terai region of Nepal. Out of the total vegetable produced, 60% is marketed, and the rest is used for domestic consumption (MoAD, 2018). Fresh vegetables are an important component of the Nepalese diet, and their per capita consumption of vegetables has doubled over the last 2 decades particularly in urban areas because of an increase in awareness level of the health benefit of vegetables, household income, and increased availability. Hence vegetable farming has become a profitable venture, and the area under cultivation has steeply increased in the Kathmandu valley and periurban areas of Nepal. Mountain and hill areas of Nepal are becoming major regions of vegetable production because of climatic and soil suitability, and these regions contribute about half of the total vegetable production nationally. Much of the vegetables are consumed in the Kathmandu valley; consequently, periurban districts such as Kavrepalanchowk, Nuwakot, and Dhading have developed into commercial vegetable growing areas as these areas have suitable agroclimatic condition for year-round vegetable production. These areas also have good road access to Kathmandu. The commercial vegetable farming practices has increased the per unit production over the years. However, it has also resulted in increased use of inputs such as fertilizer, improved seeds, and growth regulators, including pesticides. In recent years, pesticide use in vegetables has increased sharply in these areas. This, in part, is because of the lack of knowledge and science-based recommendations on pesticide use and an increasing trend of their use for cosmetic purposes of the produce. Most of the recommendations on pesticide use seem to be coming from pesticide dealers instead of research and extension agencies. Consumers also seem to prefer unblemished, cosmetically attractive produce. Although pesticide use regulations exist in Nepal, they are not strictly implemented. Furthermore, there is a lack of pesticide safety education not only among growers but also among pesticide dealers (personal observations Shrestha and Neupane).

Unsafe use of pesticides can be of more concern in fragile mountain ecosystems with high levels of biodiversity than in other regions. Also the health of the Nepalese people is probably being endangered by the current pesticide use practices.

Pesticides use in vegetables

Though per hectare pesticide use in Nepal is low compared with the global average and to other South Asian countries. Records show that 396 g a.i. ha^{-1} chemical pesticides are being used in Nepal. However, pesticide use in vegetable farming is increasing at an alarming rate. In 2014– to 2015, Nepal imported 455 MT of pesticides which was almost 30% greater than the previous year. Of this amount, 85% was used in vegetable farming alone. Pesticide use is common in the cultivation of vegetables in peri-urban areas of Kathmandu valley such as Karepalanchowk, Nuwakot, and Dhading (Mount Digit Technology, 2014).

Some empirical findings show that the vegetable growers in the peri-urban areas are using four to five folds higher doses than the recommended label rate of the pesticides. Cost of pesticides usually amount to about 20–30% of the total cost of vegetable production. However, the producers do not seem to be taking production economics into consideration when it comes to pesticide use. This is probably mainly due to lack of established scientific guidelines or knowledge of economic threshold levels based on principles of integrated pest management. As such the growers generally do not want to risk losing any portion of their yield; and hence they seem to over apply pesticides as an insurance against crop loss. Furthermore, the pesticide application technique is quite primitive, haphazard, with minimum concern of user and consumer safety. The use of pesticides not only increases the cost of production but also endangers food safety, public health, and environmental degradation, but these issues have not been properly documented (personal observation Shrestha and Neupane).

Although farmers are generally aware of some of the health hazards of pesticides, they still consider pesticides to be an indispensable part of their vegetable cropping systems (Rijal et al., 2018). Cai (2008) estimated that without pesticides, the yield loss in vegetable crops would be approximately 54%. The primary reason that farmers continue to use pesticides, even though they may be aware of the hazards, is because of the successful control of pests which helps them in short-term economics and the reluctance to invest in high initial costs involved in switching to alternative practices (Wilson and Tisdell, 2001).

Case study 3 North central US cropping systems

A typical cropping system in the North Central United States is maize–soybean or maize–soybean–wheat (*Triticum aestivum* L.) rotation, occasionally with alfalfa (*Medicago sativa* L.)-alfalfa-alfalfa–maize systems. In each year of the rotation, crops are grown as monocrops using high level of mechanization. Most fields if not all are georeferenced, and variable rates of fertilizer are applied.

Healthy soil is fundamental to all plant and animal life; therefore proper management of soil resources is essential. Recent concerns regarding global climate change as related to soil health and crop production are increasingly driving scientific research relevant to growers in the North Central United States. Producers in the North Central United States can utilize several management options that may improve soil health and ecosystem services which include no-till soil management, maintaining crop residues, diversifying crop rotations, and establishing cover crops. A region as variable as the North Central United States requires extensive research on how to best implement these and other beneficial management practices to improve sustainability. To address these challenges, it is important to understand how soil and crop management practices directly and indirectly influence the soil–water–air environment.

Bundy et al. (1999) reported that 3.6 million tons of N fertilizer were applied annually for maize production in 12 states within the North Central United States, at a cost of $600 to 800 million. This estimate excluded N from manure and legumes used in crop rotations. The total N fertilizer used by 15 US states for maize in 2014 rose to five million tons at an estimated cost of $500 Mg^{-1} or $2.5 billion for 36.6 million ha (USDA–NASS, 2015a). Snyder (2012) documented that US maize consumes 37%–51% of the total annual fertilizer N.

The need for efficient use of N fertilizer because of environmental and economic reasons has raised concerns in modern crop production systems. Environmental concerns continue to be intertwined with the growing costs of N fertilizer production and use. Accurate N fertilizer rates, along with higher N use efficiency (NUE) remain important for maximizing returns while simultaneously protecting the environment and water quality. NUE for maize in the United States increased more than 30% over the last 20 years (Fixen and West, 2002).

Recent US experience demonstrates that significant improvements in fertilizer use efficiency are possible. For US-grown maize the ratio of crop yield per unit of applied N fertilizer (i.e., the partial factor productivity for N fertilizer) has increased by 36%, from 42 kg kg^{-1} in 1980 to 57 kg kg^{-1} in 2000 (Cassman et al., 2002). The drivers of this improvement would appear to reside in the combination of continuing yield improvement through technological change and flat N fertilization rates since the concerns over environmental consequences of excess fertilizer use emerged in the 1980s.

Case study 4 Massachusetts cranberry production system

Cranberry (*Vaccinium macrocarpon* Ait.) is a temperate, perennial, woody vine indigenous to North America. It forms terminal mixed buds on short, 5-cm to 20-cm vertical branches known as uprights, originating from axillary buds on the stolon (Eck, 1990). The vertical shoots from the stolon bear fruit. The plant is primarily cultivated in northern coastal areas in the United States (Massachusetts, New Jersey, Oregon, and Washington) and Canada (Maritime Provinces and British Columbia), as well as in central regions (Wisconsin and Quebec often depend on small).

Particularly in Massachusetts, cranberries are grown in coastal watersheds and often depend on small lakes as their water source for irrigation, harvest, and winter flooding (DeMoranville, 2015). Some of the lakes and estuaries of these watersheds are included in the state's list of impaired water bodies because of P (in inland lakes) or N (in estuaries) enrichment. Under Federal Clean Water Act, all listed waters must have total maximum daily load (TMDL) limits sets. Water conservation efforts have focused primarily on irrigation, particularly for soil water management(Pelletier et al., 2013; Jabet et al., 2017), despite irrigation using half as much water as flooding (Kennedy et al., 2017). Although its water footprint is smaller than flooding, irrigation is just as important to cranberry production. It provides vital protection of the cranberry plant from frost injury, which can cause major to complete crop loss in a matter of hours (Olswski et al., 2017).

The TMDL process identifies the acceptable load and divides it among the lands that drain into the water body. When cranberry farms are within the regulated area for a TMDL, even if cranberry farms are not the major contributor to the nutrient pollution, a limit is set on the amount of N or P that can leave the farm into the impaired water body (DeMoranville, 2015). There is little solace on the part of cranberry growers because of the *Right to farm laws* in the United States deny nuisance lawsuits against farmers who use accepted and standard farming practices and have been in prior operation even if these practices harm or bother adjacent property owners or the general public.

Research in the past 30 years have shown that cranberry requirement for N (20–70 kg N ha^{-1}), P (<20 kg P ha^{-1}) and K (40–140 kg P ha^{-1}) based on tissue testing, plant growth demand, potential for remobilization, and determination of removal by crop (DeMoranville, 2015). The fact that cranberry is an acid-loving plant adapted to sandy, hydric, and nutrient-poor soils is reflected in the low tissue nutrient concentration (Davenport et al., 2000). For actual fertilizer rates applied on Massachusetts bogs and amounts removed in drainage water see Tables 1.1 and 1.2. Because cranberry uses large volumes of water in its production, N (coastal waters) and P (inland lakes) enrichment have been implicated in degradation of water resources. Water use in cranberry can be up to 12, 330 m^3, and water permits in Massachusetts are issued on that basis (DeMoranville, 2015).

In Massachusetts, cranberry nutrient management has also become cranberry water management. Research efforts have become focused not only on the nexus of those practices and on their impact beyond crop production but also on potential water quality impacts. As newer and high yielding cranberry cultivars are planted, there will be a need to do research that identifies optimum nutrient application rates

Table 1.1 Mean annual total phosphorus (TP) export in drainage water and annual applied phosphorus (P) for six Massachusetts cranberry bogs in 2003–2004.

Bog identifier	P applied fertilizer	TP incoming in applied water	TP leaving in discharge water	Net TP export in water
	kg P ha^{-1} yr^{-1}			
EH	11	0.30	4.49	4.19
PV	22	1.80	8.50	6.69
BEN	19	1.20	3.90	2.70
WS	20	0.20	5.50	5.30
M-K	24	2.00	3.19	1.20
ASH	39	1.80	3.40	1.59

Adopted from DeMoranville (2015).

Table 1.2 Mean annual total nitrogen (TN) export in drainage water and annual applied nitrogen (N) for six Massachusetts cranberry bogs in 2003–2004.

Bog identifier	N applied fertilizer	TN incoming in applied water	TN leaving in discharge water	Net TN export in water
	$kg\ N\ ha^{-1}\ yr^{-1}$			
EH	39	6.16	13.23	7.06
PV	44	11.77	18.49	6.73
BEN	36	14.35	29.37	15.02
WS	36	4.15	9.86	5.72
M-K	49	13.90	18.05	4.15
ASH	49	12.78	17.04	4.26

Adopted from DeMoranville (2015).

that maximizes productivity and minimize off-site impacts. Future research could also focus on factors that control nutrient mobilization in drainage and flood water and the importance of soil types, site hydrology, nutrient application rates and forms, and critical water management practices.

Yield gaps

The term yield gap has often been used to describe the difference between actual yield recorded on farmer fields and yields that are possible when known technologies and best practices are used. Examples of yield gaps measured for maize cropping in Africa and for agricultural production more generally in various parts of the world are estimated to be in the 40%–80% range (World Bank, 2008; Comprehensive Assessment of Water Management in Agriculture, 2007). While yield gaps are generally large in Africa (and to a lesser degree in South Asia), they have narrowed in other parts of the world such as China and in wheat and maize production areas of western Europe and North America (World Bank, 2008). The lack of inputs, in particular irrigation and nutrients, is the conspicuous explanation for the continuing large yield gaps in Africa (World Bank, 2008). Closing the yield gap in Africa is not just transferring known technologies and agronomic recommendations but involves changes in institutional structures. Rukuni (2002) listed and discussed factors affecting performance in terms of productivity and competitiveness of smallholder agriculture. These are grouped into four main categories: (1) policy environment, (2) institutions, (3) technology, and (4) infrastructure.

Policy environment refers to aspects of (1) good governance and social and political stability; (2) macroeconomic stability, pricing, and marketing policies; and (3) land tenure and property rights. Africa's farmer support system that include extension services, research, credit and input sector is in dire straits. There is a need to a managerial transition from expatriate to indigenous scientists including from the consultative group of international agriculture to national agricultural researchers. Financial transition from dependence on financial support from colonial governments and large-scale farms to mobilizing support from governments and donors is one of the many complexes African managers have been forced to grapple simultaneously in transitioning African agricultural production systems to productivity and competitiveness. Research and development and industrial capacity are needed for the manufacture of "embodied" technology such as machinery, seeds, fertilizers, chemicals, and materials. Disembodied technology, on the other hand, refers to knowledge, techniques, and management practices that increase productivity and largely transmitted through extension and advisory services. Energy and power for smallholder farmers represent a major drawback to productivity. The continued decline in draft animals, coupled with a lack of appropriate small machinery, means that farmers yields are held back.

Climate change impact

Agricultural production is already significantly impacted by climate variability, especially in the semiarid regions of the world (Keating et al., 2010). Developing mitigative and adaptive management practices for climate variability will have immediate and future benefits in the improvement of eco-efficiency under current and future climates (Meinke et al., 2007). For example, in water-limited environments, seasonal rainfall variability is one of the most important factors for fluctuations in agricultural production and risk. Crop models can assist to quantify the season-specific and site-specific outcomes of agricultural interventions (Matthews and Stephens, 2002; Whitbread et al., 2010) and, when integrated with long-term historical weather data, allow retrospective analysis of the potential value of climate forecasts for a particular decision (Meinke and Stone, 2005; Hansen, 2002).

As discussed earlier, increasing the productivity of African agriculture is fundamental to improved food security for both rural and urban poor. Reducing high levels of poverty and hunger will require greater agricultural and rural development. Because smallholder farms account for more than 90% of Africa's agricultural production and are dominated by the poor, it also follows that strategies for growth must be centered on smallholder farm productivity and sustainability.

Conclusions

Bindraban et al. (2009) suggest new approaches to the problem of building soil fertility in SSA is to use the principles of 'Integrated Soil Fertility Management' and 'Balanced Nutrient Management' recognizing that (1) neither practices based solely on mineral fertilizers or solely on organic matter management are sufficient for sustainable management of agricultural production; (2) well-adapted, disease-resistant and pest-resistant germplasm is necessary to make efficient use of available nutrients (Vanlauwe et al., 2006); (3) good agronomic practices in terms of planting dates, planting densities, and weeding are essential to ensure efficient use of scarce nutrient resources (Tittonell et al., 2005); and (iv) the need to target nutrient resources within crop rotation cycles, going beyond recommendations for single crops (Giller et al., 2002).

Example from vegetable production in Nepal emphasizes that appropriate measures need to be taken to maintain the sustainability of such production systems in Nepal and elsewhere in the world with similar situations by providing integrated pest management options, pesticide safety education, and enacting pesticide regulations.

In the two US cropping systems discussed the need for efficient use of N fertilizer (and including P fertilizer in cranberry), because of environmental and economic reasons, has raised concerns in modern crop production systems. Environmental concerns continue to be intertwined with the growing costs of N fertilizer production and use. Accurate N fertilizer rates along with higher NUE remain important for maximizing returns while simultaneously protecting the environment and water quality. In cranberry system in particular nutrient management cannot be separated from water management.

The major challenge for the Zimbabwe smallholder cropping system is identified as being able to provide better than breakeven yields while making use of minimal external inputs including menial labor. In the market-oriented vegetable production in periurban of Nepal the major challenge is the overdependence on pesticides. Therefore there is a need to sustain this production system by providing integrated pest management options, pesticide safety education, and enacting pesticide regulations. In the two US cropping systems discussed in this chapter, there is a need for efficient use of N fertilizer (and including P fertilizer in cranberry) because of environmental and economic reasons affiliated with these modern crop production systems.

References

Abuja Declaration. https://www.afdb.org/en/topics-and-sectors/initiatives-partnerships/african-fertilizer-financing-mechanism/abuja-declaration/.

Baudron, F., Tittonell, P., Corbeels, M., Letourmy, P., Giller, K.E., 2012. Comparative performance of conservation agriculture and current smallholder farm practices in semi-arid Zimbabwe. Field Crops Research 132, 117–128.

Bänziger, M., Edmeades, G.O., Beck, D., Bellon, M., 2000. In: Breeding for Drought and Nitrogen Stress Tolerance in Maize: From Theory to Practice CIMMYT. Mexico City, Mexico.

Bindraban, P.S., Loffler, H., Rabbinge, R., 2008. How to close the ever-widening gap of Africa's agriculture. International Journal of Technology and Globalisation 4 (3), 276–295.

Bindraban, P., Bulte, E., Giller, K., Meinke, H., Mol, A., van Oort, P., Oosterveer, P., van Keulen, H., Wollni, M., 2009. Beyond competition: pathways to Africa's agricultural development. Plant Research International B.V., Wageningen, The Netherlands.

Bundy, L.G., Walters, D.T., Olness, A.E., 1999. Evaluation of soil nitrate tests for predicting maize nitrogen response in the North Central Region. North Central Reg. Res. Publ. 342. Wisconsin Agriculture Experiment Station. Univ. of Wisconsin, Madison.

Cai, D.W., 2008. Understand the role of chemical pesticides and prevent misuses of pesticides. Bulletin of Agricultural Science and Technology 1 (6), 36–38.

Cassman, K.G., Dobermann, A., Walters, D., 2002. Agroecosystems, nitrogen-use efficiency, and nitrogen management. Ambio 31, 132–140.

Commission for Africa Report. 2005. http://www.commissionforafrica.info/2005-report.

Davenport, J., DeMoranville, C., Hart, J., Roper, T., 2000. Nitrogen for Bearing Cranberries in North America. 27 March 1995. https://catalog.extension.oregonstate.edu/em8741.pdf.

de Wit, C.T., 1992. Resource use efficiency in agriculture. Agricultural Systems 40, 125–151.

DeMoranville, C., 2015. Cranberry nutrient management in southeastern Massachusetts: balancing crop production needs and water quality. HortTechnology 25, 471–474.

Dixon, J., Gulliver, A., Gibbon, D., 2001. Farming Systems and Poverty. Improving Farmers' Livelihoods in a Changing World. FAO and World Bank, Rome and Washington D.C, p. 412 pp.

Eck, P., 1990. The American Cranberry. Rutgers Univ. Press, New Brunswick, NJ.

Fixen, P.E., West, F.B., 2002. Nitrogen fertilizers: meeting contemporary challenges. Ambio 31 (2), 169–176.

Giller, K.E., 2002. Targeting management of organic resources and mineral fertilizers: can we match scientists' fantasies with farmers' realities? In: Vanlauwe, B., Diels, J., Sanginga, N., Merckx, R. (Eds.), Integrated Plant Management in Sub-Saharan Africa. CAB International, pp. 151–171.

Giller, K.E., Rowe, E.C., de Ridder, N., van Keulen, H., 2006. Resource use dynamics and interactions in the tropics: scaling up in space and time. Agricultural Systems 88, 8–27.

Giller, K.E., Witter, E., Corbeels, M., Tittonell, P., 2009. Conservation agriculture and smallholder farming in Africa: the heretics' view. Field Crops Research 114, 23–24.

Hansen, J.W., 2002. Realizing the potential benefits of climate prediction to agriculture: issues, approaches, challenges. Agricultural Systems 74, 309–330.

Hochman, Z., vanRees, H., Carberry, P.S., Hunt, J.R., McCown, R.L., Gartmann, A., Holzworth, D., van Rees, S., Dalgliesh, N.P., Long, W., Peake, A.S., Poulton, P.L., McClelland, T., 2009. Re-inventing model-based decision support: 4. Yield Prophet, an internet-enabled simulation-based system for assisting farmers to manage and monitor crops in climatically variable environments. Crop and Pasture Science 60, 1057–1070.

Jabet, T., Caron, J., Lambert, R., 2017. Payback period in cranberry associated with a wireless irrigation technology. Canadian Journal of Soil Science 97 (1), 71–81.

Jeranyama, P., Hesterman, O.B., Waddington, S.R., Harwood, R.R., 2000. Relay-intercropping of sunnhemp and cowpea into a smallholder maize system in Zimbabwe. Agronomy Journal 92, 239–244.

Jeranyama, P., Waddington, S.R., Waddington, S.R., Harwood, R.R., 2007. Nitrogen Effects on Maize Yield Following Groundnut in Rotation on Smallholder Farms in Sub-Humid Zimbabwe.

Keating, B.A., Carberry, P.S., Bindraban, P.S., Asseng, S., Meinke, H., Dixon, J., 2010. Eco-efficient agriculture: concepts, challenges, and opportunities. Crop Science 50, S109–S119.

Kennedy, C.D., Jeranyama, P., Alverson, N., 2017. Agricultural water requirements for commercial production of cranberries. Canadian Journal of Soil Science 97 (1), 38–45.

Kumwenda, J.D.T., Waddington, S.R., Snapp, S.S., Jones, R.B., Blackie, M.J., 1996. Soil Fertility Management Research for Maize Cropping Systems of Smallholders in Southern Africa: A Review. NRG Paper 96-02. CIMMYT, Mexico, D.F.

MacColl, D., 1989. Studies on maize (Zea mays L.) at Bunda, Malawi. II. Yield in short rotation with legumes. Experimental Agriculture 25, 367–374.

Mapfumo, P., Mtambanengwe, F., 1999. Nutrient mining in maize-based systems of rural Zimbabwe. In maize production technology for the future: challenges and opportunities. In: Proceedings of the Sixth Eastern and Southern Africa Regional Maize Conference, 21-25 September 1998, pp. 274–277.

Matthews, R.B., Stephens, W., 2002. Crop-soil Simulation Models: Applications in Developing Countries. CABI Publ., Wallingford, UK.

McCown, R.L., Keating, B.A., Probert, M.E., Jones, R.K., 1992. Strategies for sustainable crop production in Semi-arid Africa. Outlook on Agriculture 21, 21–31.

Meinke, H., Stone, R.C., 2005. Seasonal and inter-annual climate forecasting: the new tool for increasing preparedness to climate variability and change in agricultural planning and operations. Climatic Change 70, 221–253.

Meinke, H., Nelson, R., Kokic, P., Stone, R., Selvaraju, R., Baethgen, W., 2006. Actionable climate knowledge: from analysis to synthesis. Climate Research 33, 101–110.

Meinke, H., Sivakumar, M.V.K., Motha, R.P., Nelson, R., 2007. Preface: climate predictions for better agricultural risk management. Australian Journal of Agricultural Research 58, 935–938.

Millennium Ecosystem Assessment, 2005. Our Human Planet: Summary for Decision-Makers. Island Press, Washington, DC.

MoAD (Ministry of Agriculture Development), 2016. Statistical Information on Nepalese Agriculture. Ministry of Agriculture Development, Government of Nepal, Singha Durbar, Kathmandu, Nepal.

MoAD (Ministry of Agriculture Development), 2018. Statistical Information on Nepalese Agriculture. Ministry of Agriculture Development, Government of Nepal, Singha Durbar, Kathmandu, Nepal.

Mount Digit Technology, 2014. Study on National Pesticides Consumption Statistics in Nepal. Mount Digit Technology (P.) LTD, Kathmandu, Nepal.

Mpepereki, S., Wollum, A.G., Makonese, F., 1996. Growth temperature characteristics of indigenous Rhizobium and Bradyrhizobium isolates from Zimbabwean soils. Soil Biology and Biochemistry 28, 1537–153.

Mpepereki, S., Javaheri, F., Davis, P., Giller, K.E., 2000. Soybeans and sustainable agriculture: 'Promiscuous' soybeans in southern Africa. Field Crops Research 65, 137–149.

Mukurumbira, L.M., 1985. Effects of rate of fertilizer nitrogen and previous grain legume crop on maize yields. Zimbabwe Agricultural Journal 82, 177–179.

Nye, P.H., Greenland, D.J., 1960. The soil under shifting cultivation. Technical Communication, 51. Commonwealth Bureau of Soil, Harpenden, UK.

Olszwski, F., Jeranyama, P., Kennedy, C., DeMoranville, C., 2017. Automated cycled sprinkler irrigation for spring frost protection of cranberries. Agricultural Water Management 189, 19–26.

Pearson, C.J., 2007. Regenerative, semi-closed systems: a priority for twenty-first century agriculture. Bioscience 57, 409–418.

Pelletier, V., Gallichand, J., Caron, J., 2013. Effect of soil water potential threshold for irrigation on cranberry yield and water productivity. Transactions of the ASABE 56 (6), 1325–1332.

Raintree, J., Warner, K., 1986. Agroforestry pathways for the intensification of shifting cultivation. Agroforestry System 4 (1), 39–54.

Rijal, J., Regmi, R., Ghimire, R., Puri, K., Gyawaly, S., Poudel, S., 2018. Farmers' knowledge on pesticide safety and pest management practices: a case study of vegetable growers in Chitwan, Nepal. Agriculture 8 (1), 16. https://doi.org/10.3390/agriculture8010016.

Rohrbach, D.D., 1998. Developing more practical fertility management recommendations. In: Waddington, S.R., Murwira, H.K., Kumwenda, J.D.T., Hikwa, D., Tagwira, F. (Eds.), Soil Fertility Research for Maize-Based Farming Systems in Malawi and Zimbabwe. The Soil Fertility Network/CIMMYT, Harare, Zimbabwe, pp. 237–244.

Rosegrant, M.W., Paisner, M.S., Meijer, S., Witcover, J., 2001. Global Food Projections to 2020: Emerging Trends and Alternative Futures. 2020 Vision Food Policy Report. International Food Policy Research Institute, Washington D.C.

Rukuni, M., 2002. Africa: addressing growing threats to food security. Journal of Nutrition 132, 3443S–3448S.

Rusike, J., Sukume, C., Dorward, A., Mpepereki, S., Giller, K.E., 1999. The Economic Potential of Smallholder Soybean Production in Zimbabwe. Soil Fertility Network for Maize Based Cropping Systems in Malawi and Zimbabwe/CIMMYT-Zimbabwe. Harare.

Sanginga, N.S., Woomer, P.L., 2009. Integrated Soil Fertility Management in Africa: Principles, Practices, and Developmental Process. Free download form CIAT.

Snyder, C.S., 2012. Are mid-west maize farmers over-applying fertilizer N? Better Crops with Plant Food 96, 3–4.

Tittonell, P., Vanlauwe, B., Leffelaar, P.A., Rowe, E., Giller, K.E., 2005. Exploring diversity in soil fertility management of smallholder farms in western Kenya. I. Heterogeneity at region and farm scale. Agriculture, Ecosystems and Environment 110, 149–165.

Twomlow, S., Rohrbach, D., Dimes, J., Rusike, J., Mupangwa, W., Ncube, B., Hove, L., Moyo, M., Mashingaidze, N., Maphosa, P., 2008. Micro-dosing as a Pathway to Africa's Green Revolution: Evidence from Broad-Scale On-Farm Trials Nutr Cycling Agroecosyst 101007/s10705–008–9200–4.

Vanlauwe, B., Tittonell, P., Mukulama, J., 2006. Within-farm soil fertility gradients affect response of maize to fertilizer application in western Kenya. Nutrient Cycling in Agroecosystems 76, 171–182.

Whitbread, A., Robertson, M., Carberry, P., Dimes, J., 2010. Applying farming systems simulation to the development of more sustainable smallholder farming systems in Southern Africa. European Journal of Agronomy 32, 51–58.

Wilson, C., Tisdell, C., 2001. Why farmers continue to use pesticides despite environmental, health and sustainability costs. Ecological Economics 39, 449–462.

World Bank, 2000. Can Africa Claim the 21st Century? The International Bank for Reconstruction and Development/The World Bank, Washington, DC.

World Bank, 2008. World Development Report 2008: Agriculture for Development. The World Bank, Washington, DC.

Further reading

Ababa, A., Ethiopia: CIMMYT and Ethiopian Agricultural Research Organization, McCown, R.L., Keating, B.A., Probert, M.E., Jones, R.K., 1992. Strategies for sustainable crop production in semi-arid Africa. Outlook on Agriculture 21 (1), 21–31.

Grant, P.M., 1981. The fertility of sandy soils in peasant agriculture. Zimbabwe Agricultural Journal 78, 169–175.

Perrot, N., De Vries, H., Lutton, E., van Mil, H.G.J., Donner, M., Tonda, A., Martin, S., Alvarez, I., Bourgine, P., van der Linden, E., Axelos, M.A.V., 2016. Some remarks on computational approaches towards sustainable complex agri-food systems. Trends in Food Science and Technology 48, 88–101.

Rusinamhodzia, L., Dahlind, S., Corbeelsa, M., 2016. Living within their means: reallocation of farm resources can help smallholder farmers improve crop yields and soil fertility Agriculture. Ecosystems and Environment1 216, 125–136.

U.S. Code § 3103.Definitions. www.law.cornell.edu/uscode/text/7/3103.

CHAPTER

2 Ecosystem services in sustainable food systems: operational definition, concepts, and applications

C.C. Du Preez, PhD, C.W. van Huyssteen, PhD, E. Kotzé, PhD, J.J. van Tol, PhD

Department of Soil, Crop and Climate Sciences, University of the Free State, Bloemfontein, Free State, South Africa

Introduction

In recent years, it is often highlighted that 75% of the Earth has been disturbed by humans and that this area will increase to more than 90% by 2050. The result is that one-third of the Earth's land surface is subjected to degradation through inter alia soil erosion (Fig. 2.1) and water pollution (Fig. 2.2). It is therefore estimated that 42% of the world's total population, approximately 3.2 billion people, is now impacted by the consequence of land degradation. This human-induced influence is equivalent to 10% of the annual total global gross domestic product (UNEP, 2016).

FIGURE 2.1

Soil erosion resulting from injudicious livestock grazing near Lady Grey in the Eastern Cape Province of South Africa.

The Role of Ecosystem Services in Sustainable Food Systems. https://doi.org/10.1016/B978-0-12-816436-5.00002-0

FIGURE 2.2

Water pollution caused by excessive nutrient loading that results in a blue-green algae bloom at the Hartbeespoort Dam in the Gauteng Province of South Africa.

Continued deforestation and land conversion, excessive withdrawal of water, growing consumption of fossil fuels and increasing emission of greenhouse gases are depleting and degrading the natural ecosystems, polluting the environment, and diminishing ecosystem services (MEA, 2005). These factors lead to two contrasting viewpoints: humans are parasites living in a nonsymbiotic relationship with their host planet Earth, triggering natural disasters or humans have a natural instinct to survive by competing with other species. Proponents of the latter opinion argue that passing moral judgment on and being aggressive toward fellow humans are counterproductive and contribute very little to Earth and its inhabitants. The fact remains that all 8.7 million species have fully exploited the Earth and its ecospheres (atmosphere, biosphere, hydrosphere, lithosphere, and pedosphere) with humans being the dominant force, especially since the onset of the Anthropocene period.

However, the species which have survived and evolved through judicious adaptation may achieve greater benefit by choosing mutualism rather than parasitism. This truism dictates that humans cannot survive without positive interactions between species. Therefore it is important to live in symbiosis with other species and ensure that consumable natural resources such as soil, water, and air are improved and restored by adopting mutualism and living in harmony with nature. Only then will sustainable food systems, as described in other chapters, be realized to support the world's increasing population (MEA, 2005; FAO, 2011a; UNEP, 2016).

It is unequivocally accepted globally that all people have the right to a healthy diet. This human right is captured in the United Nations sustainable development goals in which the importance of food security, through the practice of sustainable agriculture, is emphasized (Keesstra et al., 2016). In achieving this we must change the way in which we manage our food system. In this particular context a food system includes all elements and activities that relate to the production, processing, distribution, preparation, and consumption of food and the outputs of these activities which include socioeconomic and environmental outcomes (UNEP, 2016).

The food system is dependent on inter alia a large array of natural resources. These include land, water, minerals, biodiversity, and ecosystem services, which include genetic resources and marine resources such as fish stocks. The sustainable and efficient use of these resources is essential to satisfy both current and future food demand (Stephens et al., 2018).

Agriculture is predominantly responsible for most of the global land and water use, which often leads to loss of soil biodiversity (Orgiazzi et al., 2016) and other negative impacts on ecosystems such as ecotoxicity, eutrophication, and depletion of phosphorus stocks. A substantial amount of energy is also used by intensive agriculture (Karabulut et al., 2018) and combined with the escalation of soil and biomass carbon loss can contribute to climate change. On the other hand, agriculture can also contribute to environmental solutions, e.g., by binding carbon in the soil and thereby increasing biodiversity through diverse habitats. The impacts of agriculture thus depend to a substantial degree on specific aspects of the activities and, by extension, the resource management regime (UNEP, 2016).

Currently, of the 149 million km^2 total global land area, 15 million km^2 are predominantly used for crop production and around 34 million km^2 for livestock production (FAO, 2011b). Globally, croplands produce the largest share of food, although the contribution from rangelands, fisheries, hunting, and gathering should not be underestimated as these sources provide a major part of the human diet in some traditional food systems. Within food systems however, land is predominantly used for primary production.

Against this backdrop a concise view will be given on ecosystems and the services that they deliver, including how the ecosystem services contribute to sustainable development. Next follows a thorough discussion with respect to ecosystem services and cycling processes in food systems. This is followed in turn by the changes that ecosystem services are subject to and the drivers that cause these changes. Finally some attention will be given to projected changes in ecosystems.

Ecosystem services: definitions and concepts

Biosphere including subunits

All ecospheres mentioned earlier influence food systems in one way or another. The living organisms and the diversity of terrestrial environments in which they live are collectively known as the biosphere. A key characteristic of the biosphere is the enveloping and complex interdependent network of all living things with their environment and the Earth as a whole. The biotic environment can be grouped into five identifiable units of scale, each a subset of a subsequent larger one. These units consist of individuals, populations, communities, ecosystems, and finally the biosphere (Van As et al., 2012).

Individuals: This most basic unit consists of simple free living cells, such as bacteria and certain algae or more complex groups such as colonies of bacteria and algae and multicellular organisms such as plants and animals. Individual living organisms, or species, are connected together as a food web that controls both energy flow and nutrient cycling.

Populations: A population consists of all individual members of a particular species within a geographic area.

Communities: A community consists of all individuals and all species that inhabit a particular area. It is an assemblage of populations comprising many different species interacting as a complex food web in the same geographic area.

Ecosystems: An ecosystem consists of all abiotic factors in addition to the entire community of species that exists in a particular area.

Biosphere: The Earth's biosphere is the collective grouping of all the planet's ecosystems, including the atmosphere to an altitude of several kilometers, the land to a few kilometers beneath the Earth's crust, as well as all the lakes, rivers, and oceans.

Different ecosystem services

The ecosystem services referred to earlier point to the various benefits that humans freely gain from properly functioning ecosystems. Such ecosystems include, for example, agroecosystems, rangeland ecosystems, and aquatic ecosystems which all contribute to food systems. Ecosystem services are categorized into provisioning, regulating, and cultural services that directly affect people, and the supporting services needed to maintain other services (La Notte et al., 2017). Many of the services listed below are closely interlinked. Notable examples of primary production include photosynthesis, nutrient cycling, and water cycling which all involve different aspects of the same biological processes (MEA, 2005; Liere et al., 2017).

Provisioning services

These services are the products obtained from ecosystems.

- *Food:* This includes the vast range of food products derived from plants, animals, and microbes.
- *Fiber:* Materials included here are wood, jute, cotton, hemp, silk, and wool.
- *Fuel:* Wood, dung, and other biological materials that serve as source of energy.
- *Genetic resources:* This include the genes and genetic information used for animal and plant breeding and biotechnology.

- *Biochemicals, natural medicines, and pharmaceuticals:* Medicines, biocides, food additives such as alginates, and biological materials derived from ecosystems.
- *Ornamental resources:* Animal and plant products, such as skins, shells, and flowers, are used as ornaments, and whole plants are used for landscaping and ornaments.
- *Freshwater:* Freshwater is obtained from ecosystems, and thus the supply of freshwater can be considered a provisioning service. Freshwater in rivers is also a source of energy. Although water is required for all forms of life to exist, it could also be considered a supporting ecosystem service.

Regulating services

These services are the benefits obtained from the regulation of ecosystem processes.

- *Air quality regulation:* Ecosystems both contribute chemicals to and extract chemicals from the atmosphere, influencing many aspects of air quality.
- *Climate regulation:* Ecosystems influence climate both locally and globally. At a local scale, for example, changes in land cover can affect both temperature and precipitation. At the global scale ecosystems play an important role in climate by either sequestering or emitting greenhouse gases.
- *Water regulation:* The timing and magnitude of runoff, flooding, and aquifer recharge may be strongly influenced by changes in land cover, including specific alterations that change the water storage potential of the system, such as the conversion of wetlands or the replacement of forests with croplands or croplands with urban areas.
- *Erosion regulation:* Vegetative cover plays an important role in soil retention and the prevention of landslides.
- *Water purification and waste treatment:* Ecosystems can be a source of impurities, for example, in freshwater but can also help filter out and break down organic wastes introduced into inland waters and coastal and marine ecosystems and may assimilate and detoxify compounds through soil and subsoil processes.
- *Disease regulation:* Changes in ecosystems can directly change the abundance of human pathogens, such as cholera, and can alter the abundance of disease vectors, such as mosquitoes.
- *Pest regulation:* Ecosystem changes affect the prevalence of crop and livestock pests and diseases.
- *Pollination:* Ecosystem changes affect the distribution, abundance, and effectiveness of pollinators.
- *Natural hazard regulation:* The presence of coastal ecosystems such as mangroves and coral reefs can reduce the damage caused by hurricanes or large waves.

Cultural services

These services are the nonmaterial benefits people obtain from ecosystems through spiritual enrichment, cognitive development, reflection, recreation, and esthetic experiences.

- *Cultural diversity:* The diversity of ecosystems is one factor that influences the diversity of cultures.
- *Spiritual and religious values:* Many religions attach spiritual and religious values to ecosystems or their components.
- *Knowledge systems:* Ecosystems influence both traditional and formal types of knowledge systems developed by different cultures.
- *Educational values:* Ecosystems and their components and processes provide the basis for both formal and informal education in many societies.
- *Inspiration:* Ecosystems provide a rich source of inspiration for art, folklore, national symbols, architecture, and advertising.
- *Esthetic values:* Many people find beauty or esthetic value in various aspects of ecosystems, as reflected in the support for parks, scenic drives, and the selection of housing locations.
- *Social relations:* Ecosystems influence the types of social relations that are established in particular cultures. Fishing societies, for example, differ in many respects in their social relations from nomadic herding or agricultural societies.
- *Sense of place:* Many people value the sense of place that is associated with recognized features of their environment, including aspects of the ecosystem.
- *Cultural heritage values:* Many societies value the maintenance of either historically important cultural landscapes or culturally significant species.
- *Recreation and ecotourism:* People often choose where to spend their leisure time based in part on the characteristics of the natural or cultural landscapes in a particular area.

Supporting services

Supporting services are those that are necessary for the delivery of all other ecosystem services. These differ from provisioning, regulating, and cultural services in that their human impact is often indirect or occurs over a prolonged period of time, whereas changes in the other categories have relatively direct and immediate impacts on people. Some services, such as erosion regulation, can be categorized as both a supporting and a regulating service, depending on the time scale and human impact immediacy.

- *Soil formation:* Because many provisioning services depend on soil fertility, the rate of soil formation influences human well-being in many ways.
- *Photosynthesis:* Photosynthesis produces energy and oxygen necessary for most living organisms.
- *Primary production:* The assimilation or accumulation of energy and nutrients by organisms.

- *Nutrient cycling:* Approximately 20 nutrients which are essential for life, including nitrogen and phosphorus, cycle through ecosystems and are maintained at different concentrations in different ecosystemic locations.
- *Water cycling:* Water, essential for living organisms, cycles through ecosystems.

Ecosystem services and sustainable development goals

Ecosystem services are related to the United Nations sustainable development goals (Keesstra et al., 2016). The purpose behind these goals is to serve as a guideline for governments for both developed and developing countries. Some of the goals are mainly socioeconomic in character (e.g., end poverty in all its forms everywhere; ensure inclusive and equitable education; and promote sustained, inclusive, and sustainable economic growth), while others focus clearly on the biophysical aspects (e.g., ending hunger, achieving food security, and improving nutrition by promoting sustainable agriculture; taking urgent action to combat climate change and its impacts; and protecting, restoring, and promoting sustainable use of terrestrial ecosystems by reversing land degradation).

It is tempting to draw a distinction between the socioeconomic and biophysical goals. However, together these two realms define human existence and are mutually interdependent. The achievement of socioeconomically focused goals requires the consideration of the associated dynamic behavior of ecosystems. Conversely, for achieving goals with an ecosystem focus requires the consideration of socioeconomic aspects. Environmental sustainability will depend on the actions of land users such as farmers and forest managers, although urban developments also have major effects on local land use (Keesstra et al., 2016).

Ecosystem services contribute, either directly or indirectly, to nearly all land-related sustainable development goals (Keesstra et al., 2016). We should acknowledge that services are provided by nature and that human effects should be governed by the realization that every ecosystem has its own characteristically dynamic thresholds. Sustainable development can only be achieved when processes, feedbacks, and ecosystemic thresholds are taken into account. Collaboration between scientists of various disciplines and stakeholders from outside the world of science is therefore required to confront the societal challenges of our time. Key issues with respect to food systems that are part of the sustainable development goals include food, health, water, climate, and land management. Each of these issues is addressed by invited experts in the form of short essays.

Ecosystem services and cycling processes in food systems

Space and time are two features that are fundamental to an ecosystem's structure and function (Van As et al., 2012). Ecosystems, like single organisms, are dynamic and occur in space (width, depth and height) and time (past, present and future). Although the spatial aspect of ecosystems is real, ecosystems are not discrete entities sharply delimited from one another. Instead ecosystems are often quite closely interrelated with each other.

Another feature of ecosystems is that they are open. This means that there is always an exchange of energy, materials, and organisms between different ecosystems. Nutrients and solar energy constitute an input from other ecosystems. Conversely ecosystems also have an output that contributes to other ecosystems, such as river runoff. However, the losses from one ecosystem to another must be balanced by continual input, such as rainfall into river catchment areas. Nevertheless, ecosystems are largely internally self-regulating: death, decomposition, nutrient cycling, birth, and population growth are self-sustaining unless external forces upset the balance of nature (Van As et al., 2012).

Ecosystems therefore have the following properties (Van As et al., 2012):

- Ecosystems exist independently of specific components.
- An ecosystem's components are interdependent.
- An ecosystem has functional components that operate collectively.
- An ecosystem is dynamic and may change over time and space.
- A sliding organizational scale exists: for example, two populations may independently co-exist in an area without much or any interdependence or they may be entangled in complex relationships or food webs.

An ecosystem consists of all biotic factors in addition to the entire community of species that exist in a particular ecosystem.

Two processes occur concurrently and continuously in all ecosystems, namely the unidirectional flow of energy and the cycling of nutrients (Fig. 2.3). During photosynthetic reactions, carbon dioxide, water, and essential nutrients such as nitrogen, phosphorus, sulfur, and magnesium, among others, are synthesized into carbohydrates, proteins, fats, and oils. Therefore when plants are consumed by herbivores, both energy and nutrients are ingested. Similar transfers take place at all trophic levels, finally leading to the decomposers, so that nutrient cycling must accompany energy flow and vice versa (Van As et al., 2012).

When considering energy flow in an ecosystem, it is worth remembering the first law of thermodynamics. This law states that when energy is converted from one form (e.g., solar energy) to another (e.g., chemical energy), energy is neither gained nor lost. However, the second law of thermodynamics states that every transformation results in a reduction of the free energy in the system. Put simply, this means that there is a loss of available energy accompanying its transfer from one trophic level to the next. It is important to realize, however, that although there is a progressive reduction of energy with each higher trophic level, the inorganic nutrient component is not reduced. This means that at the final decomposer level, all the nutrients are released back to the environment to become available once more for recycling by plants (Van As et al., 2012).

Ecosystem processes, including water, carbon, nitrogen, and phosphorus cycling, changed more rapidly in the second half of the 20th century than at any time in recorded human history. Human modifications of ecosystems have not only altered

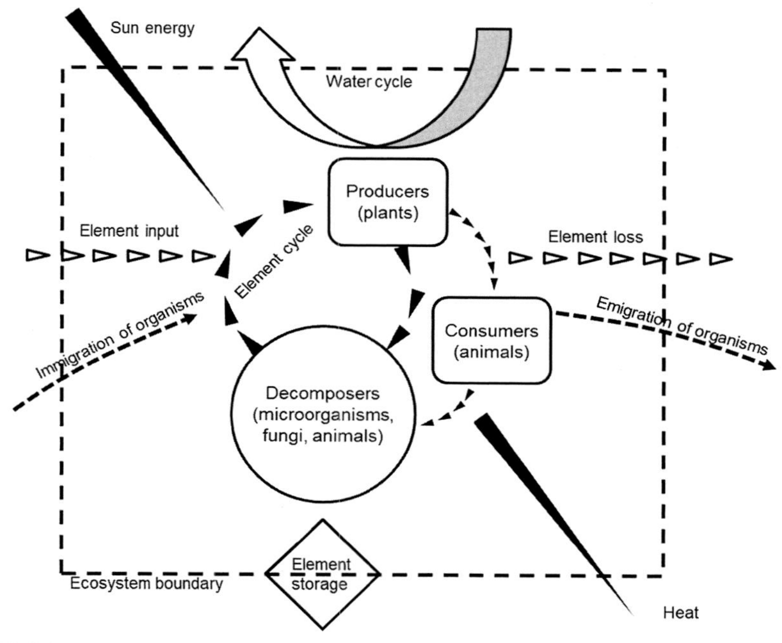

FIGURE 2.3

Energy enters an ecosystem as sunlight and ultimately leaves it as heat. The metabolical pathways essential to life, namely photosynthesis and respiration, help to recycle the chemical components.

Adapted from Van As, J., Du Preez, J., Brown, L., Smit, N. 2012. The Story of Live and the Environment: An African Perspective. Struik Nature, Cape Town.

the structure of the systems but also their processes and functions. The capacity of ecosystems to provide services is therefore derived directly from the operation of natural biogeochemical cycles, although these have been significantly modified in some cases (MEA, 2005).

Water cycle

Water withdrawals from rivers and lakes for irrigation or for urban or industrial use have doubled between 1960 and 2000. Worldwide, 70% of water use is dedicated to agriculture. Large reservoir construction has doubled, even tripled the residence time of river water, viz. the average time a drop of water takes to reach the sea. Globally humans use slightly more than 10% of the available renewable freshwater through household, agriculture, and industrial activities. In some regions such as the Middle East and North Africa however, humans use more than 100% of the renewable supplies. This excess is largely obtained through the use of groundwater supplies at rates greater than their rate or recharge.

Carbon cycle

Since 1750, the atmospheric concentration of carbon dioxide has increased by about 34%, from 280 to 376 parts per million. Approximately 60% of that increase, namely 60 parts per million, has taken place since 1959. The effect of this on terrestrial ecosystems changed during the last 50 years. In the 19th and early 20th centuries ecosystems were on average a net source of carbon dioxide. This was primarily because of deforestation with contributions from crop, pasture, and forestland degradation. Terrestrial ecosystems, however, became a net carbon sink sometime around the middle of the last century, although carbon losses from land use continue at high levels. The role played by ecosystems in carbon sequestration is influenced by afforestation, reforestation, and forest management in North America, Europe, China, and other regions; changed practices such as conservation agriculture; and the fertilizing effects of nitrogen deposition and increasing atmospheric carbon dioxide.

Nutrient cycles

In ecosystems almost 20 nutrient cycles occur, but because of spatial constraints, only the two most prominent cycles are addressed here.

Nitrogen cycle

The total amount of reactive, or biologically available, nitrogen created by human activities increased nine-fold between 1890 and 1990. Most of this increase occurred in the second half of the previous century and was associated with the increased use of fertilizers. A recent study of global human contributions to reactive nitrogen flows projected that flows will increase from approximately 165 Tg in 1999 to 270 Tg in 2050, an increase of 64% (Fig. 2.4). Synthetic nitrogen was first produced in 1913. More than half of all the synthetic nitrogen fertilizer ever used on the planet has been used since 1985. Human activities have now roughly doubled the rate of reactive nitrogen creation on land. The flux of reactive nitrogen to the oceans therefore increased by nearly 80%, from 27 Tg yr^{-1} in 1860 to 48 Tg yr^{-1} in 1990. This change is, however, not uniform over the Earth as a whole.

Phosphorus cycle

The use of phosphorus fertilizers and hence the rate of phosphorus accumulation in agricultural soils increased almost three-fold between 1960 and 1990. However, the rate of phosphorus accumulation has declined somewhat since 1990. The current flux of phosphorus to the oceans is now triple that of background rates, viz. 22 Tg P yr^{-1} versus 8 Tg P yr^{-1}.

The use of nitrogen and phosphorus in the form of fertilizers are necessitated by cropping to fulfill in the food demand of a growing population. Fertilizer application should be site-specific as illustrated by the three maize-based cropping systems in Table 2.1. Only with this kind of approach optimum crop production with minimum environmental pollution is possible (Vitousek et al., 2009).

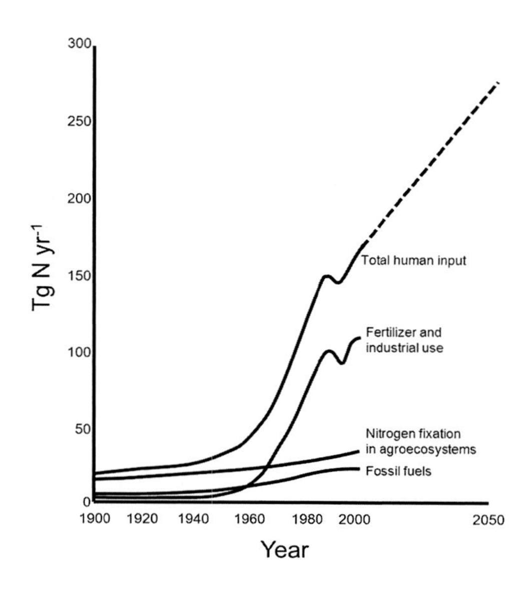

FIGURE 2.4

Global trends in the creation of active nitrogen by human activity, projected to 2050.

Adapted from MEA, 2005. Ecosystems and Human Well-Being: Synthesis. Island Press, Washington, DC.

Table 2.1 Nitrogen and phosphorus balances in three maize-based agricultural systems ($kg^{-1}\ ha^{-1}\ yr^{-1}$)

Inputs and outputs	Kenya[a]		China[b]		USA[c]	
	N	P	N	P	N	P
Fertilizer	7	8	588	92	93	14
Biological nitrogen fixation	0	0	0	0	62	0
Total agronomic inputs	7	8	588	92	155	14
Removal in grain and/or beans	23	4	361	39	145	23
Removal in other harvested products	36	3	0	0	0	0
Total agronomic outputs	56	7	361	39	145	23
Agronomic inputs minus harvest products	−52	+1	+227	+53	+10	−9

[a] *Low input maize in western Kenya.*
[b] *High input maize and wheat double-cropping in northeast China.*
[c] *Maize–soybean rotations in the upper Midwestern USA.*

Derived from Vitousek, P.M., Naylor, R., Crews, T., David, M.B., Drinkwater, L.E., Hollard, E., Johnes, P.J. Katzenberger, J., Martinelli, L.A., Matson, P.A., Nziguhela, G., Ogima, D., Palm, C.A., Robertson, G.P., Sanchez, P.A., Townsend, A.R., Zhang, F.S., 2009. Nutrient imbalances in agricultural development. Science 324, 1519–1520.

An elaboration on each of the four cycles is impossible to pursue here because of limited space. However, more detailed information on the water (e.g., Verhoef and Egea, 2013; Weil and Brady, 2017), carbon (e.g., Stevenson and Cole, 1999; Weil and Brady, 2017), nitrogen (e.g., Haygarth et al., 2013; Havlin et al., 2014), and phosphorus (e.g., Haygarth et al., 2013; Havlin et al., 2014) cycles, especially with respect to soil-plant systems, is available.

Ecosystem services and changes

The trends in the use of ecosystem services and the enhancement or degradation of a service around the year 2000 are summarized in Table 2.2. This table shows that the quantities of provisioning such as food, water, and timber increased rapidly. The increase during the second half of the 20th century was often more rapid than the rate of population growth, but generally slower than economic growth, and is continuous. In a number of cases these provisional services are used at unsustainable rates (MEA, 2005).

The ever augmenting human use resulted from a combination of substantial increases in the absolute amount of services produced by ecosystems and an increase in the fraction used by humans. For example, the world's population doubled from 3 billion people in 1960 to 6 billion people in 2000, while the global economy increased more than six-fold. In this period food production increased 2.5 times, water use doubled, and wood harvests for pulp and paper tripled (MEA, 2005).

The sustainable use of provisioning services differs within and between localities (MEA, 2005). However, the use of several provisional services is unsustainable even at a global scale. It is now generally accepted that use of capture fisheries is unsustainable since 25% of important commercial fish stocks are over exploited or significantly depleted. From 5% to possibly 25% of global freshwater use exceeds long-term accessible supplies and is maintained only through engineered water transfers or the overdraft of groundwater supplies. An estimated 15%–35% of irrigation water withdrawals exceed supply rates and are therefore unsustainable. Current agricultural practices are also unsustainable in some instances because of their reliance on insufficient sources of water, harmful impacts caused by excessive nutrient or pesticide use, acidification, salinization, nutrient depletion, and rates of soil loss that exceed rates of soil formation (Hopkins and Gregorich, 2013).

Humans have substantially altered regulating services such as disease and climate regulation by modifying the ecosystem providing the service and, in the case of waste processing services, by exceeding the capabilities of ecosystems to provide these services. Most changes to regulating services are the inadvertent results of actions taken to enhance the supply of provisioning services. Humans have substantially modified the climate regulation service of ecosystems. Initially these modifications occurred through land use changes that contributed to increases

Table 2.2 Estimated global trends in the status of provisioning, regulating, and cultural services of ecosystems

Service	Subcategory	[a]Status
Provisioning		
Food	Crops	+
	Livestock	+
	Capture fisheries	–
	Aquaculture	+
	Wild foods	–
Fiber	Timber	+/–
	Cotton, hemp, silk	+/–
	Wood fuel	–
Genetic resources		–
Biochemicals and medicines		–
Freshwater		–
Regulating		
Air quality regulation		–
Climate regulation	Global	+
	Regional, local	–
Water regulation		+/–
Erosion regulation		–
Water purification, waste treatment		–
Disease regulation		+/–
Pest regulation		–
Pollination		–
Natural hazard regulation		–
Cultural		
Spiritual and religious values		–
Aesthetic values		–
Recreation and ecotourism		+/–

[a] *+ = enhancement and – = degradation.*

Adapted from MEA, 2005. Ecosystems and Human Well-Being: Synthesis. Island Press, Washington, DC.

in the amount of carbon dioxide and other greenhouse gases such as methane and nitrous oxide in the atmosphere. Lately climate regulation service has been altered by increasing the sequestration of carbon dioxide. Ecosystems, however, remain a net source of methane and nitrous oxide (MEA, 2005).

Modifications of ecosystems have altered patterns of disease by increasing or decreasing the habitat for certain diseases or their vectors through dams and irrigation canals that provide a habitat for schistosomiasis or bringing human populations into closer contact with various disease organisms. Changes to ecosystems have contributed to a significant rise in the number of floods and wildfires on all continents since the 1940s. Ecosystems serve an important role in detoxifying wastes introduced into the environment, but there are intrinsic limits to this waste-processing capacity. For example, aquatic ecosystems cleanse on average 80% of their global nitrogen loading, but this intrinsic self-purification capacity varies widely and is being reduced by the loss of wetlands (MEA, 2005).

The use of cultural services continued to grow, but the capability of ecosystems to provide cultural benefits diminished significantly in the past century. Human cultures are strongly influenced by ecosystems, and ecosystem change can have an impact on cultural identity and social stability. The cultures, knowledge systems, religions, heritage values, social interactions, and the limited amenity services have always been influenced and shaped by the nature of an ecosystem and its conditions. Many of these benefits are being degraded either through changes to ecosystems because of a recent rapid decline in the number of sacred graves and other such protected areas for example or through societal changes such as the loss of languages and traditional knowledge, which reduces people's recognition or appreciation of cultural benefits. Rapid loss of culturally valued ecosystems can contribute to social disruption and societal marginalization. There has also been a decline in the quantity and quality of esthetically pleasing landscapes (MEA, 2005).

Global gains in the supply of food, water, timber, and other provisioning services were often achieved in the past century despite local resource depletion and local restrictions on resource use, through the shift in production and harvesting to new underexploited regions, sometimes considerable distances away. Although human demand for ecosystem services continues to grow, the demand for particular services in specific regions is declining as substitutes are being developed. Both the supply and the resilience of ecosystem services are affected by changes in biodiversity. The modification of an ecosystem, which prompts the alteration of an ecosystem service, generally results in concomitant changes to other ecosystem services. In actual terms the net benefit gained through actions to increase the productivity or harvesting of ecosystem services are fewer than initially believed after taking the negative tradeoffs into account (Holt et al., 2016). Examples of this are as follows:

- Expansion of commercial shrimp farming has had serious impacts on ecosystems, including loss of vegetation, deterioration of water quality, decline in capture fisheries, and loss of biodiversity.
- Expansion of livestock production around the world has often led to overgrazed rangeland, wildlife habitat loss, dust storms, bush encroachment, deforestation, and greenhouse gas emissions.
- Poorly designed and executed agricultural policies led to an irreversible change in ecosystems through, for example, pollution of water resources, loss of soil fertility, and extinctions of species.
- The removal of wetlands through canalization, leveeing, and drainage led to significant reduction of flood mitigation capacity.

However, positive results can also be achieved when actions taken to conserve or enhance a particular component of an ecosystem or its services benefit other services or stakeholders (Bommarco et al., 2018). For example:

- Agroforestry can meet human needs for food and fuel, restore soils, and contribute to biodiversity conservation.
- Intercropping can increase yields, enhance biocontrol, reduce soil erosion, and reduce weed invasion.
- Protection of natural forests for biodiversity conservation can also reduce carbon emissions and protect water supplies.
- Conservation of wetlands can contribute to flood control and help to remove pollutants such as nitrogen and phosphorus from water.

Complementarity and synergy often exist among provisioning, regulating, cultural and supporting services of ecosystems (Garibaldi et al., 2018)

Drivers causing ecosystem services change

Global trends impacting ecosystem services

Natural or anthropogenic factors that cause a change in ecosystem services are referred to as drivers. These drivers may operate either indirectly or directly. A direct driver unequivocally influences ecosystem processes, while an indirect driver operates more opaquely by altering one or more direct drivers (MEA, 2005).

At a global scale, there are five indirect drivers of changes in ecosystems and their services (MEA, 2005): population change, change in economic activity, socio-political factors, cultural factors, and technological change. Collectively these indirect drivers influence the level of production and consumption of ecosystem services and the sustainability of production. Both economic growth and population growth lead to increased consumption of ecosystem services, although the harmful environmental impacts of any particular level of consumption depend on the efficiency of the technologies used in the production of the service. These factors interact in complex ways in different locations to change pressures on ecosystems and uses of ecosystem services. Driving forces are almost always multiple and interactive, so that a one-to-one linkage between particular driving forces and particular changes in ecosystems rarely exists. Even so, changes in any of these indirect drivers generally result in changes in ecosystems. The causal linkage is almost always highly mediated by other factors, thereby complicating casual statements or attempts to establish the proportional aspects of the various contributions.

Most of the direct drivers in ecosystem changes currently remain constant or are growing in intensity in most ecosystems. The most important direct drivers of change in ecosystems are habitat change (land-use change and physical modification of rivers or water withdrawal from rivers), overexploitation, invasive alien species, pollution, and climate change (MEA, 2005).

For terrestrial ecosystems, the most important direct drivers since the 1950s have been land cover change (in particular, conversion to cropland) and the application of new technologies (which have contributed significantly to the increased supply of services such as food, timber, and fiber). Between one-fifth and one-half of the land area has been transformed, largely to croplands. New technologies have resulted in a significant increase in agricultural yield. In the case of cereals, for example, from the mid-1980s to the late 1990s the global area under cereals declined by around 0.3% per year, while yields increased by about 1.2% a year (MEA, 2005).

In the case of freshwater ecosystems and their services, depending on the region, the most important direct drivers of changes since the mid-1950s include modification of water regimes, invasive species and pollution, and particularly high levels of nutrient loading. It is estimated that about 50% of inland water ecosystems (excluding large lakes and closed seas) were converted in the 20th century. For example, 60% of all reservoirs in South America and 78% of all reservoirs in Asia were built after the 1980s. The introduction of nonnative invasive species is one of the major causes of species extinction in freshwater systems. While the presence of nutrients such as nitrogen and phosphorus is necessary for biological systems, high levels of nutrient loading cause significant eutrophication of water bodies and contribute to levels of nitrate in drinking water in some locations. Here nutrient load refers to the total amount of nitrogen or phosphorus entering the water during a given time. Nonpoint pollution sources such as storm water runoff in urban areas, poor or nonexistent sanitation facilities in rural areas, and the flushing of livestock manure by rainfall and snowmelt also qualify as causes of pollution. Pollution from point sources such as mining has had devastating local and regional impacts on the biota of inland waters (MEA, 2005).

Excessive nutrient loading has emerged as one of the most important direct drivers of ecosystem change in terrestrial and freshwater ecosystems (MEA, 2005). The introduction of nutrients into ecosystems can have both beneficial effects (such as increased crop productivity) and adverse effects (such as eutrophication of inland waters). Beneficial effects will eventually reach a plateau because when more nutrients are added, crop yields do not increase correspondingly, while the harmful effects will escalate.

The synthetic production of nitrogen fertilizer has been an important driver for the remarkable increase in food production over the past decades. World consumption of nitrogenous fertilizers grew nearly eight-fold between 1960 and 2005, from 10.8 to 85.1 million tons. As much as 50% of the nitrogen fertilizer applied may be lost to the environment, depending on how well the application is managed. Since excessive nutrient loading is largely the result of applying more nutrients than crops can use, it harms both farm profits and environment quality (Keeney and Hatfield, 2001).

Excessive flows of nitrogen contribute to eutrophication of freshwater ecosystems and acidification of terrestrial and freshwater ecosystems with implications for biodiversity in these ecosystems. To some degree, nitrogen also plays a role in the creation of ground level ozone which leads to loss of agricultural and forest productivity, disintegration of the ozone layer, and increased ultraviolet radiation on Earth, causing an increased incidence of skin cancer and climate change. The resulting health effects include the consequences of ozone pollution on asthma and respiratory function, increased allergies and asthma due to increased pollen production, the risk of blue-baby syndrome, an increased risk of cancer, and other chronic diseases from nitrates in drinking water and the increased risk of a variety of pulmonary and cardiac diseases from production of fine particles in the atmosphere (Follett and Follett, 2001).

Phosphorus fertilizer application has increased three-fold since 1960, with a steady increase until 1990, followed by a leveling off at a level approximately equal to applications in the 1980s. While phosphorus concentrations have increased on phosphorus-deficient soils, the growing accumulation in soils contributes to high levels of phosphorus runoff. As with nitrogen loading the potential consequences include eutrophication of freshwater ecosystems, which can lead to a degraded habitat for fish and decreased quality of potable water for both humans and livestock (Sharpley, 2007).

Many ecosystem services are reduced when inland waters become eutrophic. Water from lakes and reservoirs that experience algal blooms is more expensive to purify for drinking or other industrial uses. Eutrophication can reduce or eliminate fish populations. Possibly the most apparent loss in services is manifested in the degradation of many cultural services provided by lakes and reservoirs. Foul odors of rotting algae, slime covered lakes and reservoirs, and toxic chemicals produced by some blue-green algae during blooms prevent people from swimming, boating, and otherwise enjoying the esthetic value of these lakes and reservoirs (MEA, 2005).

The Earth's climate system has changed since the preindustrial era, in part because of human activities, and the change is projected to continue throughout the 21st century. During the last 100 years the global mean surface temperature has increased by 0.6°C, and precipitation patterns have changed spatially and temporally. Observed changes in climate, especially warmer regional temperatures, have already affected biological systems in many parts of the world. There have been changes in species distributions, population sizes, and the timing of reproduction or migration events, as well as an increase in the frequency of pest and disease outbreaks, especially in forested systems. The growing season in Europe has, for example, lengthened over the last 50 years (MEA, 2005).

Impact of agriculture on ecosystem services

Based on the information dealt with up to this point, it is clear that there is a relation between resource use, environmental impacts, and food system activities (Fig. 2.5). All four major sets of food system activities are dependent on natural resources, while their outcomes are related to food security components, societal factors, and environmental factors. Outcomes related to societal factors feedback to social-economic drivers and the outcomes related to environmental factors feedback to natural resources. However, for food security to be met, all three major components of food security and their respective elements must be satisfied (UNEP, 2016).

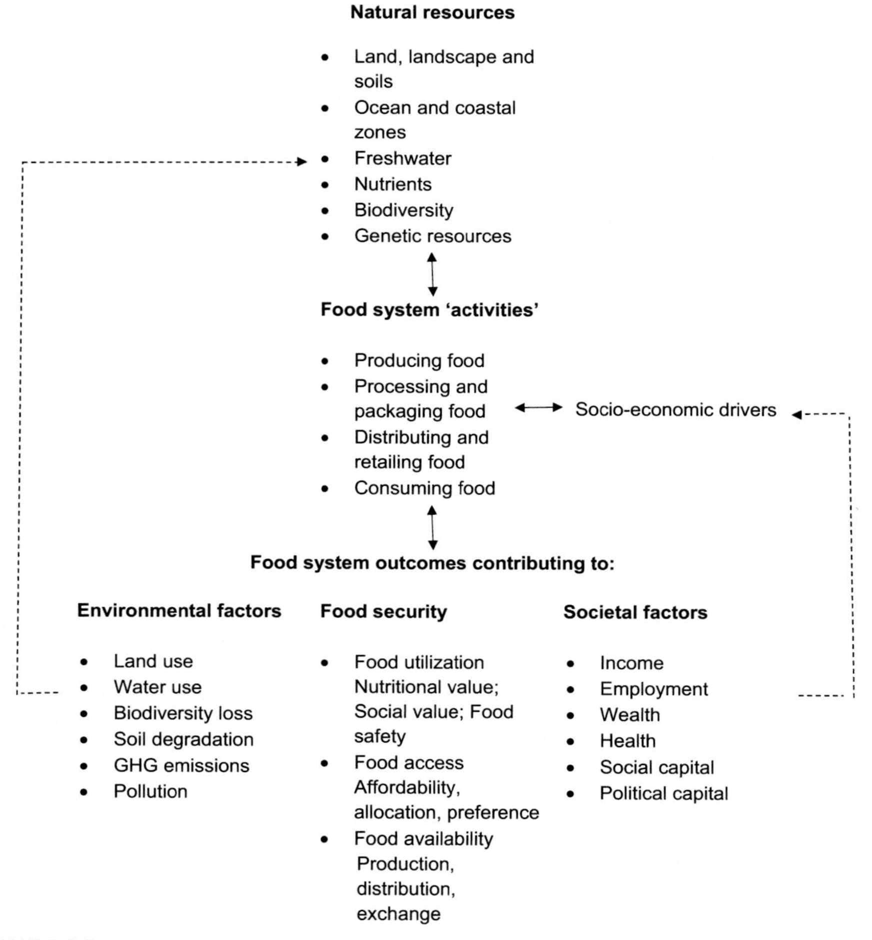

FIGURE 2.5

Relation between resource use, environmental impacts, and food system activities.

Adapted from UNEP, 2016. Food systems and natural resources. In: Westhoe, H., Ingram, J., van Berkum, S., Ozay, L., Hajer, M. (Eds.). A Report of the Working Group on Food Systems of the International Resource Panel. UNESCO, Paris.

This depiction of food systems is generic and independent of spatial scale. How it manifests in a given situation is, however, highly context dependent. Although all food systems have the same essential attributes, they vary significantly in different regions of the world and hence have different interactions with natural resources: how natural resources underpin all ecosystem activities and how these activities impact natural resources vary considerably from case to case (Rao et al., 2018).

This implies that the right questions must be addressed when the goal is the maintenance of ecosystem services at a sustainable level in agricultural landscapes (Swift et al., 2004). Several proposals were made lately on how to analyse ecosystem services in agricultural landscapes (Lee and Lautenbach, 2016; Englund et al., 2017). A strong argument is that the valuation of ecosystem services with respect to food systems will be the best prospect to provide sufficient healthy food for a growing world population (Costanza et al., 1997; Odum and Odum, 2000).

Agriculture impacts negatively on ecosystems generally (Table 2.3). These negative impacts on services of ecosystems gained momentum from the 1950s and will probably continue for the foreseeable future (Lal, 2013). Thus it is essential that ecosystem services for agriculture be used sustainably to protect natural resources and simultaneously enhance food production. However, in the long run biotechnology might reduce the need for agriculture.

Ecosystem service changes and human well-being

The structure of the world's ecosystems changed more rapidly in the second half of the 20th century than at any time in recorded human history, and virtually all of Earth's ecosystems have now been significantly transformed through human actions. The transformation of approximately 24% of Earth's terrestrial surface to cultivated systems constitutes a significant change in the structure of ecosystems. More land was converted to cropland in the 30 years after 1950 than in the 150 years between 1700 and 1850. Moreover, between 1960 and 2000, reservoir storage capacity quadrupled, and as a result the amount of water stored behind large dams is estimated to be three to six times the amount held in natural river channels. These rapid ecosystem changes now occur in developing countries, although developed countries also experienced comparable rates of change (UNEP, 2016).

Changes in ecosystem services influence all components of human well-being, including basic material needs for, inter alia, a good life and health. Even wealthy populations cannot be fully insulated from the degradation of ecosystem services. The degradation of ecosystem services influences human well-being in industrial regimes as well as wealthy populations in developing countries (MEA, 2005).

Although economic and social factors are often the primary determinants of hunger, food production remains an important factor, particularly among the rural poor. Food production is an ecosystem service in its own right, but it also depends on watershed services, pollination, pest regulation, and soil formation. Food and water are basic components for a good life and health (UNEP, 2016).

Food production needs to increase to meet the expectations of a growing human population, and at the same time the efficiency of food production (the amount produced per unit land, water and other inputs) needs to increase to reduce the harm caused to other key ecosystem services. Ecosystem condition, in particular climate, soil degradation, and water availability, influences progress toward this goal through its influence on crop yields as well as through impacts on the availability of wild sources of food (Poppy et al., 2018).

Table 2.3 Environmental impacts of agriculture

Clearing vegetation
• Regional hydrology and probably local climate altered. • Habitat damage leading to wildlife losses. • Loss of biodiversity. • Significant contribution to greenhouse effect. • Air pollution.
Tillage
• Erosion (greater than stream silt loads). • Altered groundwater (saline seeps). • Dust. • Soil structural changes. • Loss of soil organic carbon. • Altered soil flora and fauna. • Altered albedo (weather changes?).
Irrigation
• Polluted return flows (stream pollution). • Raised groundwater (waterlogging and salinity). • Soil structural changes. • Altered soil flora and fauna. • Possibility of reduced soil erosion. • Disease vector problems. • Reduced streamflow and groundwater due to extraction and evapotranspiration. • Methane generation from paddy-fields, etc. (greenhouse effect).
Altered drainage
• Pollution from land drains. • Polluted groundwater (saline seeps). • Groundwater raised (waterlogging and salinity). • Groundwater lowered (loss of soil organic carbon and subsidence). • Genetic losses as wetlands destroyed.
Agrochemicals
• Pesticides: local, regional, and global pollution possible. • Pest resistance or loss of predators. • Fertilizers: eutrophication of surface waters. • Nitrate/phosphate problems with groundwaters. • Alteration of soil structure. • Alteration of soil flora/fauna. • Greenhouse gases generated. • Agricultural lime: alteration of soil flora/fauna and changes to aquatic flora/fauna (agrochemicals may affect marine life). • Genetic losses as wildlife poisoned.

Table 2.3 Environmental impacts of agriculture—*cont'd*

Agricultural 'wastes'
• Straw/sugar cane burning: air pollution and loss of soil nutrients, especially nitrogen—much depends on when and under what environmental conditions burning occurs. • Processing crops: stream/groundwater pollution. • Livestock wastes: water pollution (genetic losses); smell; methane generation (greenhouse effect). • Silage effluent: water/air pollution. • Noise: aesthetic and wildlife damage.
Pest control/hunting
• reduces wildlife (including malaria-vector control impacts).
Pasture damage
• Over/undergrazing. • Burning to encourage sward. • Reseeding/improving pasture. • Genetic losses.
Biotechnology
• Risk of uncontrolled escape, terrorism, or military use of organisms or genetic material. • Risk of increased poverty for those who cannot adopt biotechnology leading to poor farming or use of marginal land, both leading to land degradation. • Promotion of new crops may discourage growing of traditional varieties—loss of genetic diversity. • Possibility that biotechnology might help reduce pressures on agricultural land.

Adapted from Barrow, C.J., 1995. Developing the Environment: Problems and Management. Longman, London.

Despite the growth in per capita food production over the past 4 decades, an estimated 852 million people were undernourished in 2002. Of these, nearly 95% live in developing countries. South Asia and sub-Saharan Africa, the regions with the largest numbers of undernourished people, are also the regions where growth in per capita food production has lagged most. Most notably, per capita food production has declined in sub-Saharan Africa. In this region, desertification affects the livelihoods of millions of people including a large portion of the poor in dryland ecosystems (Jones et al., 2013). Dryland ecosystems tend to have the lowest levels of human well-being.

Drylands have the lowest per capita gross domestic product and the highest infant mortality rates. Nearly 500 million people live in rural areas in arid and semi-arid lands, mostly in Asia and Africa, but also in regions of Mexico and northern Brazil. The small amount of precipitation and its high variability limit the productive potential of dryland for settled farming and nomadic pastoralism, and many ways of expanding production (such as reducing fallow periods, over grazing pasture areas, and cutting trees for fuelwood) result in environmental degradation. The combination of high variability in environmental conditions and relatively high levels of poverty leads to situations where human populations can be extremely sensitive to changes in the ecosystem (although the presence of these conditions has led to the development of very resilient land management strategies). Once rainfall in the Sahel reverted to normal low levels after 1970, following favorable rainfall from the 1950s to the mid-1960s that had attracted people to the region, an estimated 250 000 people died, along with nearly all their cattle, sheep, and goats (Stewart et al., 2006).

In other regions, however, the growth in food production and farm productivity has more than kept pace with global population growth, resulting in a significant downward pressure on the price of food. Following significant spikes in the 1970s, caused primarily by the oil crisis, there have been persistent and profound reductions in the price of foodstuffs. Over the last 40 years, food prices have dropped by around 40% in real terms because of increases in productivity. It is well established that the past increases in food production, at progressively lower unit costs, have improved the health and well-being of billions of people, particularly the most needy, who spend the largest share of their incomes on food. Increased food production has, however, not been entirely positive. Among industrial countries, and increasingly among developing countries, diet-related risks, associated mainly with over nutrition, in combination with physical inactivity, now account for one-third of the disease burden. At present, over one billion adults are overweight, with at least 300 million considered clinically obese, up from 200 million in 1995 (MEA, 2005).

The modification of rivers and lakes through the construction of dams and diversions has increased the water available for human use in many regions of the world. However, the declining per capita availability of water has negative impacts on human well-being. Water scarcity is a globally significant and accelerating condition for roughly two billion people worldwide, leading to problems with food production, human health, and economic development. Rates of water use relative to accessible supply increased from 1960 to the present at nearly 20% per decade globally, with values of 15% to more than 50% per decade for individual continents (MEA, 2005).

The degradation of ecosystem services influences human well-being in industrial regions as well as the wealthy populations in developing countries (MEA, 2005; UNEP, 2016):

- The physical, economic, or social impacts of ecosystem service degradation may cross boundaries. Land degradation and fires in poor countries, for example, have contributed to quality degradation through dust and smoke in wealthy countries.
- Degradation of ecosystem services exacerbates poverty in developing countries which can affect neighboring industrial countries by slowing regional economic growth and contributing to the outbreak of conflicts or the migration of refugees.

- Changes in ecosystems contribute to greenhouse gas emissions resulting in global climate changes which affect all countries.
- Many countries still depend directly on ecosystem services. The collapse of fisheries, for example, has harmed many communities in industrial countries. Prospects for forest, agriculture, fishing, and ecotourism industries are all directly tied to ecosystem services.
- Wealthy populations are insulated from the harmful effects of some aspects of ecosystem degradation, but not all. For example, substitutes are typically not available when cultural services are lost.

Projected changes in ecosystems

Projected changes in ecosystems and the services they provided from 2000 to 2050 and in some instances 2100 are as follows (MEA, 2005):

- The global population will probably grow to between 8.1 and 9.6 billion in 2050. This population growth is expected to be concentrated in the poorest, urban communities in sub-Saharan Africa, South Asia, and the Middle East.
- *Per capita* income is projected to increase two-fold to four-fold. Increasing income leads to increasing per capita consumption in most parts of the world for most resources and changes the structure of consumption. For example, diets tend to be higher in animal protein as income rises.
- Land-use change due primarily to agricultural expansion is projected to continue as a major direct driver of change in terrestrial and freshwater ecosystems. At the global level and across all scenarios, land-use change is projected to remain the dominant driver of biodiversity change in terrestrial ecosystems, consistent with the pattern over the past 50 years, followed by changes in climate and nitrogen deposition. However, other drivers may be more important than land-use change in particular biomes. For example, climate change is likely to be the dominant driver of biodiversity in tundras and deserts. Species invasions and water extractions are important drivers for freshwater ecosystems.
- Nutrient loading is expected to become an increasingly severe problem, particularly in developing countries. The loading of freshwater ecosystems with nutrients already has major adverse effects. The presence of nitrogen in river water will not change in most industrial countries, while a 20%–30% increase is projected for developing countries, particularly in Asia.
- Climate change and its impacts are projected to have an increasing effect on biodiversity and ecosystem services. The global temperature is expected to increase by 1.5–2.0°C above the preindustrial level by 2050 and by 2.0–3.5°C above it by 2100. Climate change will alter ecosystem services, for example, by causing changes in productivity and the growing zones of cultivated and conciliated vegetation. It is also projected to change the frequency of extreme events, with associated risks to ecosystem services. Agricultural productivity is

projected to decrease in the tropics and subtropics. By the end of the century, climate change and its impacts may be the dominant direct driver of biodiversity loss and the change in ecosystem services globally.

Conclusions

Since the 1950s, humans have changed ecosystems and the services they provide more rapidly and extensively than in any comparable period of time in human history, largely to meet rapidly growing demands for food, fresh water, timber, fiber, and fuel. This has resulted in substantial and largely irreversible loss in the diversity of life on Earth.

The changes made to ecosystems have contributed to substantial net gains in human well-being and economic development, but these gains have been achieved at growing costs in the form of the degradation of many ecosystem services, increased risks of nonlinear changes, and the exacerbation of poverty for some groups of people. These problems, unless addressed, will substantially reduce the benefits that future generations obtain from ecosystems.

The degradation of ecosystem services could worsen significantly during the first half of this century. This will create a barrier to the achievement of a better life and health for humans globally.

The challenge of reversing the degradation of ecosystems while meeting increasing demands for their services can be partially met under some scenarios that have been considered, but these involve significant changes in policies, institutions, and practices that are not currently in place. Many options do, however, exist to conserve or enhance specific ecosystem services in ways that reduce negative trade-offs or that provide positive synergies with other ecosystem services. These options are dealt with in some of the other chapters.

References

Barrow, C.J., 1995. Developing the Environment: Problems and Management. Longman, London.

Bommarco, R., Vico, G., Hallin, S., 2018. Exploiting ecosystem services in agriculture for increased food security. Global Food Security 17, 57–63.

Costanza, R., D'Arge, R., De Groot, R., Farber, S., Grasso, M., Hannon, B., Limburg, K., Naeem, S., O'Neil, R.V., Parnelo, J., Raskin, R.G., Sutton, P., Van den Belt, M., 1997. The value of the world's ecosystem services and natural capital. Nature 287, 253–260.

Englund, O., Berndes, G., Cederberg, C., 2017. How to analyse ecosystem services in landscapes- a systematic review. Ecological Indicators 73, 492–504.

FAO, 2011a. Payments for Ecosystem Services and Food Security. FAO, Rome.

FAO, 2011b. The State of World's Land and Water Resources for Food and Agriculture: Managing Systems at Risk. FAO, Rome.

Follett, J.R., Follett, R.F., 2001. Utilization and metabolism of nitrogen by humans. In: Follett, R.F., Hatfield, J.L. (Eds.), Nitrogen in the Environment: Sources, Problems, and Management. Elsevier, Amsterdam.

Garibaldi, L.A., Anderson, G.K.S., Requier, F., Fijen, T.P.M., Hipólito, J., Kleijn, D., Perez-Mendez, N., Rollin, O., 2018. Complementarity and synergisms among ecosystem services supporting crop yield. Global Food Security 17, 38–47.

Haygarth, P.M., Bardgett, R.D., Condron, L.M., 2013. Nitrogen and phosphorus cycles and their management. In: Gregory, P.J., Nortcliff, S. (Eds.), Soil Conditions and Plant Growth. Wiley-Blackwell, Oxford.

Havlin, J.L., Tisdale, S.L., Nelson, W.L., Beaton, J.D., 2014. Soil Fertility and Fertilizers: An Introduction to Nutrient Management, eighth ed. Pearson, Upper Saddle River, New Jersey.

Holt, A.R., Alix, A., Thompson, A., Malthy, L., 2016. Food production, ecosystem services and biodiversity: we can't have it all everywhere. The Science of the Total Environment 573, 1422–1429.

Hopkins, D.W., Gregorich, E.G., 2013. Managing the soil-plant system for the delivering of ecosystem services. In: Gregory, P.J., Nortcliff, S. (Eds.), Soil Conditions and Plant Growth. Wiley- Blackwell, Oxford.

Jones, A., Breuning-Madsen, H., Brossard, M., Dampha, A., Deckers, J., Dewitte, O., Gallali, T., Hallett, S., Jones, R., Kilasara, M., Le Roux, P., Micheli, E., Montarella, L., Spaargaren, O., Thiombiano, L., Van Ranst, E., Gemefack, M., Zougmore, R. (Eds.), 2013. Soil Atlas of Africa. European Commission Publication Office of the European Union, Luxembourg.

Karabulut, A.A., Crenna, E., Sala, S., Udias, A., 2018. A proposal for integration of the ecosystem-water-food-land energy (EWFLE) nexus concept into life cycle assessment: a synthesis matrix system for food security. Journal of Cleaner Production 172, 3874–3889.

Keeney, D.R., Hatfield, J.L., 2001. The nitrogen cycle, historical perspective, and current and potential future concerns. In: Follett, R.F., Hatfield, J.L. (Eds.), Nitrogen in the Environment: Sources, Problems and Management. Elsevier, Amsterdam.

Keesstra, S.D., Bouma, J., Wallinga, J., Tittonell, P., Smith, P., Cerdá, A., Montanarella, L., Quiton, J.N., Pachepsky, Y., Van der Putten, W.H., Bardgett, R.D., Moolenaar, S., Mol, G., Jansen, B., Fresco, L.O., 2016. The significance of soils and soil science towards realization of the United Nations sustainable development goals. Soils 2, 111–128.

Lal, R., 2013. Enhancing ecosystem services with no-till. Renewable Agriculture and Food Systems 28, 102–114.

La Notte, A., D'Amato, D., Mäkinen, H., Paracchini, M.L., Liquete, C., Egoh, B., Geneletti, D., Crossman, N.D., 2017. Ecosystem services classification: a systems ecology perspective of the cascade framework. Ecological Indicators 74, 392–402.

Lee, H., Lautenbach, S., 2016. A quantitative review of relationships between ecosystem services. Ecological Indicators 66, 340–351.

Liere, H., Jha, S., Philpott, S.M., 2017. Intersection between biodiversity conservation, agroecology, and ecosystem services. Agroecology and Sustainable Food Systems 41, 723–760.

MEA, 2005. Ecosystems and Human Well-Being: Synthesis. Island Press, Washington, DC.

Odum, H.T., Odum, E.P., 2000. The energetic basis for valuation of ecosystem services. Ecosystems 3, 21–31.

Orgiazzi, A., Bardgett, R.D., Barrios, E., Behan-Pelletier, V., Briones, M.J.I., Chotte, J.-L., De Deyn, G.B., Eggleton, P., Fierer, N., Fraser, T., Hedlund, K., Jeffery, S., Johnson, N.C., Jones, A., Kandeler, E., Kaneko, N., Lavelle, P., Lemanceau, P., Miko, L., Montanarella, L., Moreira, F.M.S., Ramirez, K.S., Scheu, S., Singh, B.K., Six, J., Van der Putten, W.H., Wall, D.H. (Eds.), 2016. Global Soil Biodiversity Atlas. European Commission Publication Office of the European Union, Luxembourg.

Poppy, G.M., Chiotha, S., Eigenbrod, F., Harvey, C.A., Honzák, M., Hudson, M.D., Jarvis, A., Madise, N.J., Schreckenberg, K., Shackleton, C.M., Villa, F., Dawson, T.P., 2018. Food security in a perfect storm: using the ecosystem services framework to increase understanding. Philosophical Transactions of the Royal Society B 369, 20120288.

Rao, Y., Zhou, M., Ou, G., Dai, D., Zhang, Z., Nie, X., Yang, C., 2018. Integrating ecosystem services for sustainable land-use management in semi-arid region. Journal of Cleaner Production 186, 662—672.

Sharpley, A.N., 2007. Modelling phosphorus movement from agriculture to surface waters. In: Radcliffe, D.E., Cabrera, M.L. (Eds.), Modelling Phosphorus in the Environment. CRC Press, Boca Raton.

Stephens, E.C., Jones, A.D., Parsons, D., 2018. Agricultural systems research and global food security in the 21st century: an overview and roadmap for future opportunities. Agricultural Systems 163, 1—6.

Stevenson, F.J., Cole, M.A., 1999. Cycles of Soil: Carbon, Nitrogen, Phosphorus, Sulfur, Micronutrients, second ed. John Wiley & Sons, New York.

Stewart, B.A., Koohafkan, P., Ramamoorthy, K., 2006. Dryland agriculture defined and its importance to the world. In: Peterson, G.A., Unger, P.W., Payne, W.A. (Eds.), Dryland Agriculture, second ed. American Society of Agronomy, Madison, Wisconsin.

Swift, M.J., Izac, A.-M.N., Van Noordwijk, M., 2004. Biodiversity and ecosystem services in agricultural landscapes — are we asking the right questions? Agriculture, Ecosystems and Environment 104, 113—134.

UNEP, 2016. Food systems and natural resources. In: Westhoe, H., Ingram, J., van Berkum, S., Ozay, L., Hajer, M. (Eds.), A Report of the Working Group on Food Systems of the International Resource Panel. UNESCO, Paris.

Van As, J., Du Preez, J., Brown, L., Smit, N., 2012. The Story of Live and the Environment: An Africa Perspective. Struik Nature, Cape Town.

Verhoef, A., Egea, G., 2013. Soil water and its management. In: Gregory, P.J., Nortcliff, S. (Eds.), Soil Conditions and Plant Growth. Wiley-Blackwell, Oxford.

Vitousek, P.M., Naylor, R., Crews, T., David, M.B., Drinkwater, L.E., Hollard, E., Johnes, P.J., Katzenberger, J., Martinelli, L.A., Matson, P.A., Nziguhela, G., Ogima, D., Palm, C.A., Robertson, G.P., Sanchez, P.A., Townsend, A.R., Zhang, F.S., 2009. Nutrient imbalances in agricultural development. Science 324, 1519—1520.

Weil, R.R., Brady, N.C., 2017. The Nature and Properties of Soils, fifteenth ed. Pearson Prentice Hall, Upper Saddle River, New Jersey.

CHAPTER

Indices to identify and quantify ecosystem services in sustainable food systems

3

Paramu L. Mafongoya, BSc, MSc, PhD[1], Gudeta W. Sileshi, BSc, MSc, PhD[2,3]

[1]*Professor, School of Agricultural, Earth and Environmental Sciences, University of KwaZulu-Natal, Pietermaritzburg, KwaZulu-Natal, South Africa;* [2]*Plot 1244 Ibex Meanwood, Lusaka, Zambia;* [3]*Honorary Research Fellow, School of Agricultural, Earth and Environmental Sciences, University of KwaZulu-Natal, Pietermaritzburg, KwaZulu-Natal, South Africa*

Introduction

There is increasing demand for nutritious, safe, and healthy food to satisfy the increasing global population, which is expected to reach 9 billion people by 2050. Against this demand, there is also a need to maintain biodiversity and sustainable use of natural resources such as water, soils, forest, and genetic resources. Hence food systems need to be designed to improve human health and contribute to other multiple benefits such as healthy ecosystems. Food systems which provide dietary diversity have social, cultural, economic, and environmental benefits (Sukhdev et al., 2016; Hunter et al., 2016; Nair et al., 2016). According to Sukhdev et al. (2016) the ideal sustainable food systems (SFSs) has the following attributes: (1) offers adequate nutrition and health; (2) creates biodiversity and avoids negative ecological and environmental impact; (3) ensures livelihoods for farmers; and (4) ensures equitable access to land, water, seeds, and other inputs. A food system cannot be considered sustainable unless its underlying resource base is sustained, and it has neutral or positive impacts on important ecosystem services (ESs) needed outside food systems (Gustafson et al., 2016).

Adequate metrics and indicators are needed to track progress toward achieving these targets and inform management and policy. Likewise the commitment to shaping a more sustainable, inclusive, and equitable food supply requires the development of adequate metrics, indicators, and indices to better understand the impact of food systems on the environment, broadly defined as land, water, and air, as well as biological ecosystems (Gustafson et al., 2016). Therefore the objective of this chapter is to outline the indicators and indices used to identify and quantify ESs and to provide a comprehensive list that covers most of SFSs options.

The Role of Ecosystem Services in Sustainable Food Systems. https://doi.org/10.1016/B978-0-12-816436-5.00003-2

Methods, definitions, and circumscription

A review of the literature was undertaken to identify metrics, indicators, and indices of ESs in SFSs. Journal articles, book chapters, and scientific reports were identified through a comprehensive literature search of published and unpublished work on ES in SFS. In the following synthesis, we follow the Common International Classification of Ecosystem Services (CICES) developed from the work on environmental accounting undertaken by the European Environment Agency (Haines-Young and Potschin, 2017). Accordingly, we identify three categories of ES, namely, (1) provisioning services, (2) regulation and maintenance services, and (3) cultural services in SFS.

In the literature the terms metrics, indicators, and indices are often used interchangeably and without qualification. For clarity, we make distinctions between these terms throughout this chapter. We define a metric as a specific property of a farming system, cropping system, a biotic community, or household that can be directly measured. The term "indicator" is frequently used at the interface between science and policy (Heink and Kowarik, 2010). We define an indicator as a sign of the presence or absence of the concept being measured. Indicators are often used to describe, represent, monitor, assess, or model complex world processes, components, or properties, both qualitatively and quantitatively, to be used in decision-making processes. For such purposes, several complementary and meaningful metrics are needed. As such, indicators can have numerous metrics associated with them.

An index is defined here as a scaled composite quantity that aggregates multiple indicators to allow comparisons of states of the environment across time or space without ambiguity. Indices are derived through conversion of the data from their original units to normalized (or standardized) units and then aggregating the results (Ebert and Welsch, 2004). There are many well-established indices, which provide a condensed description of multidimensional environmental or social states by aggregating several indicators into a single quantity. Unlike indicators, indices aim to describe general properties of systems that allow users to compare different regions, taxa, or trophic levels. Therefore they are of fundamental importance for monitoring food supply, conservation of biodiversity, and environmental integrity. Different types of indices are also needed to aid different groups of decision-makers, including national governments, local governments, private actors, farmers, etc., each operating at different scales and resolutions. From a user perspective, indices help in measuring progress on intractable challenges. Therefore indices need to be informative, sensitive to relevant changes, actionable, and inspire communications with other end users such as policy-makers, farmers, and consumers (Remans et al., 2017).

We found a number of key publications (e.g., Chaudhary et al., 2018; Gustafson et al., 2016) and specific studies that have tried to identify indicators of SFSs. Many indicators and indices have been developed and applied to quantify ES. Our interest in this chapter is on key ES in SFS options. As such we will focus only on indicators and indices of ES envisioned as relevant to decision-making in SFS. A comprehensive discussion of the possible set of indicators and indices of ES is beyond the scope of this chapter.

Synthesis

There is a growing evidence suggesting that the current global food system is inadequate to meet the food and nutritional needs of the world population without compromising future well-being (Meybeck and Gitz, 2017). Anthropogenic changes in land used for agriculture and the global food system is now one of the major drivers of global biodiversity loss (Ceballos et al., 2015; Chaudhary et al., 2018). Food systems are now the source of 60% of terrestrial biodiversity loss, 24% of greenhouse gas emissions, 33% of soil degradation, and 61% of the depletion of commercial fish stocks (Sukhdev et al., 2016). The current global food system is also characterized by huge inefficiencies and inequities; an estimated 30%−50% of all food produced is either lost or wasted along the value-chain (FAO, 2013). On the other hand, there is a growing consensus and a call for a paradigm shift from industrial agriculture to more diversified and SFSs that can deliver various ESs (FAO, 2015; Global Panel on Agriculture and Food Systems for Nutrition, 2016; IPES-Food , 2016; UN, 2017). In the following sections we will provide a synthesis of indicators and indices to identify and quantify ESs that may accrue because of the transition to SFSs. To ensure consistency and comparability we will strictly follow the CICES scheme of classification, but for brevity we will focus on the groups and classes of ES.

Indicators and indices of provisioning services

Provisioning services (Table 3.1) cover all nutritional, nonnutritional material, and energetic outputs from SFS. As such they included genetic materials, food, medicine, feed, fiber, energy, and fresh water. In this synthesis genetic materials refers to seeds, spores, and other plant or animal materials collected for the establishment or maintenance of new stands or population of plants or animals, the use of plants and animals at the whole organism level for breeding purposes, and gene extraction (Haines-Young and Potschin, 2017) in SFS. It also refers to individual genes extracted from plants or animals for the design and construction of new biological entities. However, our focus will be on human food and animal feed from cultivated terrestrial plants, as well as plants cultivated by in-situ aquaculture (including higher plants, fungi, algae), domestic animals, and animals reared by in-situ aquaculture, apiculture, and sericulture. Provisioning services also cover fibers and energy sources from cultivated plants, fungi, algae, bacteria, and reared animals or animals produced by in-situ aquaculture (Haines-Young and Potschin, 2017), apiculture, and sericulture. In the following sections we briefly describe these services and provide a summary of indicators and indices (where available) useful for comparing SFS options.

Provision of genetic materials

Provision of genetic materials is one of the key ESs in SFSs. Out of an estimated total of 300,000 flowering plant species, 10,000 species have been used for human food since the origin of agriculture (FAO, 2010). Out of these, only 150 to 200 species have been commercially cultivated with rice, wheat, maize, and potato supplying 50% of the world's food energy needs and 30 crops providing 90% of the world's food energy intake. A number of food crops are also used for medicinal purposes, production of cosmetic products, fibers, and bioenergy. Intensification of agricultural systems has led to a substantial reduction in the genetic diversity of domesticated plants. This is the case with large farms (>50 ha) focusing on monocultures of one or a few food crops. According to recent analyses (e.g., Herrero et al., 2017) most of the commodities produced on small farms come from diverse landscapes, producing a variety of horticulture, roots, tubers, fish, and livestock. Such production systems are expected to maintain a diversity of plant and animal genetic materials useful for multiplication, breeding, and extraction of genes for the design and construction of new biological entities to improve the quantity of food, dietary diversity, or nutritional value.

Key indicators of genetic materials mentioned in the literature include yields and area under crops used for food, medicinal, and cosmetic purposes (Table 3.1). Crop diversity has often been applied as a landscape indicator at the regional level. However, the use of crop diversity as a proxy for biodiversity at the farm level can be motivated by the fact that the number of different habitats is likely to increase with crop diversity. In conventional farming a monoculture may be successful, whereas a diversified production system may provide higher diversity of genetic materials for future generations. Typically the focus is on species diversity, but other forms of diversity, such as genetic and chemical diversity, are also important and informative in the context of SFS. A number of indices have been used in assessing diversity, but classic diversity indices include Simpson's D and E and Shannon's H and E indices. These measures both the number of observed species and the evenness of species' distribution. Shannon's diversity (H) and Simpson's diversity (D1) indices differ in their theoretical foundation and interpretation (Magurran, 2004). Shannon diversity index is sometimes referred to as Shannon-Wiener or Shannon-Weaver) index.

Shannon's H′ index has its foundations in information theory and represents the uncertainty about the identity of an unknown individual. It is calculated using the following formula:

$$H = -\sum_{i=1}^{N}(P_i \times ln(P_i)) \qquad 3.1$$

Table 1 List of indicators of provisioning services in sustainable food systems

Group/ class	Indicators	Indices
Genetic materials	- Quantity of seeds of local varieties - Yields of crops used for medicinal and cosmetic purposes (ton/ha) - Area of crops used for medicinal and cosmetic purposes (ha) - Diversity of seed sources - Diversity of suitable habitats for local varieties	- Shannon diversity index - Habitat suitability index - Index of biotic integrity
Food provision	- Yields of food crops (t/ha) - Food supply (per capita calories/day) - Milk, meat, eggs, honey - Livestock biomass (ton/ha/year) - Dietary diversity	- Crop production index - Shannon diversity index - Livestock production index - Consumer price index - Food sustainability index - Global hunger index
Feed provision	- Yields of feed crops (t/ha) - Area of feed crop (ha) - Grazing/grassland area (ha) - Volume of agro-industry by-products	NA
Fibres and other materials	- Yields of fibre crops (ton/ha) - Fibre crop area (ha) - Mass of crop by-products	NA
Energy	- Yields of crops (t/ha) - Energy crop area (ha) - Biofuel, biodiesel, bioethanol (EJ/Yr) - Energy from manure (EJ/Yr)	NA

NA = Not available
EJ = exajoule (10^{18} joules)

where N is the number of species, P_i is the proportion of species i in the sample, and ln is the natural logarithm. H equals zero when there is only one species, indicating no diversity, but the value of H increases with the number of species, and it reaches its maximum when the species occur in equal shares. A value near 4.6 would indicate that the number of individuals is evenly distributed between all the species. Shannon's equitability (E_H) can be calculated as $H/H_{\max}$, where $H_{\max} = \ln(S)$, and S is the total number of species. Equitability assumes a value between 0 and 1, with one being complete evenness. In this synthesis the H index is proposed to approximate the diversity of genetic materials maintained in SFS options.

Historically the Shannon index has also been used to measure the effects of habitat quality such as effects of pollution, land use change, etc. Recently this index has fallen out of favor because it doesn't take into account habitat-specific parameters required by specific species. Newer indices have incorporated such parameters; one such index is the habitat suitability index (HSI), which is a process-based approach meant to characterize the carrying capacity of a habitat for one particular species. On a species-by-species basis HSI relates habitat quality to observable biophysical characteristics derived from the scientific literature. For example, tree density or canopy cover may be related to a habitat's species-specific carrying capacity. For aquatic species, indicators of things such as dissolved oxygen, salinity, water depth, substrate type, and toxicity have been combined in an HSI score. With a lower demand for data compared with species diversity models, HSI models provide flexible cost-effective decision support tools for natural resource management, and ecosystem restoration.

Another useful index with potential application to genetic materials in SFS is the index of biotic integrity (IBI). Biotic integrity is defined as the ability to support and maintain a balanced, integrated, adaptive community of organisms having a species composition, diversity, and functional organization comparable to that of the natural habitat of the region. The IBI is most commonly employed indicator system in government decision-making. It has been widely used to assess aquatic resources, and the method has spawned a variety of derivative approaches geared toward terrestrial systems (Canterbury et al., 2000). The system is composed of 12 indicators, six relating to the composition and richness of species and six relating to "ecological factors" (Karr, 1991).

Provision of food

Provision of food is one of the main ESs of food systems, and it is an important goal in food security of households, regions, and nations. However, food security is an inherently multidimensional concept, which is highly debated. Availability, access, utilization, and stability are generally recognized as the dimensions to include and analyse in any index of food security. Availability refers to the amount of food that is present in a country or area through all forms of domestic production, imports, food stocks, and food aid (WFP, 2009). Thus availability represents the first necessary element to achieve food security. Availability tends to be applied to food available at a regional or national level rather than at the household level. On the other hand, food access is defined as a household's ability to access adequate amounts of food regularly through a combination of purchases, barter, borrowings, food assistance, or gifts (WFP, 2009). Food access itself consists of three elements, namely physical, economic, and sociocultural. The physical dimension can be illustrated by a situation where food is being produced in one part of a country, but an inefficient or nonexistent transport infrastructure means that food cannot be delivered to another part suffering from a lack of food.

Utilization refers to safe and nutritious food which meets the dietary needs of consumers. Utilization therefore covers a range of aspects that hinge on the consumer's understanding of what foods to select and how to prepare and store them. According to the World Food Summit stability must be present "at all times" in terms of availability, access and utilization for food security to exist.

Food security is difficult to measure and monitor because the underlying data are fraught with uncertainties due to large variations in diets and biophysical conditions. Therefore manifestations of food insecurity that can be observed and verified are often used as proxy indicators than monitoring food security itself (Kline et al., 2017). Food insecurity exists when people suffer or are at risk of suffering from inadequate consumption to meet nutritional requirements. The literature distinguishes between chronic (long term), acute (transitory), cyclical, or critical food insecurity (Kline et al., 2017). Food insecurity is typically measured via multiple indicators of malnutrition.

Sustainable diets and SFSs are recognized by the international community and called upon to orient action toward the eradication of hunger and malnutrition and the fulfilment of sustainable development goals (Meybeck and Gitz, 2017). Therefore the measurement and tracking of progress is crucial to transitioning to a sustainable and nutrition-sensitive food system, and this requires identification of appropriate indicators and indices. Among the indicators of food provision mentioned in the literature yield is by far the most common one especially in the sustainable intensification literature (Smith et al., 2015). In cropping systems yield refers to the production of crops per unit of land area. This can be measured either in kg ha^{-1} grain (Rai et al., 2011) or as value of crop produced ($/ha). In livestock systems, yield can be measured as the production of animal products (milk, meat, or eggs) per livestock animal per day or the production of milk per animal per lactation period (Descheemaeker et al., 2011). Yield variability (e.g., coefficient of variation) can also be employed as an indicator of economic and productive sustainability. While yield variability shares many similarities with risk and resilience, it is distinguished by its concrete, quantitative definition. Other indicators included food supply (per capita calories/day), dietary diversity, and GDP (Gross domestic product) per capita (Table 3.1).

Because no single indicator (such as yield) can account for the many dimensions of food and nutrition security, recent efforts to measure food and nutrition security have progressively led to the development of a variety of different indicators. The Voices of the Hungry project of FAO began to develop a global indicator that could be implemented by all member countries (Cafiero et al., 2018). Currently the Voices of the Hungry measures food insecurity worldwide, using an experience-based food insecurity scale module called the food insecurity experience scale (FIES), which is used as a common metric for measuring food insecurity at several levels of severity, across different geographic areas and cultures. FIES is a robust and cost-effective indicator to measure people's access to food (Cafiero et al., 2018). As such FIES has been endorsed by the UN as the official indicator to monitor SGD2 target 2.1.

Among the indices commonly mentioned in the food security literature are the crop production index (CPI), the livestock production index (LPI), and the Crop production index (CPPI). We believe these indices may be used to characterize provision of food by a given SFS option. The CPI is calculated as agricultural production for each year normalized to the base period 2004–2006. CPI is based on the sum of price-weighted quantities of different agricultural commodities produced after deductions of quantities used as seed and feed weighted in a similar manner. The LPI is calculated as in CPI, but this includes meat and milk from all sources, dairy products such as cheese, eggs, honey, raw silk, wool, hides and skins. The resulting aggregate represents, therefore, disposable production for any use except as seed and feed. The agricultural production indices are prepared by the Food and Agriculture Organization of the United Nations (FAO). Since the FAO indices are based on the concept of agriculture as a single enterprise, amounts of seed and feed are subtracted from the production data to avoid double counting.

The CPI is a complex measure of the average change over time in the prices of consumer items—goods and services that people buy for day-to-day living. It combines economic theory with sampling and other statistical techniques and uses data from several surveys to produce a timely and precise measure of average price change.

Recently, diversity measures have been proposed to offer a new way to monitor nutritional diversity in the design and evaluation of SFS policies that better meet the nutritional needs of healthy populations (Herrero et al., 2017; Remans et al., 2014). Among the indices proposed for this purpose are the modified functional attribute diversity and Shannon index (H). Functional diversity reflects the diversity in nutrients provided by the different food items based on the nutritional composition, and amount of each food item present (Herrero et al., 2017; Remans et al., 2014) applied a modified version of the functional attribute index (MFAD) calculated as follows:

$$MFAD = \frac{\sum_{i=1}^{n}\sum_{j=1}^{n} d_{ij}}{N} \qquad 3.2$$

where N is the number of food items and d is the dissimilarity between food items i and j as defined by multiple traits—or nutritional components—measured using some distance algorithm, such as Euclidean distance. N is the number of functional units, such that different items that are identical in their trait composition are considered the same functional unit. For example, if there are two food items with the same nutritional composition, then they are not counted twice (Remans et al., 2014).

The Shannon Index (Eq. 3.1) is a measure of food item diversity in the national food production or supply. This index weights the richness of food items by the evenness of their distribution. As such it is a measure of the relative abundance of each food item within a country or a specific area. For example, Herrero et al. (2017) calculated the Shannon diversity index to represent how many different types of foods are produced in a pixel and how evenly these different types are distributed. The limitation of this index is that it measures dietary diversity without explicit consideration of their nutrients.

More recently additional indices have been developed (or under development) for use at the global level. These include the global hunger index (GHI) and the food sustainability index (FSI). The GHI regards hunger as multidimensional problem and uses three indicators: (1) undernourishment (percentage of population with insufficient caloric intake), (2) proportion of underweight children under the age of five years, and (3) mortality rate in children under the age of five years (von Grebmer et al., 2014). An average of the 3% rates is taken, and countries are then classified in the index as serious, alarming, or extremely alarming. The index's great strength is its inclusion of three different aspects of hunger; however, because they are so closely related, distortion due to double-counting can arise. Another disadvantage is that the index fails to pick up changes to outcome distribution and reacts poorly to short-term food and health shocks. The index's strong points are that the data on the whole are reliable, it can be applied to any country, and is useful for comparing different countries. The index also provides a useful accountability instrument when dealing with governments. IFPRI (International Food Policy Research Institute) has published an annual GHI since 2006, and it was set up to assess hunger globally, monitor the progress of the MDGs (Millennium Development Goals), and interpret trends within causal models (Wiesmann, 2006).

The FSI, developed by the Economist Intelligence Unit (2017) with the Barilla Center for Food & Nutrition, measures the sustainability of food systems. The three primary categories in the index—food loss and waste, sustainable agriculture, and nutritional challenges—were defined in the Milan Protocol. The index contains 35 indicators and over 55 subindicators, organized across these three categories. Each category receives a score, calculated from a weighted mean of the underlying indicator scores, and scores are scaled from 0 to 100, where 100 is the highest sustainability and greatest progress toward meeting environmental, societal, and economic objectives (The Economist Intelligence Unit, 2017).

Provision of animal feed

One of the biggest challenges in providing food for a growing population is feeding the animals needed to satisfy the increasing demand for meat, dairy, and fish products. Grazing systems in semiarid areas offer only limited potential for intensification, and livestock production is becoming increasingly crop-based. As a result, a large proportion of animal feed comes from food systems. For example, about a third of global cereal production is fed to animals (Schader et al., 2015).

The tremendous role of crop residues in livestock production has been widely documented in the literature. The common crop residues used in feeding animals include millet and maize stover, rice straw and groundnut hay, husk, and pods of common pulses. In addition, fodder trees grown in agroforestry arrangement (Chakeredza et al., 2007), concentrate feed derived from human food (Schader et al., 2015; Wadhwa and Bakshi, 2013), food waste, and by-products of food production–consumption chains such as brans, whey, and oil-cakes (Ajila et al., 2012) have also been widely used as animal feed. Because of their high protein content, fodder trees produced in agroforestry systems can be used alone or as additives in conventional animal feeds (Chakeredza et al., 2007, 2008). Agroindustrial wastes used as animal feed are of various types and can be classified into different categories, such as crop residues, by-products from fruit and vegetable processing industry, sugar industry, starch and confectionary industry, grain and legume milling, oil industry, breweries, and distilleries (Ajila et al., 2012). By-products of the milling industry include bran, waste flour, wastes resulting from grain-cleaning processes, wheat, corn and rye germs, and hulls of some seeds. The main by-products of the oil industry include soybean cake, and products formed during refining of oil from rape, sunflower, flax, and groundnut. By-products of the starch industry include beet pulp, molasses, deface-saturation residues, potato pulp, potato cell juice, and other residues of seeds after starch extraction. The major wastes from the sugarcane industry are bagasse, molasses, and sugarcane press mud. The key by-products from the winemaking industry and distillery industries are grape pomace and yeast sludge, while the spent grain is the most abundant brewery by-product. As such the agricultural and food-industry residues constitute almost 30% of worldwide agricultural production (Ajila et al., 2012), and therefore provision of animal feed is one of the key ESs provided by food systems.

Different indicators may be used to identify and quantify the ESs of providing animal feed in SFSs (Table 3.1). Yields of agricultural by-products may be determined by estimating the amount of crop which was processed in a country or a region. Country-specific data for amounts of concentrate feed and by-products may also be derived from FAOSTAT food balance sheets.

Provision of fiber

Fiber crops are plants that are deliberately grown for the production of fiber for textile (clothes), cordage (e.g., ropes), and filling (e.g., stuffing upholstery and mattresses). Cotton, jute, kenaf, industrial hemp, sun hemp, and flax are among the well-known fiber crops. Some of these also have a promising future as agricultural biomass that can be converted to ethanol. In addition, agricultural waste also produces large amounts of biomass classified as natural fibers, which are used for building materials, as a decorative product and as a versatile raw product. Lignocellulosic agricultural by-products such as stalks, stems, straws, hulls, and cobs are a cheap source for cellulose fibers Bouf (2017). These can be used as a starting material to produce highly valuable cellulose nanofibrils, renewable fibers which have low density and are biodegradable. This new class of products have been receiving increased attention because of their potential use in nanocomposites, papermaking, packaging, biomedicine, and automotive parts (Bouf, 2017; Nechyporchuk et al., 2016). As such provision of fiber is likely to be one of the key ESs in SFSs.

Provision of energy

Energy produced from organic nonfossil material of biological origin (referred to as bioenergy) is promoted as a substitute for nonrenewable (e.g., fossil) energy to reduce greenhouse gas emissions and dependency on energy imports (Haberl et al., 2010). According to recent estimates, global bioenergy uses at present amounts to 50 EJ yr^{-1} or about 10% of humanity's primary energy supply (Haberl et al., 2010). This consists of solid, liquid, and gaseous fuels from primary (harvest), as well as secondary and tertiary sources (waste and residue utilization). The future bioenergy potentials range is 30–1000 EJ yr^{-1} for 2050 (Erb et al., 2012). The categories of biomass include (1) conventional crops for nonfood use such as starch crops (maize, wheat, and barley), oil crops (rapeseed and sunflower), and sugar crops (sugar beet and sweet sorghum); (2) dedicated crops, involving short rotation forestry and herbaceous (grasses); (3) forestry by-products, including logging residues and thinnings; (4) agricultural by-products such as straw and animal manure; (5) industrial by-products, including residues from food-based and wood-based industries; and (6) biomass waste, including demolition wood waste, sewage sludge, and organic fraction of municipal solid waste (Cigolotti, 2012).

In this review our focus is on those sources that are directly linked to food systems, namely, agriculture, agroforestry, and the food industry. The current energy potentials from agroforestry, agricultural residues, manures, and other wastes is substantial (Haberl et al., 2010). As food production expands to feed growing populations, it will induce more organic residues, both on the field and in processing, which can be used in bioenergy generation (Erb et al., 2012). While a large portion of crop residues is typically used for soil management, as animal feed and bedding, the remainder along with nearly all processing residues can be removed for bioenergy production. For example, the technical potential in 2050 is estimated at 15–70 EJ yr^{-1} from agricultural residues, 5–55 EJ yr^{-1} from dung, and 5–50 EJ yr^{-1} from organic wastes (Cigolotti, 2012).

At the household level, biogas can be cheaply produced from agricultural residues and liquid manure. In recent years increasing awareness that anaerobic digesters can help control waste odor and disposal has stimulated renewed interest in the development of technology. The application of anaerobic digesters in the treatment of animal manure is a very promising option for a sustainable management of manure, which is a source of greenhouse gases. SFSs should integrate such technology and produce energy, while also mitigating greenhouse gas emissions.

Two types of residues are associated with crop production, namely, field (primary) and processing (secondary) residues. Recoverable energy potentials of both types of residues can be estimated from annual crop production using a number of factors such as the recoverable fraction of residue production, residue to product (or crop) ratio, and gross heating value (Haberl et al., 2010). Among the relevant indicators that can be used to identify and quantify the ESs of providing energy include yields and area of crops, energy crop area, and amount of energy produced from manure (Table 3.1).

Indicators and indices of regulation and maintenance services

Regulation and maintenance services (Table 3.2) constitute all the ways in which SFS options mediate the physical, chemical, and biological conditions that affects human health, safety, or comfort. These included pollination, seed dispersal, control of crop pests, human and animal diseases, regulation of soil quality, water conditions, atmospheric composition, habitat and gene pool protection, the regulation

Table 3.2 List of indicators of regulation and maintenance services in sustainable food systems.

Group/class	Indicators	Indices
Pollination and seed dispersal	- Pollinator species richness - Pollinator species diversity - Number of beehives - Areal coverage of vegetation features supporting pollination (hedgerows, flower strips, high nature value ffarmland, etc.)	- Shannon diversity index - Habitat suitability index
Control of pest and diseases	- Diversity of natural enemies - Abundance of predators and parasitoids	- Shannon diversity index - Habitat suitability index
Control of human diseases	- Nutritional diversity - Prevalence of obesity - Prevalence of cardiovascular disease	- Shannon diversity index
Physical soil quality	- Soil hardiness - Aggregate stability - Available water capacity - Effective rooting depth	Soil structural stability index
Chemical soil quality	- Soil pH - Cation exchange capacity - Carbon to nitrogen ratio - Mineralization rate - Nutrient availability - Electrical conductivity	NA
Biological soil quality	- Soil organic carbon content - Active carbon - Soil protein - Soil respiration rate - Diversity of soil fauna - Diversity of soil flora	- Soil protein index - Shannon diversity index (H)
Decomposition and fixation	- Area of N fixing crops - Gross nitrogen balance	NA
Global climate regulation	- Biomass carbon by permanent crops - SOC in cropland/rangeland (Mg ha^{-1})	NA
Microclimate regulation	- Humidity index	- Humidity index
Filtration/sequestration/ storage/accumulation by ecosystems	- Concentration of pollutants in soil - Concentration of nutrient elements (C, N, P, K, Ca, Mg, S) in agricultural soils	NA
Mass stabilization and control of erosion rates	- Soil cover (%) - Area under agroforestry (ha) - Area under conservation farming (ha)	NA

NA = Not available.

of baseline flows (e.g., erosion, wind, and fire), and extreme events (Haines-Young and Potschin, 2017).

Pollination services

Crop pollination is an essential ES that increases the yield, quality, and stability. Approximately 65% of plant species require pollination by animals, and an analysis of data from 200 countries indicated that 75% of crop species of global significance for food production rely on animal pollination, primarily by insects (Klein et al., 2007). The global value of pollination service has been estimated at $200 billion worldwide (Gallai et al., 2009). All crop pollinators depend on diverse plant species often provided by natural and seminatural habitats, and few species are found in abundance far away from natural habitat (Senapathi et al., 2015). Diverse wild-bee communities potentially provide both enhanced stability, quality, and quantity of pollination services over space and time, compared with single, managed species (Lonsdorf et al., 2009). One of the main reasons for such dependence on diverse natural habitat is the provision of the diverse set of nesting resources, which are typically unavailable within intensively managed crop fields (Senapathi et al., 2015).

Evidence is mounting on the negative impact of agricultural practices (that increase crop monocultures, land use changes, habitat loss, etc.) and inputs (e.g., insecticides) on honey bees and other insect communities that provide pollination services. Agricultural intensification may affect functionally important pollinator species disproportionately (Kremen et al., 2002). The decline of pollination services with agricultural intensification results from significant reductions in both diversity and total abundance of native bees. If the alarming declines in honeybee populations continue, wild pollinators may become increasingly important to farmers. Therefore maintaining pollinator habitats and pollinator diversity within agricultural landscapes may be essential to ensure food production, quality, and security (Lonsdorf et al., 2009). Indeed SFSs can provide pollination services. To monitor pollination services in SFS, a number of indicators and indices may be used. These may include pollinator species richness and diversity (e.g., Shannon index), number of beehives, areal coverage of vegetation features, and HSI (Table 3.2).

Control of crop pests and diseases

For sustainable agriculture, self-regulation is considered to be crucial in preventing pest and disease outbreaks. This is usually achieved through natural control of pest insects in agroecosystems (Power et al., 2010). Pests and diseases that affect crops or rangelands or insects that transmit livestock or other disease are commonly kept in check by other organisms in the food web. However, the presence of these other organisms largely depends on the availability of appropriate habitats and food sources. Managing agricultural landscapes to allow this regulation can be an important way to deliver this service (Swinton et al., 2007). Diversified agricultural systems can provide diverse food resources for arthropod predators (spiders, wasps, ladybird beetles, carabid beetles, etc.) and parasitoids (parasitic insects), insectivorous birds, and bats. Such systems can also provide habitat for a diversity of microbial pathogens that act as natural enemies to agricultural pests and provide biological control services.

The abundance and diversity of predatory and parasitic could be used as indicators of pest control services in SFS. Assessment of a selection of high-level taxa (e.g., family) or guilds can provide enough detailed information to permit evaluation of the sustainability of a system in comparison with others (Paoletti, 1999). Among the commonly used indices are the Shannon index (Table 3.2).

Control of human diseases

In the past the primary focus of agricultural research, policy, and practice has been on increasing yields with little attention paid to improving the nutrient output of farming systems. Current food systems are dominated by processed foods that are high in added sugar, fat, and sodium or processes that remove nutrient dense fractions from whole foods. Unhealthy diets have now become key drivers of increasing prevalence of obesity and noncommunicable diseases such as cancer, cardiovascular disease, and type 2 diabetes (Lim et al., 2012; WHO, 2015). Consequently the world is now facing multiple burdens of malnutrition; while over 800 million people go hungry, more than two billion are obese (Ng et al., 2014) and over two billion suffer from micronutrient deficiencies (FAO, 2013; Meybeck and Gitz, 2017; Tulchinsky, 2010).

Intensive industrial agriculture does not appear to be sustainable and does not contribute to a healthy human diet (Dwivedi et al., 2017). The concept of sustainable diets was defined in 2010 combining two totally different perspectives: a nutrition perspective focused on individuals, and a global sustainability perspective, in all its dimensions: environmental, economic, and social (Meybeck and Gitz, 2017). Diverse and healthy diets, largely based on plant-derived food, may reduce diet-related illnesses (Dwivedi et al., 2017).

Adoption of SFSs can have positive nutritional impact by improving the quality of food in terms of diversity, nutrient content, and safety (Ruel and Alderman, 2013). Nutrition-sensitive agriculture can also boost food species diversity, which in turn is favorable for increasing the micronutrient content of the foods produced. Diverse and healthy diets which are largely based on plants may reduce noncommunicable diseases. Mixed production systems generate more diversity of key nutrients such as zinc, iron, vitamins A and B12, and folate, which are essential for human health. A recent analysis demonstrates that most global micronutrients (53%–81%) and protein (57%) are produced on more diverse (H-index>1.5) agricultural landscapes (Herrero et al., 2017). In that regard relevant indicators and indices may include nutritional diversity, prevalence of obesity, and cardiovascular disease (Table 3.2).

Soil quality

Soils provide a number of ESs that fulfill human needs (Fig. 3.1). Soil provides three types of services: provisioning, regulating, and cultural services. Through the process of enabling plants to grow, soils provide a service to humans. Soils on the surface of earth provide a physical base on which, animals, human, and infrastructure

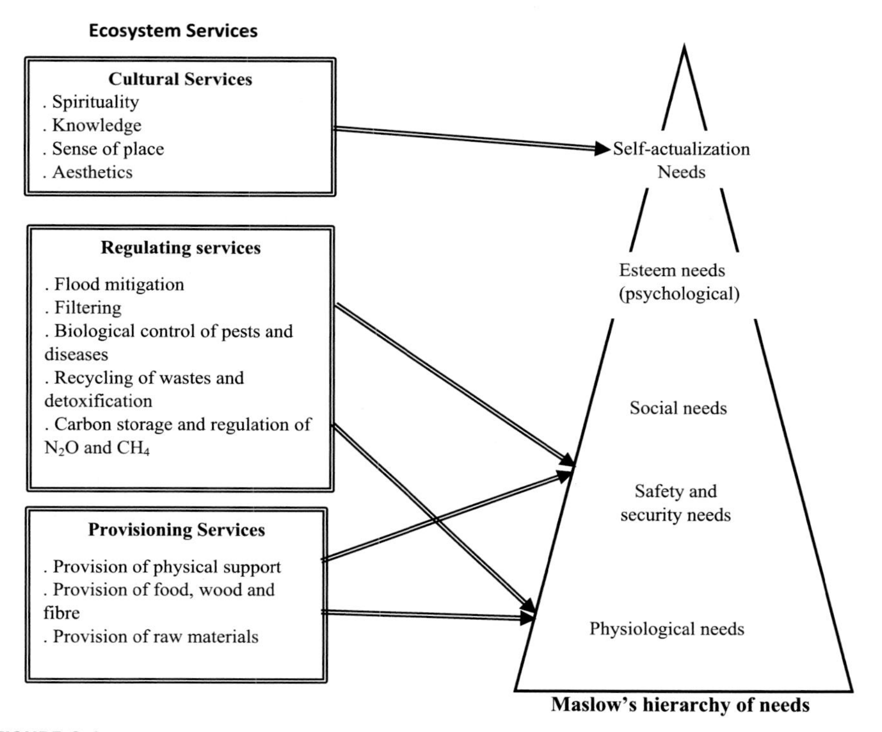

FIGURE 3.1

Framework for the provision of ecosystem services from soil natural capital.

Adapted from (Dominati et al., 2010).

stand, and this is a provision of physical support. The strength, intactness, and resilience of soil structure is a critical national capital for these services. Soils also provide raw materials; for example, clay for potting and sand for building structures and houses. The regulating services provided by soils (Fig. 3.1) enable humans to live in a stable, healthy, and resilient environment. Soil regulating services include flood mitigation, biological control of pests and diseases, recycling wastes, and detoxification. Soils and the capacity to store and return quantities of water can therefore mitigate the impacts of extreme climatic events and limit flooding. Soil structure, microorosity, infiltration, and drainage will impact on these services. Soils can filter nutrients. The presence of nitrates and phosphates in soil in large quantities, if leached, can become a contaminant in the aquatic systems (nitrification) and a threat to human health (nitrate in drinking water). Soils have ability to absorb and return nutrients, therefore avoiding their release into water. By providing habitat to beneficial species, soils can support plant growth (rhizobium and mychoryzae) and control proliferation of pests and diseases. Soil properties which control soil habitat are soil moisture and temperature, and these select the type of organisms present in the soil. This service depends on soil properties and biological processes driving interactions (symbiosis, competition). Soil can self-detoxify and recycle waste. Soil biota degrade and decompose organic matter and release nutrients that plants and other organisms can reuse. Soils also can adsorb harmful chemicals to humans. This service depends on processes such as mineralization and immobilization. Soils play an important role in regulating many atmospheric constituencies, therefore impacting on air quality. Soils can store carbon as soil organic matter (SOM). This service is mainly based on organic matter stocks and processes of driving them and also on soil conditions (e.g., moisture and temperature), which regulate soil biotic activity and thereby regulate the production of greenhouse gasses such as nitrous oxide and methane. Cultural services are also provided by soils in the form of esthetic experiences, spiritual enrichment, and recreation. Soils have various cultural uses across the world, from being a place to bury the dead, a material to build houses, or places to store or cook food.

Looking at soils through Maslow's hierarchy of needs makes it easy to figure out and thereby enables us to point out that ESs relate to human needs on two differential levels (Fig. 3.1). At the physical level, provisioning services provide useful goods for fulfillment of some physiological needs, food, fiber for clothing, source of energy, and support for infrastructure (Fig. 3.1). Regulating services fulfill some physiological needs such as clean air and clean water by regulating greenhouse gas emissions and filtering water. Provisioning and regulating services fulfill safety and security needs by ensuring stability of human habitat through soil structure stability, flood mitigation, control of pests and diseases, and recycling wastes (Fig. 3.1). At none physical level, soil ESs provide esthetics and spiritual and cultural benefits through cultural services, thereby fulfilling self-actualization needs.

Soil quality is one of the three components of environmental quality besides water and air quality (Andrews et al., 2002). Soil quality is defined as the capacity of a soil to function within ecosystems and land use boundaries to sustain biological productivity, maintain environmental quality, and promote plant, animal, and human health (Doran and Parkin, 1996). The concept of soil quality transcends the productivity of soils to encompass ecosystem sustainability as the basis for the benefit that humans derive from soils as well as intrinsic values of soils as being irreplaceable and unique. The term soil health originates in the observation that soil quality influences the health of animals and humans via quality of crops. Soil health can be defined as continued capacity of soil to function as a vital ecosystem that sustains plants, animals, and humans (Moebius-Clune et al., 2016).

The conceptual difference between soil health and soil quality is that soil quality comprises both inherent and dynamic properties, whereas soil health is on the dynamics of properties. The concept of soil quality and soil health can go beyond the reductionist approach of measuring single indicators, which remains important from a practical point of view (Kibblewhite et al., 2008). Such an integrated view would include the capacity for emerging systems properties such as the self-organization of soils, i.e., bidirectional feedbacks between soil organisms and soil structures (Lavelle, 2006) and the adaptability for changing conditions. The term soil quality is preferred by researchers, while soil health is often preferred by farmers.

In a global context, soil quality affects not only agricultural productivity but it is also a significant factor governing environmental quality, human and animal health, and food safety and quality. An optimal combination of soil physical, chemical, and biological properties affects agronomic productivity, use efficiency of water, nutrients and other inputs, and sustainability of management systems (Lal, 2015). Soil quality change over time because of natural events or anthropogenic impacts. Soil quality is enhanced by management and land-use decisions that weigh the multiple functions of soil, but it is impaired by decisions which focus only on single functions, such as crop productivity.

According to Glenk et al. (2012) soil functions are defined as "bundles of soil processes that are providing input into the delivery of valued final ESs". Soil functions are defined by seven functions of the soils as (1) biomass production; (2) storing, filtering, and transforming nutrients substances and water; (3) biodiversity pool; (4) physical and cultural environment for humans; (5) source of raw materials; (6) and carbon pools.

It is critically important to establish a conceptual linkage between soil quality indicators/indices and soil functions. Different frameworks have been developed on how to link soil-based ESs and soil functions (Haygarth and Ritz, 2009; Tóth et al., 2013). None of these frameworks include soil threats. New frames are being developed, which ESs are linked to soil threats. More recently multifunctionality has been emphasized even more strongly (Bone et al., 2010). This means soil indicators/indices should be applicable to multiple soil functions. This requires inclusion of soil ecological indicators (Bennet et al., 2010). It is an important requirement that soil-based ESs must be quantified (Schulte et al., 2014) and directly related to soil properties (Adhikari and Hartemink, 2016). Rutgers et al. (2012) identified parameters which could serve as proxies of soil ESs. The most universal parameter was soil organic matter content which reflects 9/10 ESs. In this chapter we will attempt to strengthen the conceptual linkages between soil quality indicators/indices and the multiple functions of soil.

There are several frameworks developed for quantifying and valuing ESs provided by soils (Robinson et al., 2009; Mea, 2005). In general all these frameworks contain a number of common limitations. These include the following: (1) they do not inform in detail the part played by soils in provision of ESs; (2) they do not link ESs back to natural capital stocks; and (3) they are difficult to implement practically for resource management (Dominati et al., 2010). However, recently new frameworks have been developed to address some of these limitations. The framework for soil, natural capital, and ESs of Dominati et al. (2010) has addressed some of these limitations and provided a pathway of how ecological economics concepts can be integrated with soil science in the quantification and economic valuation of ESs. This framework provides a broader and holistic approach compared with previous frameworks which are tempted to identify soil ESs by linking soil services to soil natural capital. This framework shows how external drivers impact on processes that underpin soil natural capital and ESs and how ESs contribute to human well-being.

Past management of agriculture and other ecosystems to meet the needs of increasing populations has taxed the resiliency of soils and natural processes to maintain global balances of energy and matter (FAO and ITPS. 2015). Indicators of soil quality are intimately related to physical, chemical, and biological properties of the soil (Lal, 2005; Moebius-Clune et al., 2016), and those relevant to SFS are briefly discussed in this sentence.

Physical quality

Soil physical indicators are properties that can be determined by the senses of sight and touch. Relevant indicators of soil physical quality include soil hardness, aggregate stability, available water capacity, water transmission (infiltration rate and amount), and effective rooting depth (Lal, 2015; Moebius-Clune et al., 2016).

Surface and subsurface hardness are indicators of the soil compaction status, measured as field penetration resistance in pounds per square inch (psi) using a penetrometer in the field (Moebius-Clune et al., 2016). Compaction is the result of changes in soil porosity because of intensification and long-term impacts of agricultural practices, which reduce the total pore space and increases bulk density.

Aggregate stability is a measure of how well soil aggregates resist disintegration when hit by rain drops. It is measured using a standardized simulated rainfall event on a sieve containing soil aggregates between 0.25 and 2.0 mm. The fraction of soil that remains on the sieve determines the percent aggregate stability (Moebius-Clune et al., 2016). This is an important indicator of the structural stability of soil and therefore its resistance to erosion. Stable aggregates are built by biological activity, as aggregates are largely "stuck" together by fungal hyphae, microbial colonies, and plant and microbial exudates. This means plentiful fresh and diverse organic materials (such as green manures, cover crops with vigorous fine roots, animal manures, and mulches) are needed to sustain soil biota, so that they can stabilize soil aggregates (Moebius-Clune et al., 2016).

Available water capacity reflects the quantity of water that a disturbed sample of soil can store for plant use. Available water capacity is determined from measuring water content at field capacity and permanent wilting point in the lab and calculating the difference (Moebius-Clune et al., 2016). Soils with low storage capacity have greater risk of drought stress. Available water capacity can be improved in the short term by large additions of stable organic materials (e.g., composts, biochar, and mulches), and in the long-term, building organic matter and aggregation will build porosity for storing water. This can be accomplished by reducing tillage, long-term cover cropping, mulching, rotating annual crops with diverse perennials, and generally keeping actively growing roots in the system. As they increase soil organic matter and hydraulic conductivity and reduce runoff losses, agroforestry practices with fertilizer trees can also improve available water capacity (Sileshi et al., 2014).

Effective rooting depth relates to the depth to which the roots of plants can reach without encountering physical and chemical limitations, and this is one of most important properties that determine the agricultural potential of soil. The total crop water availability is also dependent on rooting depth, which is linked to surface and subsurface hardness.

Chemical quality

Relevant indicators of soil chemical quality include pH, cation exchange capacity (CEC), carbon to nitrogen ratio (C/N), carbon and nitrogen mineralization rates, nutrient availability, electrical conductivity, salinity, sodicity, toxicity, or deficiency (Lal, 2015). Soil pH, measure of the acidity or alkalinity of soils, has a direct influence on physical, chemical (e.g., nutrient availability), and biological (e.g., microbial activity) characteristics that influence crop growth. CEC is a measure of the ability of the soil to adsorb cations, and it is also an indirect measure of negative charges in the soil, associated with the organic matter and clay content.

Salinity is a measure of the soluble salt concentration in soil and is measured via electrical conductivity. Sodicity is a calculation of the sodium absorption ratio and is measured using ICP (inductive coupled plasma) spectrometry to determine Na^{+}, Ca^{2+}, and Mg^{2+} concentrations and using an equation to calculate the absorption ratio.

The concentration of heavy metals can also be used as an indicator, as it represents the levels of metals of possible concern to human or plant health, and they are measured by digesting the soil with concentrated acid at high temperature. Heavy metals of concern include arsenic, barium, cadmium, chromium, copper, lead, nickel, and zinc. Elevated concentrations of heavy metals in soil may suppress natural microbial processes and decomposition of organic matter.

Biological quality

Relevant indicators of soil biological quality are soil organic matter, active carbon, soil protein, soil respiration, the abundance and diversity of soil fauna and flora, and absence of pathogens and pests as indicated by a soil's disease-suppressive attributes (Lal, 2015; Moebius-Clune et al., 2016).

Soil organic matter (SOM) is a measure of all carbonaceous material that is derived from living organisms (Moebius-Clune et al., 2016). SOM or the soil organic carbon pool is a key indicator of soil quality and an important driver of agricultural sustainability. Indeed, soil organic carbon pool is the most reliable indicator of monitoring soil degradation (Lal, 2015). Active carbon is a measure of the portion of the organic matter that can serve as an easily available food source for soil microbes, thus helping fuel and maintain a healthy soil food web. It is measured by quantifying potassium permanganate oxidation with a spectrophotometer (Moebius-Clune et al., 2016). Reducing tillage and increasing organic matter additions from various sources will increase the active carbon and will feed, expand, and balance the microbial community, thus increasing total organic matter over the long term (Moebius-Clune et al., 2016). Efforts to increase SOM should consider both the quantity and quality of organic inputs. The ratio of lignin + polyphenol to nitrogen is a useful quality index (Mafongoya et al., 1998).

Soil protein index is an indicator of the amount of protein-like substances that are present in the soil organic matter. The autoclaved citrate extractable index represents the large pool of organically bound nitrogen (N) in the soil organic matter, which microbial activity can mineralize and make available for plant uptake (Moebius-Clune et al., 2016). Protein content is well associated with overall soil health status because of its indication of biological and chemical soil health, in particular, the quality of the soil organic matter (Moebius-Clune et al., 2016).

Soil respiration is a direct biological activity measurement, integrating abundance and activity of microbial life. Thus it is an indicator of the biological status of the soil community, which can give insight into the ability of the soil's microbial community to accept and use residues or amendments to mineralize and make nutrients available from them to plants and other organisms, to store nutrients and buffer their availability over time, and to develop good soil structure (Moebius-Clune et al., 2016). It is measured by capturing and quantifying carbon dioxide released from a re-wetted sample of air-dried soil held in an airtight jar for 4 days (Moebius-Clune et al., 2016). Greater carbon dioxide release is indicative of a larger, more active soil microbial community.

The biological qualities of soil are also related to soil biological, which is a function of the abundance and diversity of organisms such as earthworms, termites, ants, fungi, bacteria, etc. Beneficial soil biological activity tends to decrease with increasing soil disturbance such as tillage, heavy traffic, soil compaction, as well as with extremes in low or high pH or contamination by heavy metals or salts (Moebius-Clune et al., 2016). As such the diversity of soil fauna and flora can be used as an integrative measure of soil health. The Shannon diversity index can be used a good index for tracking progress in the improvement of soil biological quality (Table 3.2).

Indices may also be constructed by a combination of physical, chemical, or biological indicators. For example, the soil structural stability index (SSI) is an integrative measure of the ability of soils to resist structural degradation, which is calculated from soil organic matter (SOM in %), clay content (%), and silt content (%) as proposed by Pieri (1992):

$$SSI = 100 \times \frac{SOM}{Clay + Silt} \tag{3.3}$$

SSI $\leq$5% indicates a structurally degraded soil because of extensive loss of SOC (soil organic carbon); 5% $<$ SSI $<$ 7% indicates a high risk of structural degradation; and SSI $>$ 7% indicates low risk (Pieri, 1992).

Decomposition and fixation

The contribution of crops and trees on farm land (e.g., agroforestry) to fixation mainly comes from nitrogen inputs via biological nitrogen fixation (BNF). BNF accounts for 60% of nitrogen production (Zahran, 1999), and it accounts for 16% of the current global nitrogen input (Liu et al., 2010). On the other hand, in Africa and South America, BNF is the single largest N source accounts for 32%–34% of the

N input (Liu et al., 2010). Therefore its further use would, at the least, ease the pressure for land through the rehabilitation of degraded areas (Herridge et al., 2008). Atmospheric N fixation by symbiotic bacteria (*Rhizobium*) in root nodules is common in 340 species of leguminous plants (Sileshi et al., 2014). N fixation also occurs in over 200 nonleguminous plants species in 25 genera of eight families associated with Actinomycetes in the genus *Frankia* (Franche et al., 2009). The broader definition of "fertilizer" trees used here includes the legume *Rhizobium,* as well as nonlegume *Frankia* symbioses. BNF can play a greater role in sustainable agriculture as it increases N recovery rates in addition to reducing the need for synthetic fertilizers.

Climate regulation

Ecosystems regulate global and regional climate (1) by providing sources or sinks of greenhouse gases (affecting global warming) and sources of aerosols (affecting temperature and cloud formation); (2) by enhancing evapotranspiration and thereby cloud formation and rainfall; and (3) by affecting surface albedo and thereby radiative forcing and temperature (Smith et al., 2013). Ecosystems can also affect the microclimate locally, through the provision of shade and shelter and the regulation of humidity and temperature. This regulation of microclimate can have a noticeable impact on human well-being (Smith et al., 2013).

Improved agricultural practices can have significant climate regulation and mitigation benefits through both biogeochemical processes that control greenhouse gas emissions and biophysical mechanisms that regulate microclimate. Among agricultural mitigation options, soil C sequestration is one of the few strategies that could be applied at large scales and at lower cost (Paustian et al., 2016; Smith, 2012). The "4 per 1000" aspiration proposed during the 21st Conference of Parties (COP21) in Paris was aimed at making agriculture a solution to address climate change while also advancing food and nutritional security (Lal, 2016; Minasny et al., 2017). While "4 per 1000" target is intended to highlight that even small increases in SOC can play a crucial role in improving soil fertility and productivity and achieving the long-term objective of limiting global warming to 1.5°C (http://4p1000.org). As such it is an important aspirational target to promotion of sustainable land management and achieving food security at the same time (Lal, 2016; Chabbi et al., 2017; Minasny et al., 2017). There is a growing body of empirical evidence suggesting that wide spread adoption of best management practices, including agroforestry, balanced application of plant nutrients, composting, conservation agriculture, and green water management on smallholder farms can enhance C sequestration and produce adequate food for the growing world population (Nath et al., 2018). In that sense the amount of SOC stored in cropland/rangeland and biomass carbon stored in permanent crops and agroforestry are relevant indicators.

Mass stabilization and erosion control

Accelerated soil erosion is often driven by anthropogenic activities such as conversion of natural ecosystems to agricultural land and mechanical tillage (Lal, 2014). Soil erosion has numerous on-site and off-site impacts. On-site impacts result in decline in soil quality because of the loss of key soil constituents (e.g., SOC, clay, and silt), reduction in available water capacity and nutrient reserves, shallowing of topsoil depth, decline of use-efficiency of inherent and applied resources, and depletion of the ecosystem carbon pool. Among off-site impacts are burial of topsoil, run-on of agricultural chemicals, contamination and eutrophication of water bodies, and inundation and emission of greenhouse gases (Lal, 2014).

Diverse plant communities boost root biomass, with knock-on positive effects for soil stabilization (Ford et al., 2016) and reduction of soil erosion. Above-ground plant parts, laterally connected rhizomes or stolons and roots combine to protect the soil against erosion by physically sheltering and fixing soils, offering resistance to rain, run-off, and attack by waves and currents (Ford et al., 2016). Fine roots physically bind together soil particles, particularly clay and silt. Soil stabilization and root biomass are positively associated with plant diversity (Ford et al., 2016). Diverse plants also lead to improvement in soil structure, which increases drainage. For example, land under agroforestry has been shown to increase drainage by 88% −900%, effectively reducing runoff and soil erosion compared with continuous sole maize (Sileshi et al., 2014). Conservation farming and other soil and water conservation measures can also stabilize soil and minimize erosion rates. Relevant indicators for quantifying these ESs in SFSs include the percentage of soil cover, area under conservation farming, and agroforestry (Table 3.2).

Indicators and indices of cultural services

Agriculture provides ESs that extend well beyond provisioning regulation/maintenance services (Swintona et al., 2007). It also provides cultural services (Fig. 3.1), which include nonmaterial and nonconsumptive benefits of SFS that affect physical and mental states of people through spiritual enrichment, cognitive development, reflection, recreation, and esthetic experience. Cultural services are primarily regarded as the environmental settings, locations, or situations that give rise to changes in the physical or mental states of people. These include characteristics of living systems that enable activities promoting health, recuperation, or enjoyment through passive or observational interactions, scientific investigation, traditional ecological knowledge, educational, entertainment, or esthetic experiences (Haines-Young and Potschin, 2017). Cultural serves also included elements of living systems that have sacred or religious meaning, existence, heritage, or bequest value (Haines-Young and Potschin, 2017).

With the increasing global integration and economic consolidation, the current food system has led to the disintegration of the social and spiritual fabric—critical connections—that are part of a community's food system (Feenstra, 2002). The importance of cultural services is not currently recognized in landscape planning and management (Tengberg et al., 2012).

Conversion of native ecosystems to industrial agriculture often results in profound changes in cultural landscapes, disappearance of religious or heritage sites, and loss of indigenous knowledge systems. Sustainable agricultural landscapes should be able to conserve these and provide various cultural services. These are

Table 3.3 List of indicators of cultural services in sustainable food systems.

Class	Indicators	Indices
Experiential use of plants, animals, and landscapes/ seascapes	- Number of visitors in agricultural areas - Number of rural enterprises offering tourism-related services - Walking and biking trails - Number of birdwatchers - Expenditures related to hunting	NA
Scientific/educational	- Amount of scientific studies on agroecosystems - Number of didactic farms	NA
Heritage/cultural	- Number of monuments in agricultural areas - Number of certified products that require traditional landscape management	NA
Esthetic	- Number of visitors in agricultural areas - Number of agricultural landscape photos on web portals	NA
Symbolic, sacred, and/or religious	- Remarkable trees - Symbolic species - Religious monuments, pilgrim paths - Sacred groves	NA
Existence/bequest	- Willingness to pay for landscape measures in cropland or rangeland	NA

NA = Not available.

context specific, and their valuation is especially difficult. An assessment of cultural ESs in food systems needs to include indicators (Table 3.3) and possible proxies of the experiential and educational use, heritage, religious, and bequest values of plants, animals, and landscapes and seascapes. In Table 3.3 we provide potential indicators. However, these indicators are neither exhaustive nor specific.

The value of ESs has been estimated in various ways, and the framework has three main parts: (1) measuring the provision of ESs; (2) determining the monetary value of ESs; and (3) designing policy tools for managing ESs (Power, 2010). Valuation of ESs is beyond the scope of this chapter. Interested readers should refer to more specific literature in this area.

Future research

This chapter provided a synthesis of the indicators and indices to be used in assessing ESs in SFSs. Some indices such as the biotic integrity indices and habitat suitability indices focus on a single species and thus have narrow utility. Others (e.g.,

Shannon diversity index) have broad utility as they use data acquired locally or nationally and thus provide a convenient measure for use at the farm or local or national levels. Yet, others such as the global food security index and GHI use national datasets, aggregate well-established indicators and mainly target national governments and multinational agencies. However, there is no fully developed set of indices for ES that combine data derived from a standard set of methods, across ecosystems and regions in SFS. There are also not many studies that evaluate food systems using multiple indicators of ESs. In conclusion there is

- Need for research to link soils ESs to soil properties
- How these ESs are understood by end-users and resource managers
- Future research must rely on interdisciplinary and transdisciplinary research approaches of soil functions and ESs. This should involve soil scientists, agronomists, economists, plant breeders, sociologists, and policymakers
- The need to test the applicability of different frameworks of quantifications and evaluation of ESs develop in different settings.

References

Adhikari, K., Hartemink, A.E., 2016. Linking soils to ecosystem services — a global review. Geoderma 262, 101—111.

Ajila, C.M., Brar, S.K., Verma, M., Tyagi, R.D., Godbout, S., Valéro, J.R., 2012. Bioprocessing of agro-byproducts to animal feed. Critical Reviews in Biotechnology 32, 382—400.

Andrews, S.S., Karlen, D.L., Mitchell, J.P., 2002. A comparison of soil quality indexing methods for vegetable production systems in northern California. Agriculture, Ecosystems and Environment 90, 25—45.

Bennett, L.T., Mele, P.M., Annett, S., Kasel, S., 2010. Examining links between soil management, soil health, and public benefits in agricultural landscapes: an Australian perspective. Agriculture, Ecosystems and Environment 139, 1—12.

Bone, J., Head, M., Barraclough, D., Archer, M., Scheib, C., Flight, D., Voulvoulis, N., 2010. Soil quality assessment under emerging regulatory requirements. Environment International 36, 609—622.

Bouf, M., 2017. Agricultural crop residue as a source for the production of cellulose nanofibrils. In: Cellulose-Reinforced Nanofibre Composites: Production, Properties and Applications. Woodhead Publishing, pp. 129—152.

Cafiero, C., Viviani, S., Nord, M., 2018. Food security measurement in a global context: the food insecurity experience scale. Measurement 116, 146—152.

Canterbury, G.E., Thomas, E.M., Daniel, R.P., Lisa, J.P., David, F.B., 2000. Bird communities and habitat as ecological indicators of forest condition in regional monitoring. Conservation Biology 14, 544—558.

Ceballos, G., Ehrlich, P.R., Barnosky, A.D., García, A., Pringle, R.M., Palmer, T.M., 2015. Accelerated modern human—induced species losses: entering the sixth mass extinction. Sci. Adv. 1, e1400253.

Chabbi, A., Lehmann, J., Ciais, P., Loescher, H.W., Cotrufo, M.F., Don, A., SanClements, M., Schipper, L., Six, J., Smith, P., Rumpel, C., 2017. Aligning agriculture and climate policy. Nature Climate Change 7, 307–309.

Chakeredza, S., Hove, L., Akinnifesi, F.A., Franzel, S., Ajayi, O., Sileshi, G., 2007. Managing fodder trees as a solution to human-livestock food conflicts and their contribution to income generation for smallholder farmers in southern Africa. Natural Resources Forum 31, 286–291.

Chakeredza, S., Akinnifesi, F., Ajayi, O.C., Sileshi, G., Mngomba, S., Gondwe, F.M.T., 2008. A simple method of formulating least-cost diets for smallholder dairy production in sub-Saharan Africa. African Journal of Biotechnology 7, 2925–2933.

Chaudhary, A., Gustafson, D., Mathys, A., 2018. Multi-indicator sustainability assessment of global food systems. Nature Communications 9, 848.

Cigolotti, V., 2012. Biomass and waste as sustainable resources. In: McPhail, S.J., et al. (Eds.), *Green Energy and Technology*, Fuel Cells in the Waste-to-Energy Chain. Springer-Verlag, London, pp. 23–44.

Descheemaeker, K., Amede, T., Haileslassie, A., Bossio, D., 2011. Analysis of gaps and possible interventions for improving water productivity in crop livestock systems of Ethiopia. Experimental Agriculture 47, 21–38.

Dominati, E., Patterson, M., Mackay, A., 2010. A framework for classifying and quantifying the natural capital and ecosystem services of soils. Ecological Economics 69, 858–1868.

Doran, J.W., Parkin, T.B., 1996. Quantitative indicators of soil quality: a minimum data set. In: Doran, J.W., Jones, A.J. (Eds.), Methods for Assessing Soil Quality. Soil Science Society of America, pp. 25–37.

Dwivedi, S.L., van Bueren, E.T.L., Ceccarelli, S., Grando, S., Upadhyaya, H.D., Orti, R., 2017. Diversifying food systems in the pursuit of sustainable food production and healthy diets. Trends in Plant Science 22, 842–856.

Ebert, U., Welsch, H., 2004. Meaningful environmental indices: a social choice approach. Journal of Environmental Economics and Management 47, 270–283.

Erb, K.-H., Haberl, H., Plutzar, C., 2012. Dependency of global primary bioenergy crop potentials in 2050 on food systems, yields, biodiversity conservation and political stability. Energy Policy 47, 260–269.

FAO, 2010. The Commission on Genetic Resources for Food and Agriculture (CGRFA) Second Report on the State of the World's Plant Genetic Resources for Food and Agriculture (PGRFA). Food and Agriculture Organization of the United Nations (FAO), Rome, Italy.

FAO, 2013. Key Recommendations For Improving Nutrition Through Agriculture and Food Systems. Available at: www.fao.org/nutrition.

FAO and ITPS, 2015. Status of the World's Soil Resources (SWSR) – Main Report. Food and Agriculture Organization of the United Nations and Intergovernmental Technical Panel on Soils, Rome, Italy.

FAO, 2015. The State of Food Insecurity in the World 2015 Food and Agricultural Organization (FAO).

Feenstra, G., 2002. Creating space for sustainable food systems: lessons from the field. Agriculture and Human Values 19, 99–106.

Ford, H., Garbutt, A., Ladd, C., Malarkey, J., Skov, M.W., 2016. Soil stabilization linked to plant diversity and environmental context in coastal wetlands. Journal of Vegetation Science 27, 259–268.

Franche, C., Lindström, K., Elmerich, C., 2009. Nitrogen-fixing bacteria associated with leguminous and non-leguminous plants. Plant and Soil 321, 35–59.

Gallai, N., Vaissiére, B.E., Potts, S.G., Salles, J., Kareiva, P., Daily, G., Ricketts, T., 2009. Assessing the monetary value of global crop pollination services. In: Tallis, H., Polasky, S. (Eds.), The Theory and Practice of Ecosystem Service Valuation in Conservation. Oxford University Press, Oxford, pp. 169–170.

Glenk, K., McVittie, A., Moran, D., 2012. Deliverable D3.1: Soil and Soil Organic Carbon Within an Ecosystem Service Approach Linking Biophysical and Economic Data. Available at: http://smartsoil.eu/smartsoil-toolbox/project-deliverables/.

Global Panel on Agriculture and Food Systems for Nutrition, 2016. Food Systems and Diets: Facing the Challenges of the 21st Century, p. 14. London, UK.

Gustafson, D., Gutman, A., Leet, L., Drewnowski, A., Fanzo, J., Ingram, J., 2016. Seven food system metrics of sustainable nutrition security. Sustainability 8, 196. https://doi.org/10.3390/su8030196.

Haines-Young, R., Potschin, M.B., 2017. *Common International Classification of Ecosystem Services (CICES)* V5.1 and Guidance on the Application of the Revised Structure. Available from: www.cices.eu.

Haberl, H., Beringer, T., Bhattacharya, S.C., Erb, K.-H., Hoogwijk, M., 2010. The global technical potential of bio-energy in 2050 considering sustainability constraints. Current Opinion in Environmental Sustainability 2, 394–403.

Haygarth, P.M., Ritz, K., 2009. The future of soils and land use in the UK: soil systems for the provision of land-based ecosystem services. Land Use Policy 26, S187–S197.

Heink, U., Kowarik, I., 2010. What are indicators? On the definition of indicators in ecology and environmental planning. Ecological Indicators 10, 584–593.

Herrero, M., Thornton, P.K., Power, B., Bogard, J.R., Remans, R., Fritz, S., Gerber, J.S., Nelson, G., See, L., Waha, K., Watson, R.A., West, P.C., Samberg, L.H., van de Steeg, J., Stephenson, E., van Wijk, M., Havlík, P., 2017. Farming and the geography of nutrient production for human use: a transdisciplinary analysis. Lancet Planet Health 1, e33–42.

Herridge, D., Peoples, M.B., Boddey, R.M., 2008. Global inputs of biological nitrogen fixation in agricultural systems. Plant and Soil 311, 1–18.

Hunter, D., Özkan, I., Beltrame, D.M.O., Samarasinghe, W.L.G., Wasike, V.W., Charrondière, U.R., Borelli, T., Sokolow, J., 2016. Enabled or disabled: is the environment right for using biodiversity to improve nutrition? Frontiers in Nutrition 3, 14.

IPES-Food (International Panel of Experts on Sustainable Food Systems), 2016. From uniformity to diversity: a paradigm shift from industrial agriculture to diversified agroecological systems. International Panel of Experts on Sustainable Food systems. www.ipes-food.org.

Karr, J., 1991. Biological integrity: A long-neglected aspect of water resource management. Ecological Applications 1, 66.

Kibblewhite, M.G., Jones, R.J.A., Montanarella, L., Baritz, R., Huber, S., Arrouays, D., Micheli, E., Stephens, M., 2008. Environmental Assessment of Soil for Monitoring Volume VI. Soil Monitoring System for Europe.

Klein, A.-M., Vaissiere, B.E., Cane, J.H., Steffan-Dewenter, I., Cunningham, S.A., Kremen, C., Tscharntke, T., 2007. Importance of pollinators in changing landscapes for world crops. Proceedings of the Royal Society of London Series B: Biological Science 274, 303–313.

Kline, K.L., Msangi, S., Dale, V.H., Woods, J., Souza, G.M., Osseweijer, P., Clancy, J.S., Hilbert, J.A., Johnson, F.X., Mcdonnell, P.C., Mugera, H.K., 2017. Reconciling food security and bioenergy: priorities for action. GCB Bioenergy 9, 557–576.

Kremen, C., Williams, N.M., Thorp, R.W., 2002. Crop pollination from native bees at risk from agricultural intensification. Proceedings of the National Academy of Sciences USA 99, 16812–16816.

Lal, R., 2005. Encyclopedia of Soil Science, second ed. CRC Press, p. 2060.

Lal, R., 2014. Soil conservation and ecosystem services. International Soil and Water Conservation Research 2, 36–47.

Lal, R., 2015. Restoring soil quality to mitigate soil degradation. Sustainability 7, 5875–5895.

Lal, R., 2016. Potential and challenges of conservation agriculture in sequestration of atmospheric CO_2 for enhancing climate-resilience and improving productivity of soil of small landholder farms. CAB Reviews 11, 1–16.

Lavelle, P., Decaens, T., Aubert, M., Barot, S., Blouin, M., Bureau, F., 2006. Soi linvertebrates and ecosystem services. European Journal of Soil Biology 42, S3–S15.

Lim, S., Vos, T., Flaxman, A., Dana, G., et al., 2012. A comparative risk assessment of burden of disease and injury attributable to 67 risk factors and risk factor clusters in 21 regions, 1990–2010: a systematic analysis for the global burden of disease study 2010. Lancet 380, 2224–2260.

Liu, J., You, L., Amini, M., et al., 2010. A high-resolution assessment on global nitrogen flows in cropland. Proceedings of the National Academy of Sciences 107, 8035–8040.

Lonsdorf, E., Kremen, C., Ricketts, T., Winfree, R., Williams, N., Greenleaf, S., 2009. Modelling pollination services across agricultural landscapes. Annals of Botany 103, 1589–1600.

Magurran, A.E., 2004. Measuring Biological Diversity. Blackwell Science, Oxford.

Mafongoya, P.L., Giller, K.E., Palm, C., 1998. Decomposition and nitrogen release patterns from tree prunings and litter. Agroforestry Systems 38, 77–97.

MEA, 2005. Millennium Ecosystem Assessment: Ecosystems and Human Well-being: Synthesis. Island Press, Washington, DC.

Meybeck, A., Gitz, V., 2017. Sustainable diets within sustainable food systems. Proceedings of the Nutrition Society 76, 1–11.

Minasny, B., Malone, B.P., McBratney, A.B., Angers, D.A., Arrouays, D., Chambers, A., Chaplot, V., Chen, Z.S., Cheng, K., Das, B.S., Field, D.J., Gimona, A., Hedley, C.B., Hong, S.Y., Mandal, B., Marchant, B.P., Martin, M., McConkey, B.G., Mulder, V.L., O'Rourke, S., Richer-de-Forges, A.C., Odeh, I., Padarian, J., Paustian, K., Pan, G., Poggio, L., Savin, I., Stolbovoy, V., Stockmann, U., Sulaeman, Y., Tsui, C.-C., Vågen, T.-G., van Wesemael, B., Winowiecki, L., 2017. Soil carbon 4 per mille. Geoderma 292, 59–86.

Moebius-Clune, B.N., Moebius-Clune, D.J., Gugino, B.K., Idowu, O.J., Schindelbeck, R.R., Ristow, A.J., Van Es, H.M., Thies, J.E., Shayler, H.A., McBride, M.B., Wolfe, D.W., Abawi, G.S., 2016. Comprehensive assessment of soil health, third ed. In: The Cornell Framework Manual, Edition 3.0. Cornell University, Geneva, NY, p. 123.

Nair, M.K., Augustine, L.F., Konapur, A., 2016. Food-based intervention to modify diet quality and diversity to address multiple micronutrient deficiency. Frontiers in Public Health 3, 277.

Nath, A.J., Lal, R., Sileshi, G.W., Das, A.K., 2018. Managing India's small landholder farms for food security and achieving the "4 per thousand" target. The Science of the Total Environment 634, 1024–1033.

Nechyporchuk, O., Belgacem, M.N., Bras, J., 2016. Production of cellulose nanofibrils: a review of recent advances. Industrial Crops and Products 25, 2–25.

Ng, M., et al., 2014. Global, regional, and national prevalence of overweight and obesity in children and adults during 1980—2013: a systematic analysis for the Global Burden of Disease Study 2013. The Lancet 384, 766—781.

Paustian, K., Lehmann, J., Ogle, S., Reay, D., Robertson, G.P., Smith, P., 2016. Climate-smart soils. Nature 532, 4 9—57.

Paoletti, M.G., 1999. Using bioindicators based on biodiversity to assess landscape sustainability. Agriculture, Ecosystems and Environment 74, 1—18.

Pieri, C., 1992. Fertility of Soils: A Future for Farming in the West African Savannah. Springer, Berlin, p. 348.

Power, A.G., 2010. Ecosystem services and agriculture: tradeoffs and synergies. Philosophical Transactions of the Royal Society B 365, 2959—2971.

Rai, M., Reeves, T.G., Pandey, S., Collette, L., 2011. Save and Grow: A Policymaker's Guide to Sustainable Intensification of Smallholder Crop Production. Food and Agriculture Organization of the United Nations, Rome.

Remans, R., Wood, S.A., Saha, N., Anderman, T., DeFries, R.S., 2014. Measuring nutritional diversity of national food supplies. Global Food Security 3, 174—182.

Remans, R., Attwood, S., Bailey, A., Weise, S., 2017. Towards an agrobiodiversity index for sustainable food systems. In: Bioversity International (2017) Mainstreaming Agrobiodiversity in Sustainable Food Systems. Scientific Foundations for an Agrobiodiversity Index, pp. 141—152.

Ruel, M.T., Alderman, H., 2013. Nutrition-sensitive interventions and programmes: how can they help to accelerate progress in improving maternal and child nutrition? Lancet 382, 536—551.

Rutgers, M., van Wijnen, H.J., Schouten, A.J., Mulder, C., Kuiten, A.M.P., Brussaard, L., 2012. A method to assess ecosystem services developed from soil attributes with stakeholders and data of four arable farms. Science of the Total Environment 415, 39—48.

Robinson, D.A., Lebron, I., Vereecken, H., 2009. On the definition of the natural capital of soils: a framework for description, evaluation and monitoring. Soil Science Society of America Journal 73, 1904—1911.

Schader, C., Muller, A., Scialabba, N.E.-H., Hecht, J., Isensee, A., Erb, K.-H., Smith, P., Makkar, H.P.S., Klocke, P., Leiber, F., Schwegler, P., Stolze, M., Niggli, U., 2015. Impacts of feeding less food-competing feedstuffs to livestock on global food system sustainability. Journal of The Royal Society Interface 12, 20150891.

Schulte, R.P.O., Creamer, R.E., Donnellan, T., Farrelly, N., Fealy, R., O'Donoghue, C., 2014. Functional and management: a framework for managing soil-based ecosystem services for the sustainable intensification of agriculture. Environmental Science and Policy 38, 45—58.

Senapathi, D., Biesmeijer, J.C., Breeze, T.D., Kleijn, D., Potts, S.G., Carvalheiro, L.G., 2015. Pollinator conservation — the difference between managing for pollination services and preserving pollinator diversity. Current Opinion in Insect Science 12, 93—101.

Sileshi, G.W., Mafongoya, P.L., Akinnifesi, F.K., Phiri, E., Chirwa, P., Beedy, T., Makumba, W., Nyamadzawo, G., Njoloma, J., Wuta, M., Nyamugafata, P., Jiri, O., 2014. Fertilizer trees. In: Encyclopedia of Agriculture and Food Systems, vol. 1. Elsevier, San Diego, pp. 222—234.

Smith, P., 2012. Soils and climate change. Current Opinion in Environmental Sustainability 4, 539—544.

Smith, A., Thorne, A., Snapp, S., 2015. Measuring Sustainable Intensification in Smallholder Agroecosystems: A Review. International Livestock Research Institute, Nairobi.

Smith, P., Ashmore, M.R., Black, H.I.J., Burgess, P.J., Evans, C.D., Quine, T.A., Thomson, A.M., Hicks, K., Orr, H.G., 2013. The role of ecosystems and their management in regulating climate, and soil, water and air quality. Journal of Applied Ecology 50, 812–829.

Sukhdev, P., May, P., Müller, A., 2016. Fix food metrics. Nature 540, 33–34.

Swintona, S.M., Lupia, F., Robertson, G.P., Hamiltond, S.K., 2007. Ecosystem services and agriculture: cultivating agricultural ecosystems for diverse benefits. Ecological Economics 64, 245–252.

Tengberg, A., Fredholm, S., Eliasson, I., Knez, I., Saltzman, K., Wetterberg, O., 2012. Cultural ecosystem services provided by landscapes: assessment of heritage values and identity. Ecosystem Services 2, 14–26.

The Economist Intelligence Unit, 2017. Food Sustainability Index. http://foodsustainability.eiu.com/.

Tóth, G., Gardi, C., Bódis, K., Ivits, E., Aksoy, E., Jones, A., Jeffrey, S., Petursdottir, T., Montanarella, L., 2013. Continental-scale Assessment of provisioning soil functions in Europe. Ecological Processes 32, 1–18.

Tulchinsky, T.H., 2010. Micronutrient deficiency conditions: global health issues. Public Health Reviews 32, 243.

UN, 2017. The UN Decade of Action on Nutrition (2016–2025) Work Programme. Available at: http://www.who.int/nutrition/decade-of-action/workprogramme-doa2016to2025-en.pdf?ua=1.

von Grebmer, K., Saltzman, A., Birol, E., et al., 2014. Global Hunger Index: The Challenge of Hidden Hunger. Welthungerhilfe, International Food Policy Research Institute, and Concern Worldwide, Bonn, Washington, DC, and Dublin.

Wadhwa, M., Bakshi, M.P.S., 2013. In: Makkar, H.P. (Ed.), Utilization of Fruit and Vegetable Wastes as Livestock Feed and as Substrates for Generation of Other Value-Added Products. FAO, Rome.

WFP, 2009. Emergency Food Security Assessment Handbook. World Food Programme (WFP).

WHO, 2015. Obesity and overweight. Fact sheet N°311. WHO, Geneva.

Wiesmann, D.A., 2006. Global Hunger Index: Measurement Concept, Ranking of Countries, and Trends. International Food Consumption and Nutrition Division Paper 212. IFPRI, Washington, DC.

Zahran, H.H., 1999. *Rhizobium*-legume symbiosis and nitrogen fixation under severe conditions and in an arid climate. Microbiology and Molecular Biology Reviews 63, 968–989.

CHAPTER

Harnessing ecosystem services from biological nitrogen fixation

4

Sipho T. Maseko PhD [1], Mpelang P. Maredi BSc [1], Cherian Mathews PhD [2], Felix D. Dakora PhD [2]

[1]*Department of Crop Sciences, Tshwane University of Technology, Pretoria, South Africa;*
[2]*Professor, Chemistry Department, Tshwane University of Technology, Pretoria, South Africa*

Introduction

Nitrogen is the most important mineral element needed by plants for growth and functioning. However, only a tiny fraction of planetary N is available to plants in the pedosphere. Much of atmospheric N_2 (78%) is unavailable for direct use by plants that lack mechanisms to fix it. One process that converts atmospheric N_2 into inorganic N fertilizer for use by plants is the Bosch-Haber process (Smil, 2001). When applied in amounts not toxic to the environment, N fertilizers have contributed hugely to increased food production and nutritional security. There are however two main challenges with the frequent use of inorganic N sources: (1) they are expensive and unaffordable to small-scale farmers and (2) they are susceptible to leaching and can thus pollute the environment and the atmosphere. Affordability, access, and distribution of N fertilizers differ within, between, and among countries. For example, across sub-Saharan Africa, access to and usage of N fertilizers is increasing given the subsidized support by some governments (Sanchez, 2015; Moloto et al., 2018). In contrast, there is a decrease in the use of N fertilizers in other regions such as East Africa and China (Mueller et al., 2017). Overall the application of N fertilizers for increased crop production is associated with some inefficiency as only about 50% of the applied N gets absorbed by plants (Erisman et al., 2011). The remainder either remains in the soil or is lost through runoff, volatilization, denitrification, and/or leaching (Russo et al., 2017; Van der Laan et al., 2017; Tang et al., 2018). These factors explain why N fertilizers are adopted with caution, despite their contribution to food security. Given their high cost, N fertilizers generally do not play a significant role in low-input agricultural systems.

Members of the family Leguminoseae are plants with the ability to convert atmospheric N_2 into NH_3, a useable form by plants—a process called biological nitrogen fixation (BNF). Part of this legume-fixed N is added to soils in the form of organic matter and is less susceptible to leaching and volatilization when compared with nitrogenous fertilizers (Reckling et al., 2016). This attribute makes atmospheric

The Role of Ecosystem Services in Sustainable Food Systems. https://doi.org/10.1016/B978-0-12-816436-5.00004-4

N_2 fixation by legumes critical to the sustainable N nutrition of plants, animals, and soil microbes. The N_2 fixation process is particularly important to low-input, resource-poor smallholder farmers who cannot afford enough quantities of nitrogen fertilizer, but it can also be used by resourceful commercial farming communities. Nitrogen fixation is achieved through a legume-bacteria symbiotic interaction, and such bacteria are collectively called rhizobia. The benefits to resource-poor, low-input systems are huge as majority of farmers in that category can hardly afford the much needed chemical N fertilizers for increasing crop yields. Under low-input systems, legumes are used in intercropping and rotations while the commercial agriculture sector include legume crops largely for rotational purposes. Whether in smallholder and commercial farms, legumes are of economic benefit in reducing N applications. Incorporation of legume residues into soil as green manure is estimated to reduce by 20%—30% the amount of N fertilizer required to obtain same cereal grain yields and biomass when grown in a fallow-cereal system (Yang et al., 2014).

Apart from their ability to contribute N to natural and agricultural ecosystems, legumes can induce the solubilization and accumulation of other mineral nutrients in the rhizosphere and aboveground plant organs, whether planted as sole or intercropped with nonlegumes (Dakora and Philips, 2002; Xue et al., 2016; Darch et al., 2018). The enhanced availability and accumulation of mineral nutrients is largely caused by changes in rhizosphere pH and exudation of Kreb cycle intermediates by legume roots. An increase or decrease in the rhizosphere pH of legume plants, whether grown sole, intercropped with nonlegumes, or in rotation with other crops, can result in the solubilization of a variety of nutrient elements (Xue et al., 2016; Stagnari et al., 2017). Mobilization of mineral nutrients in the rhizosphere often leads to increased concentrations and ultimately enhanced uptake and accumulation in plant organs. However, where legumes are planted in rotation with nonlegumes, it is not always the case that there is increased nutrient availability or accumulation, as this is influenced largely by how the legume residue is managed (Reddy et al., 2013; Chen et al., 2018).

Legumes also play a crucial role in maintaining the physical, chemical, and biological niche of soils against extreme environmental events (Gao et al., 2017; Sun et al., 2018). This feature of legumes is highly desirable, given the current changing climate which creates unpredictable and extreme weather patterns with increased frequency and intensity of high temperatures, rainfall, and drought (Lakhraj-Govender et al., 2017; Maluleke and Mokwena, 2017; Sibanda et al., 2018; Hoffmann et al., 2018). The inclusion of legumes in cropping systems is known to increase the resilience of rhizosphere communities (Sun et al., 2018) largely because of the greater and more complex microbial community structure associated with legumes compared with cereals and grass species (Zhou et al., 2017). As a result, climatic events such as drought have had negative effect on cereal crops such as maize compared with legumes (Wagner et al., 2018), and this resilience justifies the inclusion of legumes in cropping systems.

Although the land area under cereal cultivation is larger than that under legumes (Foyer et al., 2016), the gradual increase in acreage under legume production underscores their contribution to a balanced ecosystem functioning and use by humans. For example, legumes serve as good sources of food and feed for humans and animals, cover crops for erosion control, biofertilizers for soil mineral enrichment, suppressors of weeds and diseases, enhancers of soil physicochemical and biological properties, promoters of microbial community resilience after extreme climatic events, reducers of greenhouse gas emissions, producers of biofuels, as well as esthetics in ecotourism and medicinal plants for treating human disorders and diseases (Stagnari et al., 2017). As a source of food, legumes are not only lower than cereals in area under cultivation but also in overall dietary intake on per capita consumption (McDermott and Wyatt, 2017), despite the fact that pulses are a better source of energy, protein, vitamins, minerals, dietary fiber, phytochemicals, antioxidants, and bioactive compounds than the highly consumed cereals (Amarowicz and Pegg, 2008; Singh et al., 2017).

This review is aimed at underlining the valuable role played by legume species in agricultural and natural ecosystems. This includes their ability to contribute N especially to low inputs systems through nitrogen fixation to increase the solubility, uptake, and accumulation of mineral nutrient elements in plant organs and in the rhizosphere soil and benefits associated with cereal-legume rotation, including increased biomass production, suppression of diseases, increased soil and tissue mineral nutrients, increased and diverse microbial populations, and increased resistance of physicochemical and biological properties to disturbances.

N contribution to soil

Symbiotic nitrogen fixation in legumes can occur in both natural and agricultural ecosystems and contribute substantial N that is cheap, sustainable, and environmentally friendly, in that it is less prone to leaching and volatilization and hence to environmental pollution. BNF is therefore an alternative to the use of N fertilizers which are costly and inaccessible to resource-poor farmers. Although legumes such as common bean contributed more N to the soil when supplemented with low levels of fertilizer P ($20\ kg\ ha^{-1}$) (Samago et al., 2018), there are reports where legumes produced substantial N without fertilization or incorporation of biomass (Pule-Meulenberg and Dakora, 2009; Nyemba and Dakora, 2010; Belane et al., 2011; Mohale et al., 2014, Table 4.1). For example, cowpea genotypes assessed for BNF in South Africa could contribute $31-131\ kg\ ha^{-1}$ of N without mineral supplements or rhizobial inoculation (Belane et al., 2011, Table 4.1). Measurements of BNF in farmers' fields without any fertilizer inputs showed $4-200\ kg\ N\ ha^{-1}$ contribution by Bambara groundnut (Pule-Meulenberg and Dakora, 2009; Nyemba and Dakora, 2010; Mohale et al., 2014). Groundnut can potentially contribute $58-188\ kg\ ha^{-1}$ of N (Mokgehle et al., 2014) and mungbean about $31-111\ kg\ ha^{-1}$ of N (Mokobane, 2013). There is also N that is contributed to ecosystems by tree legumes growing in natural settings. These include that shown by Tye and Drake (2012) who reported that *Acacia mearnsii* depends on atmospheric

Table 4.1 Effect of inoculant and crop species on dry matter and symbiotic performance of the cowpea component planted in sole and intrahole with sorghum and maize.

Treatment	Shoot DM ($g\ plant^{-1}$)	N (%)	N content ($g\ plant^{-1}$)	δ15N (‰)	Ndfa (%	N fixed ($kg\ ha^{-1}$)	Soil N uptake ($kg\ ha^{-1}$)
TUv 546 + R	78.89 ± 4.46a	1.11 ± 0.84b	0.85 ± 0.08a	−1.36 ± 0.15 ab	85.48 ± 3.63 ab	61.94 ± 5.81a	9.18 ± 1.97abc
Pan 311 + R	82.74 ± 6.73a	0.93 ± 0.78b	0.77 ± 0.07 ab	−1.43 ± 0.10 ab	79.52 ± 5.18 ab	50.40 ± 6.24abc	13.39 ± 3.88abc
TUv 546 − R	47.14 ± 6.83b	2.01 ± 1.50a	0.82 ± 0.04a	0.11 ± 0.64a	64.97 ± 9.66b	46.70 ± 8.18abcd	21.55 ± 5.46a
Pan 311 − R	28.08 ± 2.70de	2.01 ± 1.87a	0.56 ± 0.05cd	−0.32 ± 0.55a	71.47 ± 8.39b	31.05 ± 2.95d	15.71 ± 5.46abc
TUv 546 + ZM 521 + R	28.38 ± 1.27de	2.02 ± 1.83a	0.57 ± 0.02cd	−1.09 ± 0.23 ab	83.66 ± 7.68 ab	39.28 ± 3.47bcd	8.07 ± 3.67abc
Pan 311 + ZM 521 + R	25.50 ± 1.23e	1.96 ± 1.80a	0.50 ± 0.03d	−1.30 ± 0.21 ab	86.41 ± 3.18 ab	36.01 ± 2.63cd	5.76 ± 1.44bc
TUv 546 + ZM 521 − R	31.18 ± 3.91cde	2.13 ± 1.90a	0.64 ± 0.07bcd	−1.33 ± 0.24 ab	86.74 ± 3.68 ab	47.72 ± 6.71abcd	5.90 ± 1.71bc
Pan 311 + ZM 521 − R	34.84 ± 1.58cde	1.85 ± 1.65a	0.65 ± 0.04bcd	−2.07 ± 0.46b	97.97 ± 6.97a	52.97 ± 5.85abc	0.98 ± 4.27c
TUv 546 + M 48 + R	35.03 ± 3.95cde	2.16 ± 1.87a	0.73 ± 0.07abc	−0.06 ± 0.97a	67.46 ± 14.65b	40.72 ± 9.25bcd	20.23 ± 8.95 ab
Pan 311 + M 48 + R	32.72 ± 2.32cde	1.96 ± 1.73a	0.63 ± 0.04bcd	−0.94 ± 0.29 ab	80.84 ± 4.35 ab	41.73 ± 2.33bcd	10.99 ± 2.73abc
TUv 546 + M 48 − R	38.99 ± 2.23bcd	2.05 ± 1.91a	0.80 ± 0.05 ab	−1.34 ± 0.37 ab	86.97 ± 5.55 ab	57.76 ± 5.26 ab	8.59 ± 4.01abc
Pan 311 + M 48 − R	41.47 ± 1.91bc	1.82 ± 1.72a	0.76 ± 0.04 ab	−0.29 ± 0.60a	70.95 ± 9.12b	43.96 ± 5.84abcd	19.06 ± 6.27 ab
F-statistic	25.79**	13.95***	4.43***	2.02*	1.72*	2.41*	1.94*

*Values (Mean ± SE) followed by dissimilar letters in a column for each treatment is significantly different, * ($P \leq 0.05$), ** ($P \leq .01$), *** ($P \leq .001$). Crop species: Cowpea = PAN 311; TUv 546. M 48 = Sorghum, ZM 521 = maize, +R = with rhizobium inoculation, −R = without rhizobium inoculation.*

N_2 fixation 22% more than local species of acacia. Studies by Spriggs and Dakora (2008), Kanu and Dakora (2012), and Maseko and Dakora (2015) have shown that *Cyclopia*, *Aspalathus,* and *Psoralea* species display a high dependence on N_2 fixation for their N nutrition. *Aspalathus linearis* plants can also obtain over 100 kg N ha^{-1} from symbiotic fixation for their N nutrition (Muofhe and Dakora, 1999).

Where the aboveground biomass of legumes is not incorporated into the soil, symbiotic N contribution to the soil from the decomposition of legume roots and nodules is small. However, where legume is incorporated, significantly higher N is contributed to the soil. Adu-Gyamfi et al. (2007), for example, found a significantly large N balance where shoot biomass (minus grain) was incorporated into soil of farmer-managed cropping fields in Malawi and Tanzania. However, the N balance of soil following incorporation of legume residues can decrease or increase depending on the legume species. For example, a groundnut-maize rotation was more likely to yield a negative N balance compared with a velvet bean-maize sequence for a longer term N benefit (Okito et al., 2004). Thus the addition of residue from different legumes can contribute varying amounts of soil N as evidenced by mungbean which contributed 112 kg N ha^{-1} with residue incorporation versus 74 kg N ha^{-1} without incorporation, yielding an N balance of 64 kg N ha^{-1} and 9 kg N ha^{-1}, respectively (Shah et al., 2003). N balance from lentils also recorded 27 kg N ha^{-1} versus 16 kg N ha^{-1} with and without residue incorporation, respectively. The amount of N contributed by legumes to ecosystems is also influenced by soil fertility status. Kerman et al. (2018), for example, showed that legumes planted in nutrient-poor soils contributed significantly less N (15–123 kg N ha^{-1}) than those from fertile fields (16–145 kg N ha^{-1}), a finding similar to Mohale et al. (2014) who observed that N contribution by Bambara groundnut in farmers' fields was greater at sites with relatively high pH, Ca, P, and Fe concentration.

As a result the growth, grain yield, N_2 fixation, and N contribution by legumes is affected by the application of organic and inorganic fertilizers, as well as biofertilizers, individually and in combination (Table 4.1; Miheretu and Sarkodie-Addo, 2017; Manzeke et al., 2017; Htwe et al., 2019). Phosphorus in particular has been found to stimulate N_2 fixation. For example, the application of 40 kg P ha^{-1} to cowpea, groundnut, and Bambara groundnut increased N_2 fixation and N contribution when compared with control plants (Yakubu et al., 2010). There is therefore ample evidence to support the view that growth, grain yield, N_2 fixation, and N contribution by legumes is enhanced by the exogenous supply of minerals such as P, Mg, S, Mo, Zn, Se, and even biofertilizers (Moswatsi et al., 2013; Kyei-Boahen et al., 2017; Van Vugt et al., 2018; Getachew et al., 2017; Silva et al., 2018; Egamberdieva et al., 2018; Da Silva et al., 2017; Schütz et al., 2018).

Symbiotic N can also contribute to the N nutrition of companion crops through belowground transfer of the fixed N. This transfer of N can occur through (1) decomposition of belowground organs (e.g., roots and nodules) and incorporation of aboveground biomass into soil, (2) exudation of N compounds by roots of legumes (Dakora and Philips, 2002), and (3) N uptake mediated by mycorrhizal hyphae (He et al., 2003; Coskun et al., 2017). Exudation of N compounds can lead to the transfer of N from a legume to another plant as shown in a study by Dahlin and Stenberg (2010) which demonstrated N transfer between legumes and grasses.

N contribution by legume cover crops

The inclusion of legumes as cover crops in cropping systems can accrue multiple benefits which include soil water retention, control of pests and weeds, reduced runoff and soil erosion, retention, and reduced leaching of N, improved P supply and cycling, increased soil organic carbon, and enhanced soil aggregate stability (Zheng et al., 2018; Hallama et al., 2019). A hairy vetch-wheat cover crop system combined with N fertilization markedly reduced N loss in conservation agriculture without affecting yield (Shelton et al., 2018). A contribution by *Crotalaria ochroleusa* was not significant when used as a cover crop with biomass incorporation into the soil (Souza et al., 2018). Based on 65 studies in the USA and Canada, Marcillo and Miguez (2017) legumes used as winter cover crops increased maize yields through provision of sufficient N to crops grown in rotation, thus reducing inputs such as chemical fertilizers (Magdoff and Weil, 2004), as a result of increased N in soils (Wayman et al., 2015). Despite these marked positive effects of leguminous cover crops in cropping systems, in Southern Africa small-scale farmers are unable to tap these benefits because of uncontrolled bush fires, as well as the combined practice of crop and livestock farming, where livestock are allowed into farmers' fields for feed.

Factors affecting N contribution by legumes

The amount of N that a legume can contribute is affected by a variety of factors, including inoculation and the tillage system. The use of rhizobial inoculants alone could provide more than 50% of N fertilizer required for crop production in most marginal lands (Chianu et al., 2012). Cowpea inoculated with strain BR3267 showed grain productivity similar to the plants receiving 50 kg N ha^{-1} in the dryland areas of Brazil (Martins et al., 2003; Dwivedi et al., 2015). In Table 4.1, bacterial-inoculation of cowpea genotypes TUv 546 and Pan 311 contributed the highest N when grown as sole crops (61.94 and 50.40 kg ha^{-1}) compared with when intercropped with maize (39.28 and 36.01 kg ha^{-1}) or sorghum (40.72 and 41.73 kg ha^{-1}). By contrast, when TUv 546 was raised without rhizobial inoculation, the sole-planted grain legume contributed N that was similar to that when it was intercropped with maize (46.70 and 47.72 kg ha^{-1}) but exhibited the lowest compared with when intercropped with sorghum (57.76 kg ha^{-1}). Uninoculated sole-planted Pan 311 on the other hand contributed the lowest N (31.05) compared with when intercropped with maize (52.97) and sorghum (43.96).

Soil mineral enrichment by symbiotic legumes

High productivity legumes can contribute significantly to soil carbon enrichment. Low carbon in soils can be improved through the inclusion of symbiotic legumes in cropping systems. A maize-oat-alfalfa rotation resulted in higher soil organic carbon when compared with monoculture (Gregorich et al., 2000). Koné et al. (2008) also found an increase in soil organic carbon where *Mucuna pruriens*, *Pueraria phaseoloides,* and *Lablab purpureus* were grown as mixed culture. N-fixing tree species such as *Acacia auriculiformis* recorded higher soil organic matter and N concentration than nonfixing trees in a degraded field (Wang et al., 2010). Rotation of legumes with cereals increased soil organic carbon, as well as its stabilization in a no-till cropping system (Vieira et al., 2009; Ghosh et al., 2012).

Soil mineral enrichment by symbiotic legumes can occur through root-induced changes in rhizosphere pH. The rhizosphere is a carbon-rich zone usually colonized by a wide variety of microbial communities (Sasse et al., 2018). Changes in rhizosphere pH can be induced by cation/anion imbalance, root respiration, and organic acid exudation (Rengel, 2015). During N_2 fixation, legumes typically take up excess cations compared with anions, thus creating a cation/anion imbalance in the rhizosphere (Tang et al., 1997; Dakora and Philips, 2002; Shen et al., 2011; Betencourt et al., 2012; Lalati et al., 2013). The protons released as a result of the imbalance in cation and anion uptake leads to acidification of the rhizosphere, and a net decrease in rhizosphere pH which in turn promotes the dissolution of mineral elements such as P, Mn, Zn, Fe, and Cu in the rhizosphere (Treeby et al., 1989; Neumann and Römheld, 2002; Richardson et al., 2009).

Natural stands of legumes have also been reported to influence the chemistry of rhizosphere soils (Maseko and Dakora, 2017), for example, reported greater accumulation of P, K, Ca, Mg, Cu, and Mn in the rhizosphere of *Cyclopia genistoides* and *Cyclopia subternata* compared with the bulk soil as a result of a decreased rhizosphere pH. There was also an increase in rhizosphere acid and alkaline phosphatase activity of *Cyclopia* and *genistoides* and *C. subternata*, which resulted in greater availability and supply of P to plants (Maseko and Dakora, 2013). Muofhe and Dakora (2000) also reported an increase in the rhizosphere pH of inoculated *A. linearis* plants when compared with uninoculated NO_3-fed plants which presumably led to better plant nodulation from increased rhizobial survival. Lower rhizosphere pH combined with higher organic acid exudation and greater phosphatase activity also resulted in enhanced P uptake by legumes (Maltais-Landry, 2015).

The intercropping of legumes with nonlegumes also enriches the mineral nutrition, especially in low-nutrient soils (Singh and Singh, 2017). In low-nutrient soils the intercropping of legumes with cereals has generally resulted in increased concentrations of N, P, Fe, Na, S, Ca, and Mg (Betencourt et al., 2012; Makoi et al., 2014; Wang et al., 2017; Li et al., 2018). Interestingly, there are a few reports that showed no change or a decrease in nutrient concentration in soils even with the intercropping of legumes and cereals (Ndakidemi, 2005). In the case of N contribution the amount of N that is contributed by an intercropped legume can be lower or greater than that which it fixes when grown as a sole crop (Table 4.1). For example, where cowpea was planted in monoculture, it fixed higher N compared with when grown as a mixed culture (Makoi et al., 2010). On the other hand, increased P concentration associated with the intercropping of legumes and cereals is affected by rhizobial inoculation. For example, the results in Table 4.2 show that rhizobia-inoculated sole stands of cowpea exhibited the highest Pi in their rhizosphere and recorded about two-fold greater P in shoots compared with their uninoculated counterparts. In fact the markedly greater concentration of P in the rhizosphere is associated with higher phosphatase activity in the rhizosphere of grain legumes (Table 4.2; Makoi et al., 2010). Although phosphatase activity is higher in legumes compared with nonlegumes (Table 4.2; Venterink, 2011), the intercropping of legumes and cereal crops can also increase phosphatase activity than that of sole-cropped cereals (Table 4.2).

The increase in mineral nutrition promoted by legume/cereal intercropping improves nutrient accumulation in parts of legume and nonlegume crops. For example, there is vast literature that has demonstrated that intercropping of groundnut and maize improves the Fe nutrition of peanut (Inal et al., 2007; Zuo and Zhang, 2008; Guo et al., 2014; Dai et al., 2018). The significantly increased Fe nutrition could be caused by an improved jasmonate signaling which reportedly weakens ethylene signaling in the roots of groundnut (Dai et al., 2018). Besides Fe, intercropping maize and groundnut also significantly increased K, Fe, and Zn in shoots of the groundnut and K in shoot material of maize (Inal et al., 2007). In a field experiment where sweet potato was intercropped with soybean or groundnut without application of inorganic fertilizers, it resulted in significantly greater Fe and Zn contents in the sweet potato storage roots compared with any other treatment (Munda, 2017). Finally, for symbiotic grain legumes, higher N fixation is largely related to greater contribution of mineral nutrients (Belane et al., 2014). For example, cowpea genotypes that showed higher N_2 fixation consistently exhibited greater accumulation of P, K, Mg, S, Na, Fe, Cu, Zn, Mn, and B in leaves compared with low N_2 fixers which accumulated lesser amounts (Belane et al., 2014). Overall the soil mineral enrichment by symbiotic legumes especially under field conditions of smallholder-managed cropping systems depend on the fertility of the soil and availability of rainfall (Houshmandfr et al., 2019).

Growth and grain yield associated with legume/cereal rotation

Literature from research conducted on the benefit of rotation of legumes with cereals especially under field conditions in the African continent largely involve grain legumes and maize. When maize is grown in succession with maize and/or other cereals, including wheat, pearl millet, and sorghum, largely, its aboveground yield, particular the grain yield, is reduced or remain unchanged. By contrast the

Table 4.2 Effect of inoculant and crop species on phosphatase activity in the rhizosphere, inorganic and organic P concentration in rhizosphere soil, P concentration in plant shoots, and plant growth of sole and intrahole planted cowpea, sorghum, and maize.

Treatment crop	Rhizosphere phosphatase activity		Rhizosphere Pi concentration ($mg.kg^{-1}$)	Rhizosphere Po concentration ($g.kg^{-1}$)	Shoot P concentration		Rhizosphere pH
	Acid	Alkaline			Legume	Cereal	
	(μg p-nitrophenol.g^{-1} soil/h^{-1})				(%)		
TUv 546 + R	186.36 ± 6.09b	148.05 ± 6.30bc	9.33 ± 1.03ab	1542.49 ± 112.50a	13.05 ± 0.68a		6.16 ± 0.14bcd
Pan 311 + R	232.75 ± 9.82a	215.43 ± 3.87a	9.33 ± 0.94ab	1549.49 ± 31.50a	12.05 ± 1.03a		6.20 ± 0.16bcd
TUv 546 − R	155.61 ± 16.70cd	209.07 ± 12.98a	7.25 ± 0.63bcd	1646.49 ± 78.50a	6.83 ± 1.05bcd		6.34 ± 0.06bcd
Pan 311 − R	181.74 ± 11.62bc	227.08 ± 1.68a	10.00 ± 1.08a	1567.49 ± 85.50a	4.75 ± 0.40d		6.28 ± 0.18bcd
TUv 546 + ZM 521 + R	106.76 ± 1.60f	119.94 ± 0.80cde	4.63 ± 0.43e	1595.75 ± 114.25a	5.60 ± 1.15d	3.29 ± 0.48de	6.10 ± 0.16bcd
Pan 311 + ZM 521 + R	120.84 ± 4.03ef	101.82 ± 12.75ef	8.13 ± 0.80abc	1554.99 ± 8.00a	4.83 ± 0.33d	3.44 ± 0.52de	6.57 ± 0.26bc
TUv 546 + ZM 521 − R	162.06 ± 8.84bcd	158.35 ± 22.31b	9.75 ± 0.60ab	1574.24 ± 72.25a	8.54 ± 0.96b	6.64 ± 0.96bc	6.42 ± 0.08bcd
Pan 311 + ZM 521 − R	107.52 ± 2.53f	138.10 ± 6.03bcd	6.70 ± 0.31cde	1539.99 ± 156.25a	5.63 ± 0.14d	5.05 ± 0.23cd	6.13 ± 0.11bcd
TUv 546 + M 48 + R	123.79 ± 5.85ef	110.11 ± 9.06de	8.88 ± 0.94abc	1538.74 ± 40.25a	6.64 ± 0.92bcd	3.27 ± 0.23de	6.32 ± 0.29bcd
Pan 311 + M 48 + R	110.58 ± 12.17f	142.60 ± 9.43bcd	4.50 ± 0.20e	1687.50 ± 173.00a	4.35 ± 0.55d	3.87 ± 0.22de	5.97 ± 0.11d
TUv 546 + M 48 − R	230.84 ± 14.37a	102.14 ± 8.52ef	7.25 ± 0.48bcd	1487.24 ± 113.25a	6.06 ± 0.59cd	2.52 ± 0.26de	6.33 ± 0.19bcd
Pan 311 + M 48 − R	146.94 ± 10.01de	156.94 ± 10.43b	5.13 ± 0.83de	1458.24 ± 138.25a	8.31 ± 0.72bc	2.94 ± 0.45e	6.05 ± 0.13cd
ZM 521	122.52 ± 6.10ef	141.50 ± 17.79bcd	8.25 ± 1.11abc	1456.99 ± 142.00a		7.74 ± 1.10a	7.07 ± 0.05a
M 48	161.13 ± 7.82bcd	71.86 ± 5.35f	9.00 ± 0.82abc	1515.49 ± 58.50a		10.83 ± 1.47a	6.62 ± 0.15ab
F-statistic	20.33***	17.99***	5.82***	0.37 ns	13.29***	13.67***	3.06**

*Values (Mean ± SE) followed by followed by dissimilar letters in a column for each treatment is significantly different, * ($P \leq .05$), ** ($P \leq .01$), *** ($P \leq .001$), ns = not significant. Crop species: Cowpea = PAN 311; TUv 546. M 48 = Sorghum, ZM 521 = maize, +R = with* rhizobium *inoculation, −R = without* rhizobium *inoculation* cultivation of the cereal in rotation with grain legumes, including faba bean, velvet bean, groundnut, cowpea, soybean, as well as climbing and bush common beans, increased its growth and grain yield (Nebiyu et al., 2014; Okpara and Igwe, 2014; Uzoh et al., 2017; Niyuhire et al., 2017; Franke et al., 2018; Van Vugt et al., 2018). There are reports on significant increase in grain and total dry matter yield of other cereals such as pearl millet and sorghum when planted in rotation with grain legumes (Bagayoko et al., 2000a,b). Of noteworthy is that greater increases in the yield of the subsequent cereal to a precursor legume crop is recorded where the residue of the legume crop is incorporated into the soil (Uzoh et al., 2017). Furthermore the application of organic and inorganic fertilizers increased the yield parameters of maize (Van Vugt et al., 2018). Also, where the precursor grain legume is inoculated with biofertilizers, alone or in combination with fertilizers, these resulted in markedly higher yield parameters of maize (Rurangwa et al., 2018).

Although rotating cereals with legumes has had beneficial effects on the intercropped plants and soils, there is evidence that shows lack of and/or negative impact (Reddy et al., 2013; Chen et al., 2018). Lack of and/or negative effects of cereal-legume rotations include cases where the residues of the legume is burnt. For some farmers, especially those that do not practice crop and livestock farming, burning of the crop residue is common practice and part of land management (Punyalue et al., 2018). Through burning residue, N and other mineral nutrients contained in the residue are lost through volatilization. In other cases the stover from a preceding legume crop can be used as animal feed during the dry season or as household fuel (Smil, 1999). Largely these practices which result in the removal of the stover from cropping fields, the potential N, and mineral nutrients likely to have been contributed by the legume is lost, resulting in negative or no positive effect of cereal-legume rotation (Mohammad et al., 2012).

Across the world, intercropping of cereals and legumes is largely reported to benefit soils and crops. Where intercropping has had positive effects of cropping systems, there is no single factor that can be considered to be solely responsible for these positive effects of legumes grown in rotation with cereals. Generally, legumes grown as precursor to cereals in a legume-cereal rotation system contribute enhanced growth of the cereal, mineral nutrients, suppress weeds and diseases, promote diversity of microbes, and improves the physicochemical and biological properties of soils (Correia et al., 2014; Lal, 2017; Bainard et al., 2017). However, the benefits of a legume-cereal rotation are enhanced by the application of fertilizers and/or biofertilizers (Uzoh et al., 2017; Niyuhire et al., 2017; Schütz et al., 2018).

Effect of legume/cereal rotation on the fertility of soils

The rotation of cereal and legume crops has multiple benefits to soils, including increased quality and fertility (Soltani et al., 2014). According to literature, long-term rotations that include a diversity of cereal and legumes result in a markedly greater benefit compared with short-term set-ups using single cereal and a legume (Zuber et al., 2018; Maiga et al., 2019a,b). Among other parameters the quality and fertility of soils are determined largely by the concentration of especially essential elements that are contributed through increased soil organic carbon (Acosta-Martinez et al., 2008). For example, a rotation of grain legumes with maize increases the concentration of main macronutrients (Bagayoko et al., 2000a,b; Zingore et al., 2008; Yusuf et al., 2009; Okpara and Igwe, 2014), secondary macronutrients, and micronutrients (Manzeke et al., 2012). Improved soil fertility and quality through increased levels of soil organic carbon is affected by the incorporation of residues into the soil and application of fertilizers (Mohammad et al., 2012). In general the incorporation of plant residues into the soil increases the concentration of mineral nutrients (Yang et al., 2014).

The mineral nutrient benefit legume and cereal rotation has also been revealed in aboveground plant organs. This benefit is attributed partly to the ability of legumes to produce organic acid into soil solution. Interestingly, majority of literature on nutritional benefit of legume-cereal rotation on aboveground is on zinc. In 2009 Zn was ranked 17 in a list of leading risk factors responsible for deaths across the world (WHO, 2009). Thompson and Cohen (2015) suggested that to address nutrition-related challenges such as that of the deficiency of zinc in food crops or cropping fields of low-input households, there ought to be an approach that include empowering poor farmers and food consumers. Perhaps cropping systems such as that on rotating cereals and legumes could empower low-input farmers address the widespread deficiency of zinc. Where wheat was grown after a preceding clover, it had higher accumulation of zinc in the grain and shoots compared with when it was planted after a fallow treatment (Soltani et al., 2014). Interestingly the preceding clover resulted in the highest increase of shoot Zn concentration in the successive wheat compared with sunflower, Sudan grass, and safflower (Soltani et al., 2014). A rotation of clover and wheat where the legume was grown as a preceding crop resulted in significantly enhanced accumulation of grain Zn and N in the succeeding wheat (Khoshgoftarmanesh et al., 2017). Rotation of dry bean (*Phaseolus vulgaris* L.) and wheat with the legume planted as a preceding crop where at the end of the cropping, residues of the common bean were incorporated into the soil, and it had a significant effect on wheat through increasing its grain Zn compared with wheat residues and control treatment (Arani et al., 2018). In a field experiment where clover was grown as a preceding crop in a clover-wheat rotation, it had the highest effect on transforming carbonates-bound zinc form to exchangeable fraction and organically bound zinc labile forms and also enhanced the tissue Zn concentration in wheat (Norouzi et al., 2014). Incorporation of soybean plants into the soil as green manure increased the concentration of N and Zn in the grain of wheat in a soybean-wheat rotation experiment (Yang et al., 2014).

Effect of legume/cereal rotation on suppression of plant diseases

A cereal-legume rotation has positive effects, including increased soil fertility, enhanced microbial diversity, and improved disease suppressive capacity. The former is reportedly through fostering microbial communities that enhances the resistance against and/or that inhibit pathogens (Latz et al., 2012), or the increased abundance of disease suppressive microorganisms that would compete with pathogens for space or nutrients or inhibit pathogens through production of antibiotics (Latz et al., 2015). Studies that reported suppression of diseases by cereal-legume rotation include that by Bagayoko et al. (2000a,b) who showed markedly lower populations of nematodes where sorghum was rotated with groundnut; however, this was specific to plant species as the benefit was not observed in the pearl millet/cowpea rotation. A maize-soybean-wheat rotation resulted in markedly increased abundance of the *prnD* gene responsible for producing antifungal compound PRN, an indication of increased disease suppressive compared with monocultures (Peralta et al., 2018).

Effect of legume/cereal rotation on microbial populations in the soil

The presence of a diverse and richness of soil microorganisms in cropping systems is of great significance. These are largely responsible for residue decomposition, nutrient cycling, organic matter transformations, degradation of agrochemicals, and building soil tilth and structure (Lupwani and Kennedy, 2007). Biederbeck et al. (2005) reported that where legumes are incorporated as green manure, these resulted in greater bacteria, fungi, microbial biomass-carbon, microbial biomass-N, and enzyme activities (dehydrogenase, phosphatase, and arylsulfatase). Gonzalez-Chavez et al. (2010) showed that where wheat and soybean were rotated in a no-till cropping system, it resulted in markedly greater richness and biodiversity, as well as enhanced microbial biomass and chemical properties of the soil. Paungfoo-Lonhienne et al. (2017) reported an increase in the abundances of bacteria and archea where groundnut and soybean were planted as break crops to a continuous sugarcane plantation. The activities of acid and alkaline phosphatase were increased in the 20 cm topsoil of a cotton-groundnut rotation (Acosta-Martinez et al., 2008). The use of Chinese vetch as a winter crop to rice increased the NH_4^+ content and the populations of soil microbes (Xie et al., 2017). Intercropping of *Zanthoxylum bungeanum* and *Glycine max* resulted in higher density, biomass, and richness of macrofauna compared with that where *Z. bungeanum* was intercropped with sweet pepper and prickly ash monoculture (Wang et al., 2018). Interestingly microbial populations stimulated through cereal-legume rotations are decreased in soils that are low in fertility and/or without fertilizers. For example, Ai et al. (2018) found that cereal-legume rotations showed the lowest impact on bacterial microbiota relative to application of organic and inorganic fertilizers.

References

Acosta-Martinez, V., Acosta-Mercado, D., Sotomayor-Ramírez, D., Cruz-Rodríguez, L., 2008. Microbial communities and enzymatic activities under different management in semiarid soils. Applied Soil Ecology 38, 249−260.

Adu-Gyamfi, J.J., Myaka, F.A., Sakala, W.D., Odgaard, R., Versterager, J.M., Hong-Jensen, H., 2007. Biological nitrogen fixation and nitrogen and phosphorus budgets in farmer-managed intercrops of maize-pigeonpea in semi-arid southern and eastern Africa. Plant and Soil 295, 127−136.

Ai, C., Zhang, S., Zhang, X., Guo, D., Zhou, W., Huang, S., 2018. Distinct responses of soil bacterial and fungal communities to changes in fertilization regime and crop rotation. Geoderma 319, 156−166.

Amarowicz, R., Pegg, R.B., 2008. Legumes as a source of natural antioxidants. European Journal of Lipid Science and Technology 110, 865−878.

Arani, A.B., Namdari, A., Nazarli, H., 2018. Wheat grain enrichment with zinc through using zinc fertiliser preceding plant residues incorporation. Agriculture 64, 80−86.

Bagayoko, M., Buerkert, A., Lung, G., Bationo, A., Römheld, V., 2000a. Cereal/legume rotation effects on cereal growth in Sudano-Sahelian West Africa: soil mineral nitrogen, mycorrhizae and nematodes. Plant and Soil 218, 103−116.

Bagayoko, M., George, E., Römheld, V., Buerkert, A., 2000b. Effects of mycorrhizae and phosphorus on growth and nutrient uptake of millet, cowpea and sorghum on a West African soil. The Journal of Agricultural Science 135, 399−407.

Bainard, L.D., Navarro-Borrell, A., Hamel, C., Braun, K., Hanson, K., Gan, Y., 2017. Increasing the frequency of pulses in crop rotations reduces soil fungal diversity and increases the proportion of fungal pathotrophs in a semiarid agroecosystem. Agriculture, Ecosystems & Environment 240, 206−214.

Belane, A.K., Asiwe, J., Dakora, F.D., 2011. Assessment of N_2 fixation in 32 cowpea (*Vigna unguiculata* L. Walp) genotypes grown in the field at Taung in South Africa, using 15N natural abundance. African Journal of Biotechnology 10, 11450−11458.

Belane, A.K., Pule-Meulenberg, F., Makhubedu, T.I., Dakora, F.D., 2014. Nitrogen fixation and symbiosis-induced accumulation of mineral nutrients by cowpea (*Vigna unguiculata* L. Walp.). Crop & Pasture Science 65, 250−258.

Betencourt, E., Duputel, M., Colomb, B., Desclaux, D., Hinsinger, P., 2012. Intercropping promotes the ability of durum wheat and chickpea to increase rhizosphere phosphorus availability in a low P soil. Soil Biology and Biochemistry 46, 181−190.

Biederbeck, V.O., Zentener, R.P., Campbell, C.A., 2005. Soil microbial populations and activities as influenced by legume green fallow in a semiarid climate. Soil Biology and Biochemistry 37, 1775−1784.

Chen, J., Heiling, M., Resch, C., Mbaye, M., Gruber, R., Dercon, G., 2018. Does maize and legume crop residue mulch matter in soil organic carbon sequestration? Agriculture, Ecosystems & Environment 265, 123−131.

Chianu, J.N., Chianu, J.N., Mairura, F., 2012. Mineral fertilizers in the farming systems of sub-Saharan Africa. A review. Agronomy for Sustainable Development 32, 545–566.

Correia, M.V., Pereira, L.C.R., De Almeida, L., Williams, R.L., Freach, J., Nesbitt, H., Erskine, W., 2014. Maize-mucuna (*Mucuna pruriens* (L.) DC) relay intercropping in the lowland tropics of Timor-Leste. Field Crops Research 156, 272–280.

Coskun, D., Britto, D.T., Shi, W., Kronzucker, H.J., 2017. Nitrogen transformation in modern agriculture and the role of biological nitrification inhibition. Nature Plants 3, 170–174.

Dahlin, A.S., Stenberg, M., 2010. Cutting regime affects the amount and allocation of symbiotically fixed N in green manure leys. Plant and Soil 331, 401–412.

Dai, J., Qiu, W., Wang, N., Nakanishi, H., Zuo, Y., 2018. Comparative transcriptomic analysis of the roots of intercropped peanut and maize reveals novel insights into peanut iron nutrition. Plant Physiology and Biochemistry 127, 516–524.

Da Silva, J.R., Alexandre, A., Brigido, C., Oliveira, S., 2017. Can stress response genes be used to improve the symbiotic performance of rhizobia? AIMS Microbiology 3, 365–382.

Dakora, F.D., Philips, D., 2002. Root exudates as mediators of mineral acquisition in lownutrient environments. Plant and Soil 245, 35–47.

Darch, T., Giles, C.D., Blackwell, M.S.A., et al., 2018. Inter-and intra-species intercropping of barley cultivars and legume species, as affected by soil phosphorus availability. Plant and Soil 427, 125–138.

Dwivedi, A., Dev, I., Kumar, V., Yadav, R.S., Yadav, M., Gupta, D., Singh, A., Tomar, S.S., 2015. Potential role of maize-legume intercropping systems to improve soil fertility status under smallholder farming systems for sustainable agriculture in India. International Journal of Life Sciences Biotechnology and Pharma Research 4, 145.

Egamberdieva, D., Jabborova, D., Wirth, S.J., Alam, P., Alyemeni, M.N., Ahmad, P., 2018. Interactive effects of nutrients and *Bradyrhizobium japonicum* on the growth and root architecture of soybean (*Glycine max* L.). Frontiers in Microbiology. https://doi.org/10.3389/fmicb.2018.01000.

Erisman, J.W., Galloway, J.N., Seitzinger, S., Bleeker, A., Butterbach-Bahl, K., 2011. Reactive nitrogen in the environment and its effect on climate change. Current Opinion in Environmental Sustainability 3, 281–290.

Foyer, C.H., Lam, H.M., Nguyen, H.T., et al., 2016. Neglecting legumes has compromised human health and sustainable food production. Nature Plants 2. https://doi.org/10.1038/nplants.2016.112.

Franke, A.C., van den Brand, G.J., Vanlauwe, B., Giller, K.E., 2018. Sustainable intensification through rotations with grain legumes in Sub-Saharan Africa: a review. Agriculture, Ecosystems & Environment 261, 172–185.

Gao, D., Wang, X., Fu, S., Zhao, J., 2017. Legume plants enhance the resistance of soil to ecosystem disturbance. Frontiers of Plant Science 8, 1295. https://doi.org/10.3389/fpls.2017.01295.

Getachew, Z., Abera, G., Beyene, S., 2017. Rhizobium inoculation and sulphur fertilizer improved yield nutrients uptake and protein quality of soybean (*Glycine max* L.) varieties on Nitisols of Assosa area, Western Ethiopia. African Journal of Plant Science 11, 123–132.

Ghosh, P.K., Venkatesh, M.S., Hazra, K.K., Kumar, N., 2012. Long-term effect of pulses and nutrient management on soil organic carbon dynamics and sustainability on an inceptisol of Indo-Gangetic Plains of India. Experimental Agriculture 48, 473–487.

Gonzalez-Chavez, M.C., Aitkenhead-Peterson, J.A., Gentry, T.J., Zuberer, D., Hons, F., Loeppeert, R., 2010. Soil microbial community, C, N, and P responses to long-term tillage and crop rotation. Soil and Tillage Research 106, 285–293.

Gregorich, E.G., Liang, B.C., Drury, C.F., Mackenzie, A.F., McGill, W.B., 2000. Elucidation of the source and turnover of water soluble and microbial biomass carbon in agricultural soils. Soil Biology and Biochemistry 32 (5), 581–587.

Guo, X., Xiong, H., Shen, H., Qiu, W., Ji, C., Zhang, Z., Zuo, Y., 2014. Dynamics in the rhizosphere and iron-uptake gene expression in peanut induced by intercropping with maize: role in improving iron nutrition in peanut. Plant Physiology and Biochemistry 76, 36–43.

Hallama, M., Pekrun, C., Lambers, H., Kandeler, E., 2019. Hidden miners-the roles of cover crops and soil microorganisms in phosphorus cycling through agroecosystems. Plant and Soil 434, 7–45.

He, X., Critchley, C., Bledsoe, C., 2003. Nitrogen transfer within and between plants through common mycorrhizal networks (CMNs). Critical Reviews in Plant Sciences 22, 531–567.

Hoffmann, M.P., Odhiambo, J.J.O., Koch, M., Ayisi, K.K., Zhao, G., Soler, A., Rotter, R., 2018. Exploring adaptations of groundnut cropping to prevailing climate variability and extremes in Limpopo Province, South Africa. Field Crops Research 219, 1–13.

Houshmandfr, A., Ota, N., Siddique, K.H.M., Tausz, M., 2019. Crop rotation options for dryland agriculture: an assessment of grain yield response in cool-season grain legumes and canola to variation in rainfall totals. Agricultural and Forest Meteorology 275, 277–282.

Htwe, A.Z., Moh, S.M., Soe, K.M., Moe, K., Yamakawa, T., 2019. Effects of biofertilizer produced from *Bradyrhizobium* and *Streptomyces griseoflavus* on plant growth, nodulation, nitrogen fixation, nutrient uptake, and seed yield of mung bean, cowpea, and soybean. Agronomy. https://doi.org/10.3390/agronomy9020077.

Inal, A., Gunes, A., Zhang, F., Cakmak, I., 2007. Peanut/maize intercropping induced changes in rhizosphere and nutrient concentrations in shoots. Plant Physiology and Biochemistry 45, 350–356.

Kanu, S.A., Dakora, F.D., 2012. Symbiotic nitrogen contribution of root-nodule bacteria nodulating *Psoralea* species in the Cape Fynbos, South Africa. Soil Biology and Biochemistry 54, 68–76.

Kerman, M., Franke, A.C., Adjei-Nsiah, S., Ahiabor, B.D.K., Abaidoo, R.C., Giller, K.E., 2018. N_2-fixation and N contribution by grain legumes under different soil fertility status and cropping systems in the Guinea savannah of northern Ghana. Agriculture, Ecosystems & Environment 261, 201–210.

Koné, A.W., Tondoh, J.E., Angui, P.K., Bernhard-Reversat, F., Loranger-Merciris, G., Brunet, D., Brédoumi, S.T., 2008. Is soil quality improvement by legume cover crops a function of the initial soil chemical characteristics? Nutrient Cycling in Agroecosystems 82, 89–105.

Khoshgoftarmanesh, A.H., Norouzi, M., Afyuni, M., Schulin, R., 2017. Zinc biofortification of wheat through preceding crop residue incorporation into the soil. European Journal of Agronomy 89, 131–139.

Kyei-Boahen, S., Savala, C.E.N., Chikoye, D., Abaidoo, R., 2017. Growth and yield response of cowpea to inoculation and phosphorus fertilization in different environments. Frontiers of Plant Science. https://doi.org/10.3389/fpls.2017.00646.

Lakhraj-Govender, R., Grab, S., Ndebele, N.E., 2017. A homogenized long-term temperature record for the Western Cape Province in South Africa. International Journal of Climatology 37, 2337–2353.

Lal, R., 2017. Improving soil health and human protein nutrition by pulses-based cropping systems. Advances in Agronomy 145, 167–204.

Lalati, M., Pansu, M., Drevon, J.J., Ounane, S.M., 2013. Advantage of intercropping maize (*Zea mays* L.) and common bean (*Phaseolus vulgaris* L.) on yield and nitrogen uptake in Northeast Algeria. International Journal of Research in Applied Sciences 1, 1–7.

Latz, E., Eisenhauer, N., Rall, B.C., Allan, E., Roscher, C., Scheu, S., Jousset, A., 2012. Plant diversity improves protection against soil-borne pathogens by fostering antagonistic bacterial communities. Journal of Ecology 100, 597–604.

Latz, E., Eisenhauer, N., Rall, B.C., Scheu, S., Jousset, A., 2015. Plant community composition and the pathogen-suppressive potential of soils. Scientific Reports. https://doi.org/10.1038/srep23584.

Li, Q., Chen, J., Wu, L., Luo, X., Li, N., Arafat, Y., Lin, S., Lin, W., 2018. Belowground interactions impact the soil bacterial community, soil fertility and crop yield in maize/peanut intercropping systems. International Journal of Molecular Sciences. https://doi.org/10.3390/ijms19020622.

Lupwayi, N.Z., Kennedy, A.C., 2007. Grain legumes in Northern Great Plains: impacts on selected biological soil processes. Agronomy Journal 99, 1700–1709.

Magdoff, F., Weil, R.R. (Eds.), 2004. Soil Organic Matter in Sustainable Agriculture. CRC Press.

Maiga, A., Alhameid, A., Singh, S., Polat, A., Singh, J., Kumar, S., Osborne, S., 2019a. Responses of soil organic carbon, aggregate stability, carbon and nitrogen fractions to 15 and 24 years of no-till diversified crop rotations. Soil Research 57, 149–157.

Maiga, A., Alhameid, A., Singh, S., Polat, A., Singh, J., Kumar, S., Osborne, S., 2019b. Responses of soil organic carbon, aggregate stability, carbon and nitrogen fractions to 15 and 24 years of no-till diversified crop rotations. Soil Research. https://doi.org/10.1071/SR18068.

Makoi, J.H.J.R., Chimphango, S.B.M., Dakora, F.D., 2010. Elevated levels of acid and alkaline phosphatase activity in roots and rhizosphere of cowpea (*Vigna unguiculata* L. Walp) genotypes grown in mixed culture and at different densities with sorghum (*Sorghum bicolour* L. Moench). Crops and Pasture Science 61, 279–286.

Makoi, J.H., Chimphango, S.B., Dakora, F.D., 2014. Changes in rhizosphere concentration of mineral elements as affected by differences in root uptake and plant growth of five cowpea genotypes grown in mixed culture and at different densities with sorghum. American Journal of Experimental Agriculture 4, 193.

Maltais-Landry, G., 2015. Legumes have a greater effect on rhizosphere properties (pH, organic acids and enzyme activity) but a smaller impact on soil P compared to other cover crops. Plant and Soil 394, 139–154.

Maluleke, W., Mokwena, R.J., 2017. The effect of climate change on rural livestock farming: case study of Giyani policing area, Republic of South Africa. South African Journal of Agricultural Extension 45, 26–40.

Manzeke, G.M., Mapfumo, P., Mtambanengwe, F., Chikowo, R., Tendayi, T., Cakmak, I., 2012. Soil fertility management effects on maize productivity and grain zinc content in smallholder farming systems of Zimbabwe. Plant and Soil 361, 57–69.

Manzeke, M.G., Mtambanengwe, F., Nezomba, H., Watts, M.J., Broadley, M.R., Mapfumo, P., 2017. Zinc fertilization increases productivity and grain nutritional quality of cowpea (*Vigna unguiculata* L. Walp.) under integrated soil fertility management. Field Crops Research 213, 231–244.

Marcillo, G.S., Miguez, F.E., 2017. Corn yield response to winter cover crops: an updated meta-analysis. Journal of Soil and Water Conservation 72, 226–239.

Martins, L.M.V., Xavier, G.R., Rangel, F.W., Ribeiro, J.R.A., Neves, M.C.P., Morgado, L.B., Rumjanek, N.G., 2003. Contribution of biological nitrogen fixation to cowpea: a strategy for improving grain yield in the semi-arid region of Brazil. Biology and Fertility of Soils 38, 333–339.

Maseko, S.T., Dakora, F.D., 2013. Plant enzymes, root exudates, cluster roots and mycorrhizal symbiosis are the drivers of P nutrition in native legumes growing in P deficient soil of the Cape fynbos in South Africa. Journal of Agricultural Science and Technology A 3, 331–340.

Maseko, S.T., Dakora, F.D., 2017. Accumulation of mineral elements in the rhizosphere and shoots of Cyclopia and Aspalathus species under different settings of the Cape fynbos. South African Journal of Botany 110, 103–109.

Maseko, S.T., Dakora, F.D., 2015. Nitrogen nutrition, carbon accumulation and δ13C of *Cyclopia* and *Aspalathus* species in different settings of the Cape fynbos, South Africa. Journal of Plant Ecology 9, 586–595.

McDermott, J., Wyatt, A., 2017. The role of pulses in sustainable and healthy food systems. Annals of the New York Academy of Sciences 1392, 30–42.

Miheretu, A., Sarkodie-Addo, J., 2017. Response of cowpea (*Vigna unguiculata* L. Walp) varieties following application of nitrogen fertilizers and inoculation. Journal of Agriculture and Veterinary Science 10, 32–38.

Mohale, K.C., Belane, A.K., Dakora, F.D., 2014. Symbiotic N nutrition, C assimilation, and plant water use efficiency in Bambara groundnut (*Vigna subterranea* L. Verdc) grown in farmers' fields in South Africa, measured using 15N and 13C natural abundance. Biology and Fertility of Soils 50, 307–319.

Mohammad, W., Shah, S.M., Shehzadi, S., Shah, S.A., 2012. Effect of tillage, rotation and crop residues on wheat crop productivity, fertilizer nitrogen and water use efficiency and soil organic carbon status in dry area (rainfed) of north-west Pakistan. Journal of Soil Science and Plant Nutrition 12, 715–727.

Mokobane, K.F., 2013. Evaluation of symbiotic N nutrition, C accumulation, P uptake and yield of fifteen mungbean genotypes planted at two sites in South African region. Magister Technologiae, Agriculture (Crop Sciences) Dissertation. Tshwane University of Technology.

Mokgehle, S.N., Dakora, F.D., Mathews, C., 2014. Variation in N_2 fixation and N contribution by 25 groundnut (*Arachis hypogaea* L.) varieties grown in different agro-ecologies, measured using 15N natural abundance. Agriculture, Ecosystems & Environment 195, 161–172.

Moloto, R.M., Moremi, L.H., Soundy, P., Maseko, S.T., 2018. Biofortification of common bean as a complementary approach to addressing zinc deficiency in South Africans. Acta Agriculturae Scandinavica, Section B-Soil & Plant Science 68, 575–584.

Moswatsi, M.S., Kutu, F.R., Mafeo, T.P., 2013. Response of cowpea to variable rates and methods of zinc application under different field conditions. African Crop Science Conference Proceedings 11, 757–762.

Mueller, N.D., Lassaletta, L., Runck, B.C., Billen, G., Garnier, J., Gerber, J.S., 2017. Declining spatial efficiency of global cropland nitrogen allocation. Global Biogeochemical Cycles 31, 245–257.

Munda, E., 2017. Effect of Intercropping and Phosphorous Application on the Growth and Yield of Sweet Potato, Groundnut and Soybean (Doctoral Dissertation. Stellenbosch University, Stellenbosch.

Muofhe, M.L., Dakora, F.D., 2000. Modification of rhizosphere pH by the symbiotic legume *Aspalathus linearis* growing in a sandy acidic soil. Functional Plant Biology 27, 1169–1173.

Muofhe, M.L., Dakora, F.D., 1999. Nitrogen nutrition in nodulated field plants of the shrub tea legume *Aspalathus linearis* assessed using 15N natural abundance. Plant and Soil 209, 181–186.

Ndakidemi, P.A., 2005. Nutritional Characterisation of the Rhizosphere of Symbiotic Cowpea and Maize Plants in Different Cropping System (Doctoral Dissertation. Cape Peninsula University of Technology.

Nebiyu, A., Huygens, D., Upadhayay, H.R., Diels, J., Boeckx, P., 2014. Importance of correct B value determination to quantify biological N_2 fixation and N balances of faba beans (*Vicia faba* L.) via 15N natural abundance. Biology and Fertility of Soils 50, 517–525.

Neumann, G., Römheld, V., 1999. Root excretion of carboxylic acids and protons in phosphorus-deficient plants. Plant and Soil 211, 121–130.

Niyuhire, M.C., Pypers, P., Vanlauwe, B., Nziguheba, G., Roobroeck, D., Merckx, R., 2017. Profitability of diammonium phosphate use in bush and climbing bean-maize rotations in smallholder farms of Central Burundi. Field Crops Research 21, 252–260.

Norouzi, M., Khoshgoftarmanesh, A.H., Afyuni, M., 2014. Zinc fractions in soil and uptake by wheat as affected by different preceding crops. Soil Science & Plant Nutrition 60, 670–678.

Nyemba, C., Dakora, F.D., 2010. Evaluating N_2 fixation by food grain legumes in farmers' fields in three agro-ecological zones of Zambia, using 15N natural abundance. Biology and Fertility of Soils 46, 461–470.

Okito, A., Alves, B.J.R., Urquiaga, S., Boddey, R.M., 2004. Nitrogen fixation by groundnut and velvet bean and residual benefit to a subsequent maize crop. Pesquisa Agropecuária Brasileira 39, 1183–1190.

Okpara, I.M., Igwe, C.A., 2014. Soil chemical properties and legume-cereal rotation benefits in an Ultisol in Nsukka, South eastern Nigeria. African Journal of Biotechnology 13 (23).

Paungfoo-Lonhienne, C., Wang, W., Yeoh, Y.K., Halpin, N., 2017. Legume crop rotation suppressed nitrifying microbial community in a sugarcane cropping soil. Scientific Reports. https://doi.org/10.1038/s41598-017-17080-z.

Peralta, A.L., Sun, Y., McDaniel, M.D., Lennon, J.T., 2018. Crop rotational diversity increases disease suppressive capacity of soil microbiomes. Ecosphere 9, 2235.

Pule-Meulenberg, F., Dakora, F.D., 2009. Assessing the symbiotic dependency of grain and tree legumes on N_2 fixation for their N nutrition in five agro-ecological zones of Botswana. Symbiosis 48, 68–77.

Punyalue, A., Jamjod, S., Rerkasem, B., 2018. Intercropping maize with legumes for sustainable highland maize production. Mountain Research and Development 38, 35–44.

Reckling, M., Bergkvist, G., Watson, C.A., Stoddard, F.L., Zander, P.M., Walker, R.L., Pristeri, A., Toncea, I., Bachinger, J., 2016. Trade-offs between economic and environmental impacts of introducing legumes into cropping systems. Frontiers of Plant Science 7, 669. https://doi.org/10.3389/fpls.2016.00669.

Reddy, K.N., Zablotowicz, R.M., Krutz, L.J., 2013. Corn and soybean rotation under reduced tillage management: impacts on soil properties, yield, and net return. American Journal of Plant Sciences 4, 10–17.

Rengel, Z., 2015. Availability of Mn, Zn and Fe in the rhizosphere. Journal of Soil Science and Plant Nutrition 15, 397–409.

Richardson, A.E., Barea, J.M., McNeill, A.M., Prigent-Combaret, C., 2009. Acquisition of phosphorus and nitrogen in the rhizosphere and plant growth promotion by microorganisms. Plant and Soil 321, 305–339.

Rurangwa, E., Vanlauwe, B., Giller, K.E., 2018. Benefits of inoculation, P fertiliser and manure on yields of common bean and soybean also increase yield of subsequent maize. Agriculture, Ecosystems & Environment 261, 219–229.

Russo, T.A., Tully, K., Palm, C., Neill, C., 2017. Leaching losses from Kenyan maize cropland receiving different rates of nitrogen fertilizer. Nutrient Cycling and Agroecosystems 108, 195–209.

Samago, T.Y., Anniye, E.W., Dakora, F.D., 2018. Grain yield of common bean (*Phaseolus vulgaris* L.) varieties is markedly increased by rhizobial inoculation and phosphorus application in Ethiopia. Symbiosis 75, 245–255.

Sanchez, P.A., 2015. En route to plentiful food production in Africa. Nature Plants 1, 1–2.

Sasse, J., Kant, J., Cole, B.J., Klein, A.P., Arsova, B., Schlaepfer, P., Gao, J., Lewald, K., Zhalnina, K., Kosina, S., Bowen, B.P., 2018. Multi-lab EcoFAB Study Shows Highly Reproducible Physiology and Depletion of Soil Metabolites by a Model Grass, p. 435818 bioRxiv.

Schütz, L., Gattinger, A., Meier, M., Müller, A., Boller, T., Mäder, P., Mathimaran, N., 2018. Improving crop yield and nutrient use efficiency via biofertilization-A global meta-analysis. Frontiers of Plant Science 8, 2204.

Shah, Z., Shah, S.H., Peoples, M.B., Schwenke, G.D., Herridge, D.F., 2003. Crop residue and fertiliser N effects on nitrogen fixation and yields of legume-cereal rotations and soil organic fertility. Field Crops Research 83, 1–11.

Shelton, R.E., Jacobsen, K.L., McCulley, R.L., 2018. Cover crops and fertilization alter nitrogen loss in organic and conventional conservation agriculture systems. Frontiers of Plant Science 8, 2260.

Shen, J., Yuan, L., Zhang, J., Li, H., Bai, Z., Chen, X., Zhang, W., Zhang, F., 2011. Phosphorus dynamics: from soil to plant. Plant Physiology 111.

Sibanda, S., Grab, S., Ahmed, F., 2018. Spatio-temporal temperature trends and extreme hydro-climatic events in southern Zimbabwe. South African Geographical Journal 100, 210–232.

Silva, V.M., Boleta, E.H.M., Lanza, M.G.D.B., et al., 2018. Physiological, biochemical, and ultrastructural characterization of selenium toxicity in cowpea plants. Environmental and Experimental Botany 150, 172–182.

Singh, R., Singh, G.S., 2017. Traditional agriculture: a climate-smart approach for sustainable food production. Energy, Ecology and Environment 2, 296–316.

Singh, B., Singh, J.P., Shevkani, K., Singh, N., Kaur, A., 2017. Bioactive constituents in pulses and their health benefits. Journal of Food Science & Technology 54, 858–870.

Smil, V., 1999. Nitrogen in crop production: an account of global flows. Global Biogeochemical Cycles 13, 647–662.

Smil, V., 2001. Enriching the Earth: Fritz Haber, Carl Bosch and the transformation of world agriculture. Cambridge, MA.

Soltani, S., Khoshgoftarmanesh, A.H., Afyuni, M., Shrivani, M., Schulin, R., 2014. The effect of preceding crop on wheat grain zinc concentration and its relationship to total amino acids and dissolved organic carbon in rhizosphere soil solution. Biology and Fertility of Soils 50, 239–247.

Souza, G.A.V.D.S., de Souza, T.A.F., Santos, D., Rios, E.S., de Lima Souza, G.J., 2018. Agronomic evaluation of legume cover crops for sustainable agriculture. Russian Agricultural Sciences 44, 31–38.

Spriggs, A.C., Dakora, F.D., 2008. Field assessment of symbiotic N_2 fixation in wild and cultivated *Cyclopia* species in the South African fynbos by 15N natural abundance. Tree Physiology 29, 239–247.

Stagnari, F., Maggio, A., Galieni, A., Pisante, M., 2017. Multiple benefits of legumes for agriculture sustainability: an overview. Chemical and Biological Technologies in Agriculture 4, 2. https://doi.org/10.1186/s40538-16-0085-1.

Sun, F.S., Pan, K., Li, Z., Wang, S., Tariq, A., Olatunji, O.A., Sun, X., Zhang, L., Shi, W., Wu, X., 2018. Soybean supplementation increases the resilience of microbial and nematode communities in soil to extreme rainfall in an agroforestry system. The Science of the Total Environment 626, 776–784.

Tang, C., Barton, L., McLay, C.D.A., 1997. A comparison of proton excretion of twelve pasture legumes grown in nutrient solution. Australian Journal of Experimental Agriculture 37, 563–570.

Tang, Y., Li, X., Shen, W., Duan, Z., 2018. Effect of the slow-release nitrogen fertilizer oxamide on ammonia volatilization and nitrogen use efficiency in paddy soil. Agronomy 8, 53. https://doi.org/10.3390/agronomy8040053.

Thompson, B., Cohen, M., 2015. Increased concentrations of atmospheric carbon dioxide and zinc deficiency. The Lancet Global Health 3, e585–e586.

Treeby, M., Marschner, H., Römheld, V., 1989. Mobilization of iron and other micronutrient cations from a calcareous soil by plant-borne, microbial, and synthetic metal chelators. Plant and Soil 114, 217–226.

Tye, D.R.C., Drake, D.C., 2012. An exotic Australian Acacia fixes more N than a coexisting indigenous Acacia in a South African riparian zone. Plant Ecology 213, 251–257.

Uzoh, I.M., Obalum, S.E., Igwe, C.A., Abaidoo, R.C., 2017. Quantitative separation of nitrogen and non-nitrogen rotation benefits for maize following velvet bean under selected soil management practices. Agricultural Research 6, 378–388.

Van der Laan, M., Bristow, K.L., Stirzaker, R.J., Annandale, J.G., 2017. Towards ecologically sustainable crop production: a South African perspective. Agriculture, Ecosystems and Environment 236, 108–119.

Van Vugt, D., Franke, A.C., Giller, K.E., 2018. Understanding variability in the benefits of N_2-fixation in soybean-maize rotations on smallholder farmers' fields in Malawi. Agriculture, Ecosystems & Environment 261, 241–250.

Venterink, H.O., 2011. Legumes have a higher root phosphatase activity than other forbs, particularly under low inorganic P and N supply. Plant and Soil 347, 137–146.

Vieira, F.C.B., Bayer, C., Zanatta, J.A., Mielniczuk, J., Six, J., 2009. Building up organic matter in a subtropical Paleudult under legume cover-crop-based rotations. Soil Science Society of America Journal 73, 1699–1706.

Wagner, T., Richter, J., Joubert, D.F., Fischer, C., 2018. A dominance shift in arid savannah: an herbaceous legume outcomes local C4 grasses. Ecology and Evolution 2018, 1–9.

Wang, F., Li, Z., Xia, H., Zou, B., Li, N., Liu, J., Zhu, W., 2010. Effects of nitrogen-fixing and non-nitrogen-fixing tree species on soil properties and nitrogen transformation during forest restoration in southern China. Soil Science & Plant Nutrition 56, 297–306.

Wang, X., Deng, X., Pu, T., Song, C., Yong, T., Yang, F., Sun, X., Liu, W., Yan, Y., Du, J., Liu, J., Yang, W., 2017. Contribution of interspecific interactions and phosphorus application to increasing soil phosphorus availability in relay intercropping systems. Field Crops Research 204, 12–22.

Wang, S., Pan, K., Tariq, A., et al., 2018. Combined effects of cropping types and simulated extreme precipitation on the community composition and diversity of soil macrofauna in the eastern Qinghai-Tibet Plateau. Journal of Soils and Sediments. https://doi.org/10.1007/s11368-018-1998-z.

Wayman, S., Cogger, C., Benedict, C., Burke, I., Collins, D., Bary, A., 2015. The influence of cover crop variety, termination timing and termination method on mulch, weed cover and soil nitrate in reduced-tillage organic systems. Renewable Agriculture and Food Systems 30, 450–460.

World Health Organization, 2009. Global Health Risks: Mortality and Burden of Disease Attributable to Selected Major Risks. World Health Organization, Geneva.

Xie, Z., He, Y., Tu, S., Xu, C., Liu, G., Wang, H., Cao, W., Liu, H., 2017. Chinese milk vetch improves plant growth, development and 15N recovery in the rice-based rotation system of South China. Scientific Reports. https://doi.org/10.1038/s41598-017-03919-y.

Xue, Y., Xia, H., Christie, P., Zhang, Z., Li, L., Tang, C., 2016. Crop acquisition of phosphorus, iron and zinc from soil in cereal/legume intercropping systems: a critical review. Annals of Botany. https://doi.org/10.1093/aob/mcv182.

Yakubu, H., Kwari, J.D., Sandabe, M.K., 2010. Effect of phosphorus fertilizer on nitrogen fixation by some grain legume varieties in Sudano - Sahelian Zone of North Eastern Nigeria. Nigerian Journal of Basic and Applied Sciences 18, 19–26.

Yang, N., Wang, Z., Gao, Y., Zhao, H., Li, K., Li, F., Malhi, S.S., 2014. Effects of planting soybean in summer fallow on wheat grain yield, total N and Zn in grain and available N and Zn in soil on the Loess Plateau of China. European Journal of Agronomy 58, 63–72.

Yusuf, A.A., Abaidoo, R.C., Iwuafor, E.N.O., Olufajo, O.O., Sanginga, N., 2009. Rotation effects of grain legumes and fallow on maize yield, microbial biomass and chemical properties of an Alfisol in the Nigerian savanna. Agriculture, Ecosystems & Environment 129, 325–331.

Zheng, W., Gong, Q., Zhao, Z., Liu, J., Zhai, B., Wang, Z., Li, Z., 2018. Changes in the soil bacterial community structure and enzyme activities after intercrop mulch with cover crop for eight years in an orchard. European Journal of Soil Biology 86, 34–41.

Zhou, Y., Zhu, H., Fu, S., Yao, Q., 2017. Metagenomic evidence of stronger effect of stylo (legume) than bahiagrass (grass) on taxonomic and functional profiles of the soil microbial community. Scientific Reports 7, 10195. https://doi.org/10.1038/s41598-017-10613-6.

Zingore, S., Murwira, H.K., Delve, R.J., Giller, K.E., 2008. Variable grain legume yields, responses to phosphorus and rotational effects on maize across soil fertility gradients on African smallholder farms. Nutrient Cycling in Agroecosystems 80, 1–18.

Zuber, S.M., Behnke, G.D., Nafziger, E.D., Villamil, M.B., 2018. Carbon and nitrogen content of soil organic matter and microbial biomass under long-term crop rotation and tillage in Illinois, USA. Agriculture. https://doi.org/10.3390/agriculture8030037.

Zuo, Y., Zhang, F., 2008. Effect of peanut mixed cropping with gramineous species on micronutrient concentrations and iron chlorosis of peanut plants grown in a calcareous soil. Plant and Soil 306, 23–36.

Further reading

Jacoby, R., Peukert, M., Succurro, A., Koprivova, A., Kopriva, S., 2017. The role of soil microorganisms in plant mineral nutrition-current knowledge and future directions. Frontiers of Plant Science 8, 1617. https://doi.org/10.3389/fpls.2017.01617.

Lalati, M., Blavet, D., Alkama, N., Laoufi, H., Drevon, J.J., Gérard, F., Pansu, M., Ounane, S.M., 2014. The intercropping of cowpea-maize improves soil phosphorus availability and maize yields in an alkaline soil. Plant and Soil 385, 181–191.

CHAPTER

The role of synthetic fertilizers in enhancing ecosystem services in crop production systems in developing countries

Justice Nyamangara[1], Jefline Kodzwa[1], Esther Nyaradzo Masvaya[2], Gabriel Soropa[3]

[1]*Department of Environmental Sciences and Technology, Chinhoyi University of Technology, Chinhoyi, Zimbabwe;* [2]*International Crop Research Institute for the Semi-Arid Tropics, Matopos Research Station, Bulawayo, Zimbabwe;* [3]*Department of Crop Science and Post-Harvest Technology, Chinhoyi University of Technology, Chinhoyi, Zimbabwe*

Introduction

Optimal use of synthetic fertilizer, both organic and inorganic, is critical for enhancing ecosystem services of any food production system while avoiding potential negative environmental impacts. Ecosystem services relate to both direct and indirect human livelihood benefits (Fisher et al., 2008; La Notte et al., 2017) and are different and dynamic (Müller and Burkhard, 2012), and indicators for each should be clearly defined and capture the intended benefits to human population, as well as potential environmental impacts. The benefits of fertilizer use include higher crop yields which translate to improved human livelihood, food security and nutrition. Higher crop yields also contribute to preservation of wildlife habitat as it allows farmers to continuously cultivate the same piece of land for many years, eliminating the need for clearing new lands. Negative environmental impacts of fertilizer can be because of overapplication (e.g., water pollution) or suboptimal or nonuse (e.g., nutrient mining and land degradation) of the fertilizer. For most of the developing countries, especially in sub-Saharan Africa (SSA), the negative effects of suboptimal or nonuse of fertilizer are more common.

Crop production forms the major food production ecosystem in SSA, and it strongly interacts with other food ecosystems such as livestock and fisheries as a source of feed while benefiting from soil nutrient inputs (Homann-Kee Tui et al., 2015). The major food crops grown in SSA include maize (*Zea mays* L.), sorghum (*Sorgum bicolor* L.), pearl millet (*Pennisetum glaucum*), cassava (*Manihot esculenta*), groundnut (*Arachis hypogaea* L.), and soyabean (*Glycine max*). Most of the production is by smallholder farmers under various farming systems, and the

The Role of Ecosystem Services in Sustainable Food Systems. https://doi.org/10.1016/B978-0-12-816436-5.00005-6

use of synthetic fertilizer use is also variable. In Asia the major food crops are rice (*Oryza sativa*) and wheat (*Triticum aestivum*), and in South America the major cereal crops are wheat and maize.

The use of synthetic fertilizers to increase crop production in SSA is still very low (Table 5.1) because of various market, climatic, and soil-related conditions, and yet population growth rate outstrips food production (Liverpool-Tasie et al., 2016; Mwangi, 1996). Farmers in SSA use less than 10 kg nutrients per ha compared with 73 kg in South America and up to 135 kg in Asia (Marenya and Barrett, 2009; Palm et al., 2001). According to the Abuja Summit declaration the use of fertilizer in SSA should reach 50 kg of nutrients per ha by 2030 (AU/NEPAD/NPCA, 2006), a figure still low to produce enough food and to offset general nutrient mining caused by nonuse or use of suboptimal fertilizer rates over several decades in the subregion. The population of SSA is expected to increase by 2.5 fold and demand for cereals approximately triple by 2050 (Godfray and Garnett, 2014; van Ittersum et al., 2016). Therefore serious action needs to be taken to increase crop productivity through the use of modern and climate sensitive technologies, including synthetic fertilizers (Mwangi, 1996). Nitrogen (N) and phosphorus (P) are particularly deficient in the majority of soils in SSA which are characterized by low organic matter content (natural nutrient store) and high acidity in high rainfall zones and salinity in low rainfall areas.

Table 5.1 Fertilizer consumption (kg ha^{-1}/arable land) in selected countries in SSA and developing countries/regions.

Country/Region	1979/80†	1991/92†	2002‡	2015‡
Cote d'Ivoire	16.5	10.4	31.0	50.2
Ethiopia	2.7	7.1	17.0	18.5
Kenya	16.9	39.1	27.3	28.6
Malawi	19.3	44.7	29.7	30.2
Nigeria	3.6	13.3	4.5	8.3
South Africa	–	–	61.2	58.5
Zambia	11.4	11.9	26.1	55.9
Zimbabwe	44.3	52.8	35.7	22.9
Brazil	–	–	120.8	163.7
India	–	–	100.3	171.0
Sub-Saharan Africa	12.4	13.6	–	15.0
Latin America & Caribbean	–	–	89.5	122.7
Middle East & North Africa	–	–	88.6	105.3
South Asia	–	–	99.8	164.5

SSA, *Sub-Saharan Africa.*
Notes: *†Mwangi (1997), ‡World Bank Group (https://data.worldbank.org/indicator/AG.CON.FERT.ZS, consulted: 05/12/2018)*

In this chapter we explore the effect synthetic fertilizers, or lack thereof, in the ecosystem services of selected food production systems in SSA. Case studies from other developing countries in South Asia and South America will also be referred to where relevant.

The role of fertilizer in the agricultural Green Revolution of Zimbabwe

Zimbabwe agriculture has evolved over the years from the early 1960s to date because of various supporting factors in the industry. The first Green (Agricultural) Revolution in Zimbabwe (1960–80) was based on maize, the staple food crop, and built on technological foundations established in the 1900s by foreign settlers through public and private investments in agricultural development (Alumira and Rusike, 2005; Eicher, 1995). The implementation and success of the revolution initially targeted large-scale commercial farmers and later on skewed toward smallholder farmers, especially those in high potential (subhumid) areas. The four prime movers of the revolution were (1) new technology developed after long-term public and private investments in agricultural research; (2) human capital and managerial skills that were produced by investments in schools, training centers, and on-the-job experience; (3) biological and physical capital investments in improved varieties, livestock herds, and infrastructure such as roads, irrigation, and dams; and (4) investments in farmer support institutions such as marketing, credit, fertilizer, and seed distribution systems (Eicher, 1995). This resulted in increased yields for many crops, especially maize and tobacco (*Nicotiana tabacum*) during the period 1960–1980. The promotion of large-scale commercial production was based on the assumption that larger farms provide ideal conditions for economic development through higher agricultural productivity supported by economies of scale and mechanization (Musodza, 2015). However, the first Green Revolution did not manage to capture world attention because it excluded smallholder farmers, the majority whom continued to live in poverty under low productivity systems.

The second Green Revolution was led by smallholder farmers (2–4 ha land holding) soon after independence in 1980 after the new government introduced new policies and implemented supporting programs to assist marginalized smallholder farmers to increase productivity of both food and cash crops (Rukuni et al., 2006). This resulted in smallholder farmers doubling maize production from 738,000 tonnes in 1980 to 1.3 million tonnes in 1986 after adoption and use of improved maize varieties and fertilizers. About 70% of the produce in the second revolution came from smallholder farmers located in high potential areas of Mashonaland provinces which are the heart of the maize belt of the country. Owing to its success and because it was led by smallholder farmers, the Zimbabwe second Agricultural Revolution drew world attention and sent a strong message that smallholder farmers are capable of increasing crop productivity if enough support is rendered (Eicher, 1995). However, it is worth to note that preconditions that helped the first Agricultural Revolution to be successful were also important to sustain the success of the second Agricultural Revolution.

The four major components or preconditions that played a crucial role in supporting the second Zimbabwe Agricultural Revolution from 1980 to 1990 were the availability of hybrid maize varieties, improved agricultural extension services, expanded access to credit, and subsidies and higher guaranteed government maize prices. This section focusses on the role of synthetic fertilizer in the smallholder-led second agricultural Green Revolution in Zimbabwe.

The role of fertilizer in Zimbabwe's smallholder-led Agricultural Revolution

Before independence in 1980 few smallholder farmers had the capacity to purchase inorganic fertilizers because of limited access to credit for agricultural inputs, hence low adoption of hybrid maize varieties that had potential to increase yields but only when soil fertility is high (Moyo, 2014; Sserunkuuma, 2005) (Sserunkuuma, 2005; Moyo, 2014). However, that changed soon after independence as the new government improved smallholder farmers' access to credit facilities (through Agriculture Finance Corporation) and guarantee on output market prices resulting in smallholder farmers' fertilizer sales increasing from 7% to 34% from 1980 to 1986 (Eicher, 1995; Rukuni et al., 2006; ACB, 2016). Most fertilizer used by smallholder farmers was applied to maize. The increased use of fertilizer resulted in increased yields from the adopted maize hybrids by smallholder farmers, especially those in high potential areas of the country.

It is worth noting that the impact of maize hybrids among smallholder farmers was only realized after farmers had access to credit to purchase fertilizers. The adopted hybrids responded well to improved fertilizer use and hence yielded better compared with the improved open-pollinated varieties (OPVs) grown by most farmers. If grown without fertilizer, the yield of improved varieties was only slightly better than that of traditional varieties and improved OPVs, but productivity was increased several fold by fertilizer input. The high fertilizer use was skewed in favor of smallholder farmers in high potential areas compared with those in low rainfall areas (Fig. 5.1).

Fertilizer use among smallholder farmers decreased after the 1986/87 cropping season as credit loans were cut back because of high default rates. Those farmers who continued to use fertilizer applied suboptimal rates as they could no longer afford recommended rates because of increased poverty, high prices, weak access to input credit, lack of appropriate packages at nearby selling points, inadequate policies, and institutional support and low profitability (Chianu et al., 2012; Kaizzi et al., 2017; Tadele, 2017).

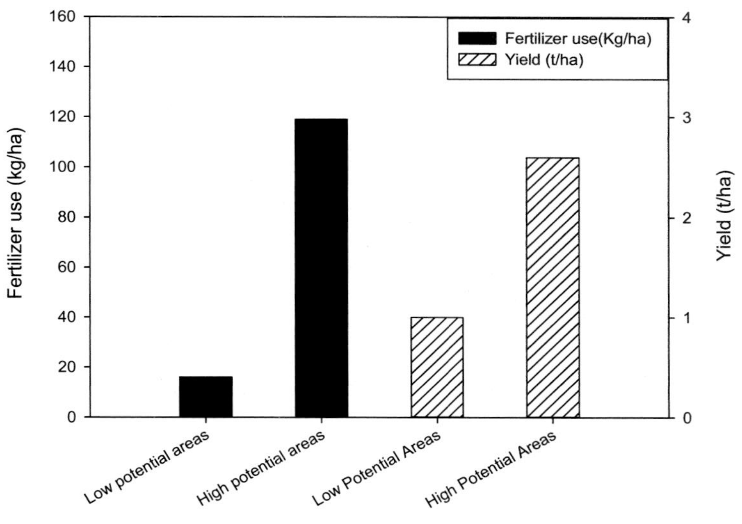

FIGURE 5.1

Average fertilizer application rate and average maize grain yield in low potential and high potential smallholder farming areas during the second agriculture revolution in Zimbabwe (Rukuni et al., 2006).

Potential environmental risk of fertilizers use

The increased fertilizer use among smallholder farmers greatly supported maize productivity during the period 1980–86. In the late 1980s the issues of environmental sustainability started to rise, and high fertilizer use was deemed to be risky and associated with potential to cause environmental problems (Rukuni et al., 2006), but this was in large-scale commercial farming subsector where high fertilizer rates were being applied to crops such as maize and tobacco. High levels of fertilizer applications were linked with environmental problems related to high levels of nitrogen (N) leaching into groundwater and phosphorus (P) deposition in surface waters though runoff and erosion.

The fertilizer application rates used by smallholder farmers were still low, and instead in many cases resulted in nutrient mining and consequent soil degradation. Henao and Baanante (2006) indicated that soil fertility in Africa was declining because of nutrient mining causing a decrease in crop yields, and in medium to long term a key source of environmental degradation. Extreme cases of nutrient mining and acidification, which are key indicators of soil chemical degradation, have been reported in the smallholder farming the located in high rainfall areas in Zimbabwe (Mugwira and Nyamangara, 1998; Zingore et al., 2008). Low soil organic matter contents (less than 1%), which limit limits crop response to fertilization, have also been reported in some smallholder areas (Mugwira and Nyamangara, 1998; Zingore et al., 2008; Du Preez, van Huyssteen & Mnkeni, 2011.

The role of fertilizer in the agricultural Green Revolution in Malawi

The economy of Malawi is agro-based and agriculture is key to improving food security, livelihoods, and nutrition of the smallholder farmers who constitute 80% of the country's population (Dorward and Kydd, 2004). The smallholder farmers depend on growing maize, the staple crop, as a primary crop under rain fed conditions on less than 1 hectare of land. The small land holding per household imply that smallholder farmers have to boost productivity to produce enough food for own consumption and possibly for sale. The first attempt to launch the agricultural Green Revolution in Malawi was in 1990, using semi-flint hybrids in 1990, and was disrupted by currency devaluations that resulted in the high cost of hybrid-seed and fertilizer and ever since the question of how to set the Green Revolution back on track dominated the policy agenda (Orr and Orr, 2002). The second launch was in 2005–06, riding off the call by the then United Nations Secretary General (Kofi Annan) for the support of African smallholder farmers with vital agricultural inputs, including fertilizer, high-yielding seed, water management, and training in what he termed the "Uniquely African Green Revolution" (Negin et al., 2009). The key components of the Green Revolution were subsidized inputs, including fertilizer, improved seed, market access, and farmer training.

Through the national input subsidy program, the Government of Malawi in 2005–06 season allocated two coupons per household to buy two 50 kg bags of fertilizer at 25% of the commercial price to grow maize on one acre (0.4 ha), 2 kg of hybrid seed, and 4.5 kg of open pollinated seed varieties at 63% subsidy (Denning et al., 2009). The subsidies and free distribution of seed and fertilizer encouraged maize production by smallholders (Chinsinga, 2011) and harvests surpassed those of previous years turning the country from a recipient of food aid into a food exporter and food aid donor to neighboring countries. From a 43% national food deficit in 2004–05, Malawi achieved a 53% surplus in 2006–07, and some of the surplus was exported to neighboring countries. The incremental effect of the fertilizer subsidy on maize production was estimated at 670,000 t for 2006–07, valued at US$117 million (Denning et al., 2009). Maize production in 2005–06, which received above average rainfall, was also greater than in the 2001–02 and 2002–03 seasons, which also received above average rainfall, by 1.02 million tonnes and 600,000 tonnes, respectively, suggesting a great impact of the subsidy program beyond better rainfall effect.

Malawi's recent experience may provide important lessons for achieving food security through smallholders in SSA. The focus was to support private agro-input dealers throughout the country, to support fertilizer use, and to curb the problem of physical inaccessibility of fertilizers. The increased fertilizer use had been attributed to stable fertilizer market policies which led to the increased expansion of private fertilizer distribution networks and greater competition among fertilizer wholesalers and importers. Overall coupon redemption was 95% for fertilizer and the private sector distributed 28% of the fertilizer (Denning et al., 2009).

Role of synthetic fertilizers in the sustainability of conservation agriculture

Conservation agriculture (CA) was first practiced in the United States of America back in the 1960s as a response to the devastating effects of the 1930s Dust Bowl (Hobbs, 2007). It is now practiced in all continents by both smallholder and large-scale farmers and in most land-based ecologies (rain fed and irrigated, organic and inorganic) (Table 5.2) and is one technology believed to increase resilience of agricultural production in the face of climate change (Rurinda et al., 2015). CA is based on three principles, namely, minimum soil disturbance, permanent soil surface mulch, and diversified cropping system that includes legume (FAO, 2014), but farmers have made various modifications to suit their agro-ecologies, farming systems, and/or market imperatives. CA systems include annual cropland systems with large-scale mechanized monocropping of maize, soyabeans, wheat, and other row crops and perennial crop systems, including orchard and plantation systems, pasture systems, mixed annual and perennial systems, agroforestry systems, and rice-based systems (Bhan and Behera, 2014; Joshi, 2011; Kassam et al., 2017).

The most current update shows that the global total CA cropland area in 2015/2016 was at least 180 Mha, corresponding to some 12.5% of the total global cropland, with the spread being equally split between the industrialized regions (52%) and the developing regions (48%) (Kassam et al., 2017). The distribution of the CA cropland area in selected continents is shown in Table 5.2.

Table 5.2 Cropland area under CA from selected regions in 2015–16.

Region	Area under CA (Mha)	% of global CA area	% of arable cropland of reporting countries
South America	69.9	39.0	63.2
North America	63.2	35.2	28.1
Asia	13.2	7.4	3.8
Africa	2.7	1.5	2.0

CA, Conservation agriculture.
Adapted from Kassam et al. (2017).

Role of synthetic fertilizer in CA systems in sub-Saharan Africa and South Asia

Agriculture in sub-Saharan Africa and South Asia faces significant challenges in coming decades to reconcile the need for increased food production to feed a growing population without significantly increasing the area under agricultural production (Stevenson et al., 2014). CA has been promoted by several scientists and institutions in these regions with the expectation that it can help reconcile these competing objectives and contribute to sustainable intensification. The mechanisms through which CA has been promoted in SSA and South Asia have been extensively reviewed (Andersson and Giller, 2012; Farooq and Siddique, 2015; Giller et al., 2009; Kassam et al., 2010). However, CA is not widely adopted in sub-Saharan Africa and South Asia as compared with the Americas (Table 5.1) owing to a lack of economic incentive for smallholder farmers—that the process of conversion to CA is not profitable over planning horizons of most farmers (Stevenson et al., 2014).

(a) sub-Saharan Africa

Vanlauwe et al. (2014) have argued that the appropriate use of fertilizer is required to define CA to enhance both crop productivity and produce sufficient crop residues to ensure soil cover under smallholder conditions in SSA. Other studies have also highlighted the importance of fertilizer in CA systems in smallholder SSA (Masvaya et al., 2017; Nyamangara et al., 2013). It has been suggested that CA systems require more N fertilizer than conventional tillage systems (Giller et al., 2009; Nyamangara et al., 2013) to offset immobilization that is brought about by the application of poor quality crop residues; however, there have not been any follow-up studies on how much more fertilizer this should be.

External inputs such as fertilizers are required to ensure that intensive systems become sustainable (Wopereis et al., 2006). Fertilizer, when applied appropriately, has been demonstrated to increase crop yields substantially and profitably under smallholder farming conditions (Table 5.3). In a study in the Eastern Cape in South Africa the use of inorganic fertilizer was the farmer preferred fertility management option with CA followed by a combination of both manure and inorganic fertilizers (Muzangwa et al., 2017). However, another study showed that with N fertilizer applications of below 100 kg N ha^{-1}, there were fewer yield advantages of CA over conventional tillage, but more yield benefits were obtained with high applications of above 100 kg N ha^{-1}. Nitrogen is often the most limiting nutrient for maize produced in the tropics (Osmond and Riha, 1996).

The use of mineral fertilizers is therefore key to production in CA systems, yet the use of fertilizers by smallholder farmers remains low as availability in rural areas remains a key constraint to their use, as well as their cost remains prohibitive (Masvaya et al., 2017; Muzangwa et al., 2017). Smallholder farmers in SSA use an average of 10 kg ha^{-1} fertilizer (IFDC, 2012). The situation is aggravated by lack of policy and institutional support, weak fertilizer markets, farmers' lack of access to credit and inputs, inappropriate fertilizer packaging sizes, deteriorating soil

Table 5.3 Effects of fertilizer application in conservation agriculture systems in southern Africa. Results are mainly from on-farm fields.

Country/ Region	Soil type	Cropping system	Tillage	Average N fertilizer applied (kg N ha^{-1})	Yield benefit from fertilizer application (%)	Reference
Zimbabwe	Various	Rainfed, continuous maize	Manual planting basins	31.9	231.0	Nyamangara et al. (2013)
Zimbabwe	Sand	Rainfed, continuous maize	Ripping	30.0	86.0	Masvaya et al. (2017)
Ethiopia	Various	Rainfed, continuous teff	Ripping	18.0	49.0	RockstrRockström_et_al_
Tanzania	Various	Rainfed, continuous maize	Ripping	50.6	74.0	
Mozambique	Sand	Rainfed, maize-pigeon pea intercropping	No tillage	40.0	0.6	Rusinamhodzi et al. (2012)
		Rainfed, maize-cowpea intercropping		40.0	2.6	
Zimbabwe	Red clay	Rainfed, continuous maize	Manual planting basins	60.0	156	Rusinamhodzi (2015)
	Sand	Rainfed, continuous maize	Manual planting basins	60.0	223	

science capacity, and weak agricultural extension (Chianu et al., 2012). Studies that involved CA and fertilizer use in SSA have focused mainly on increasing productivity as shown in Table 5.3, while not much work has been done on the subsequent impact of fertilizer use on the environment.

While crop diversification, fertilizer application, and increased duration of no-till may improve absolute maize yields under CA, we found little evidence that they improve relative adaptive capacity under conditions of enhanced heat or moisture stress (Steward et al., 2018).

(B) South Asia

In Asian countries such as India, efforts to adopt and promote CA technologies have been underway for just over a decade, and it is only recently that the technologies are finding rapid acceptance by farmers (Bhan and Behera, 2014). Efforts to develop and spread CA have been made through the combined efforts of several State Agricultural Universities, Indian Council of Agricultural Research institutes, and the Rice-Wheat Consortium for the Indo-Gangetic Plains. The spread of CA is taking place in the irrigated regions of the Indo-Gangetic plains where the rice-wheat cropping system dominates. In this region the focus of developing and promoting conservation technologies has been on zero-till seed-cum fertilizer drill for sowing of wheat in rice-wheat system. This is because the zero-till provided immediate, identifiable, and demonstrable economic benefits such as reductions in production costs, savings in fuel and labor requirements, and timely establishment of crops resulting in improved crop yields (Erenstein et al., 2012; Stevenson et al., 2014). Despite the clear benefits and increasing adoption of conservation tillage methods, most farmers, especially the small and medium scale farmers, have difficulties in following the wider basic tenets of CA, particularly year-round minimal soil disturbance, residue retention, and crop rotation (Stevenson et al., 2014).

Unlike in SSA, infrastructure, access to market, and institutional support are relatively well developed in Asia (Bhan and Behera, 2014; Lal, 2015). In CA systems in South Asia, soil nutrient deficiency correction is a prerequisite to ensure effectiveness of the system (Legoupil et al., 2015). In India the average fertilizer NPK requirement (kg ha^{-1}) in paddy rice is 81.7 N: 24.3 P_2O_5: 13.1 K, whereas in wheat is 99.6 N: 30.2 P_2O_5: 6.9 K and in maize is 41.7 N: 14.7 P_2O_5: 3.8 K. However, the NPK ratios are likely to vary with crops, cropping systems, CA practices, and soils and their reactions (Jat et al., 2011). According to Jat et al. (2011) the focus on fertilizer use in South Asia should be "feed the soil and let the soil feed the plant". The need for site-specific NPK ratios in CA-based systems is because CA has a distinct influence on soil quality and nutrient dynamics as compared with the traditional agriculture based on intensive tilled systems.

Role of synthetic fertilizer in CA success in the Americas

Agriculture in South America is highly dependent on imported N, P, and K fertilizers with the region being the world's third largest consumer of these fertilizers. The major crops grown include maize, wheat, soya bean (*Glycine* max (L.) Merr.), sugar bean (*Phaseolus vulgaris*), sugarcane (*Saccharum officinarum*), and coffee (*Coffea arabica*). Brazil, Argentina, Mexico, and Colombia constitute the major users of fertilizer in the region (FAO, 2015). However, Derpsch (2008) reported that the concepts about liming and fertilization have changed a lot in South America after shifting to the CA system. As in SSA, Derpsch (2008) suggested that N fertilizer requirements are higher under CA in South America. During the first 3 to 5 years, no-tillage may require greater N fertilizer inputs than conventional tillage systems (10−30 kg N ha^{-1} greater for cereal crops) (Derpsch et al., 2014). Applying the same amount of N for both no-tillage and conventional tillage may lead to yield reductions in no-tillage during initial years. The right amount of N must be applied in the right way to avoid volatilization losses and to avoid excessive soil disturbance when applying it.

In a study that considered CA adoption in Paraguay, Derpsch et al. (2016) put forward that medium-scale and large-scale farmers tended to succeed practicing CA, while small-scale farmers often do not. The authors cited that although fertilizer use is common with CA in South America, in Paraguay small-scale farmers tended to have more degraded soils than medium-scale and large-scale farmers, because small-scale framers generally cannot afford fertilizer and lime, or they are not available in remote areas, resulting in their infrequent application. As such disadoption rates were high among these farmers (Derpsch et al., 2016). Without fertilizer, it will be very difficult to produce the high amounts of residues needed for the restoration of soil fertility in an acceptable lapse of time under CA systems. Therefore fertilizers will need to be applied to be able to produce the biomass necessary for soil rehabilitation. In contrast, medium-scale and large-scale farmers can apply chemical fertilizers and lime and restore their soils as they can apply for credit and loan facilities, especially when they are members of a cooperative. Some studies on fertilizer use effects on crop yields in selected CA systems in comparison with conventional tillage are presented in Table 5.4.

Environmental effects of conservation agriculture

CA in the SSA, South Asia, and South America regions is very significant and widespread, but studies on the environmental impacts associated with the adoption of the technology are very few. Most studies have analyzed the relationship between the adoption of CA systems, soil properties, and crop yields. CA is reported in some studies to increase system diversity and stimulates biological processes in the soil and above the surface, as well as due to reduced erosion and leaching. In the long term the use of chemical fertilizer and pesticides, including herbicides, decrease the greenhouse gas (GHG) emission, eutrophication, and production costs (Holland, 2004). CA systems practiced in South Asia such as double no-till with retention of crop residues have been reported to reduce GHG emission equivalent nearly 13 tonnes ha^{-1} and regulate canopy temperature at grain filling stage to mitigate the

Table 5.4 Effects of fertilizer application in conservation agriculture systems in South America.

Country/Region	Soil type	Cropping system	Tillage	Average N fertilizer applied (kg N ha^{-1})	Yield difference between CA and conventional tillage (%)	Reference
Mexico	Silty loam	Continuous maize	No-till and conventional tillage	120.0	30	Govaerts et al. (2005)
		Continuous wheat	No-till and conventional tillage	120.0	10	
		Maize-wheat rotation	No-till and conventional tillage	120.0	20 (Maize) 10 (Wheat)	
Mexico	Clay	Irrigated wheat	Permanent raised beds and conventional tillage	278.0	4–7	Verhulst et al. (2011)
Southern Brazil	Clay	Maize-soyabean rotation	No-till and conventional tillage	91.0 (Maize)	6 (Maize) 5 (Soya bean)	Calegari et al. (2008)
Humid Argentina	Sand	Continuous wheat	No-till and conventional tillage	72.0 (NT) 0.0 (CT)	39	Abril et al. (2007)

terminal heat effects in wheat (Jat et al., 2011; Legoupil et al., 2015). It further helps to sequester carbon in soil at a rate ranging from about 0.2 to 1.0 t ha^{-1} $year^{-1}$ or more depending on the location and management practices (González-Sánchez et al., 2012).

However, intensifying agriculture with systems such as CA while helpful to feed the growing population, it often leaves several negative impacts on soil systems, including groundwater pollution, poor nutrient-use efficiency, waterlogging, salinization, and GHG emissions (Hobbs, 2007; Hobbs et al., 2006). The development of numerous macropores in CA soils with time could enhance NO_3^- leaching especially in sandy soils because of the low NO_3^- retention capacity in coarse soils (Daryanto et al., 2017; Kleinman et al., 2009). These negative effects of CA may also result in global warming and air pollution (Hobbs et al., 2006). Strides have been made to quantify the environmental impact of CA in the face of climate change focusing mainly on the ability of the technology to sequester carbon and minimize GHG emissions. The increase in N used in fertilizer may result in higher annual N_2O emissions in CA than in conventional tillage soils especially in the more humid South America which have been reported by several authors (e.g., Almaraz et al. (2009); Mutegi et al. (2010); Omonode and Vyn (2013). The discrepancy of the effects of tillage system on N_2O emissions is probably also related to distinct cropping system characteristics such as rainfall regime and soil type (Bayer et al., 2015). Therefore more research is needed to ascertain if the benefits of CA outweigh negative environmental impacts.

Fertilizer microdosing in semiarid areas of sub-Saharan Africa

In semiarid areas of SSA, limited access to and use of soil nutrients along with water shortages and drought substantially compromise production of food crops (Reynolds et al., 2015; Sanchez, 2002). Crop production without supplementary fertilizers on such inherently infertile soils results in cereal yields of less than 500 kg ha^{-1}, yields that always leave most households food insecure (Hove et al., 2008). On the other hand fertilizer rates recommended to smallholder farmers in the semiarid areas are often too high and ignore the unreliable nature of rainfall which is the major determinant of crop growth and yield. The small amounts applied in synchrony with rainfall amount and distribution, the so-called microdosing, can increase crop yields in semiarid areas. In fertilizer microdosing the fertilizer is applied at a rate which is considerably lower than generally recommended rates or rates required for optimal crop growth (Vandamme et al., 2018). To optimize use efficiency, fertilizer microdoses are not broadcast but applied locally in places where plants can easily access it.

Microdose fertilizer placement, specifically for N, has been used with success in widely spaced cereals such as maize, sorghum, and millet (Ncube et al., 2007; Twomlow et al., 2011). The targeted application of small quantities of fertilizer

PICTURE 5.1

Microdosing in Zimbabwe.

Photo credit: ICRISAT-Zimbabwe.

has been promoted as a sustainable "step up the ladder" of agricultural intensification (Aune and Bationo, 2008). The amount of fertilizer used under microdosing and the timing of application vary depending upon the target crop, region, planting density, and fertilizer formulation among other factors. The primary differences between microdosing and the recommended dosage are (A) the quantity-less than 6 grams of fertilizer (equivalent to a bottle-cap full or a three-finger pinch) placed at the base of each plant (Picture 1), (B) the timing microdosing requiring an earlier application after planting, and (C) the application method microdosing is placed into the soil at an optimized depth and distance from the crop (Natcher et al., 2016) (Picture 5.1).

Microdosing and crop productivity in SSA

Emerging literature continues to inform the practice of microdosing, as researchers study how a range of fertilizer quantities and application dates affect agronomic efficiency and profitability. For example, Sime and Aune (2014) investigated the effect of three separate microdosing rates of 27, 50, and 80 kg ha^{-1} of NPK fertilizer on maize in Ethiopia. Hayashi, Abdoulaye, Gerard, and Bationo (2008) investigated the effect of delayed application of microdose quantities on millet production. However, based on studies so far, microdosing at its various rates and timing has in general shown to be an effective technique in SSA for enhancing crop production and profitability while also addressing limited access to fertilizer (Hayashi et al., 2008; Tabo et al., 2011; Twomlow et al., 2011).

Hayashi et al. (2008), working in Niger, found that microdosing on millet increased yields and was more profitable than the control (no fertilizer) regardless of different application dates of the microdose quantities. Palé et al. (2009), investigating the effect of different fertilizer application rates and rainwater harvesting techniques on millet in Burkina Faso found that a modified form of fertilizer microdosing (microdose of 4g NPK per planting station + 20P ha^{-1} + 30N ha^{-1}) was effective in increasing grain yields over the recommended dosage rate. Tabo et al. (2008), working in Burkina Faso, Mali and Niger, reported that microdosing increased yields of millet and sorghum by 44%–120% in comparison with traditional farmer practices or recommended rates. In Zimbabwe, Twomlow et al. (2011) using data from broad scale on-farm trials on millet, sorghum and maize, reported that microdosing increased yield by 30%–100% compared with farmer practices.

The study by Twomlow et al. (2011) showed that 25–34 kg N ha^{-1} was optimum for maize in semiarid regions of Zimbabwe (Fig. 5.2). This study also showed a strong linear response at lower application rates, and about 11 kg of grain was produced per kg of fertilizer input, implying the economic returns to fertilizer investments at these suboptimal levels was profitable. Through a series of on-station and on-farm experiments on rice, a study in Benin demonstrated that a microdose of P consistently boosted early vigor and grain yield of dry-seeded, dibbled rice when targeted to the planting hole (Vandamme et al., 2018).

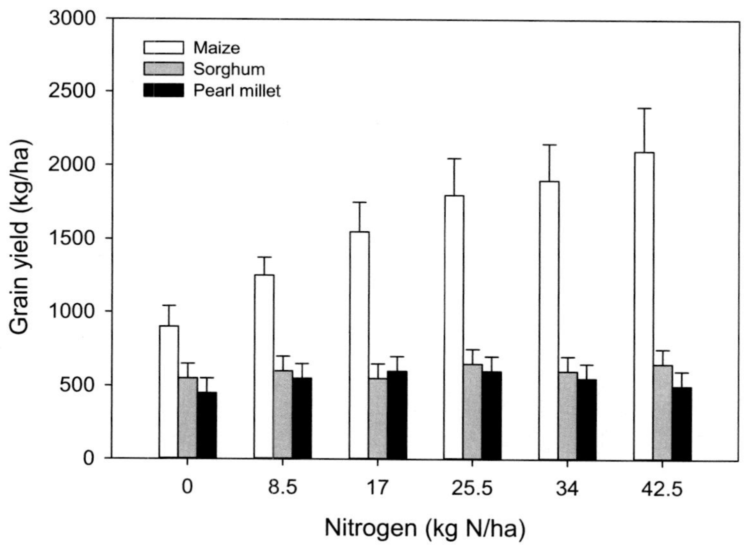

FIGURE 5.2

Grain response of maize, sorghum, and pearl millet to increasing levels of N fertilizer under farmer management. Mean of results from nine sites for three seasons for maize and two seasons for sorghum and pearl millet. Error bars represent standard errors of differences between the predicted means of the nitrogen by crop yield by season (Twomlow et al., 2011).

Table 5.5 Summary of the effects of fertilizer microdosing.

Positive effects	Negative effects
Increased yield at low cost	Limited quantities can be applied through microdosing
Efficient use of fertilizer	Nutrient mining if a small dose gives a high return
Earlier harvest	May increase labor demand if sowing and fertilizer application is undertaken in separate operations
Less risky than broadcasting of fertilizer	Depends on availability of mineral fertilizer
Can be mechanized	
Low-cost intensification	
Can stimulate the fertilizer market	
Less susceptible to parasitism by Striga species	
Possible entry point for agricultural intensification	
Requires relatively little equipment or technical skill	

Adapted from Aune and Coulibaly (2015).

However, in another study in Benin, Ibrahim, Abaidoo, Fatondji, and Opoku (2016) reported negative partial nutrient balances of -37 kg N ha^{-1}, -1 kg P ha^{-1}, and -34 kg K ha^{-1} under fertilizer microdosing in a low-input millet cropping system in Niger. These nutrient mining tendencies are likely to be more intense in maize cropping systems. Again, Singh and Ajeigbe (2007) also reported that the average use of 10 kg ha^{-1} yr^{-1} fertilizer in West Africa led to a negative balance of nutrients in the soil and continuous decline in crop yields leading to perpetuation of malnutrition, hunger, and poverty through the vicious circle of 'low input-low production-low income' and food insecurity. The effects of fertilizer microdosing, both positive and negative, are presented in Table 5.5.

Environmental effects of fertilizer microdosing

Fertilizer microdosing has been identified as one sustainable intensification practice that is also a climate smart technology (Murendo and Wollni, 2015). Fertilizer microdosing can help to improve the soil quality and fertility of highly eroded soils in Africa in a sustainable and affordable way by reducing costs spent on inputs and maximizing the efficiency of their use. However, the possibility of soil nutrient mining arising from fertilizer microdosing technology raises much concern on the sustainability of the technology. To improve the efficiency of the approach and better contribute to sustainable intensification, microdosing could be combined with use of manure or compost, improved seed, and water conservation techniques in arid regions to further increase yields and build natural capital. Most microdose fertilization studies have been based on sole crops of cereals such as maize, wheat, millet, and sorghum, but the effect of intercropping, strip cropping, or mixed cropping of these crops and microdosing of fertilizer remains largely unknown (Blessing et al., 2017). In most countries in SSA, most food crop farms are intercropped because of the predominance of smallholder farm holdings which tend to be small. Mixed cropping in combination with fertilizer microdosing and the use of organic manures may lead to sustainable crop production.

General conclusion

The use of synthetic fertilizer is generally low in developing countries, especially in SSA, and therefore most of the negative effects of cropping on the environment can be attributed to nutrient mining and consequent land degradation and not the use of synthetic fertilizer per se. As demonstrated in Malawi and Zimbabwe, increased use of synthetic fertilizer by smallholder farmers in SSA can boost yield and food security at both household and national levels. However, a sustainable model for fertilizer availability and affordability by smallholder farmers needs to be developed. Synthetic fertilizer is critical for the success of technologies, such as CA, which are meant to sustain crop productivity and food security while preserving the environment. In semiarid areas and other areas where smallholder farmers do not use fertilizer, fertilizer microdosing can be promoted as an entry point to fertilizer use to stimulate crop productivity at low risk in case of extended dry spells or droughts. However, prolonged use of the technology can result in nutrient mining and consequent soil degradation.

References

Abril, A., Baleani, D., Casado-Murillo, N., Noe, L., 2007. Effect of wheat crop fertilization on nitrogen dynamics and balance in the Humid Pampas, Argentina. Agriculture, Ecosystems & Environment 119 (1), 171–176. https://doi.org/10.1016/j.agee.2006.07.005.

ACB 2016. Zimbabwean Smallholder Support at the Crossroads: Diminishing Returns from Green Revolution Seed and Fertiliser Subsidies and the Agro-ecological Alternative. The African Centre for Biodiversity, Johannesburg. www.acbio.org.za

Almaraz, J.J., Zhou, X., Mabood, F., Madramootoo, C., Rochette, P., Ma, B.-L., Smith, D.L., 2009. Greenhouse gas fluxes associated with soybean production under two tillage systems in southwestern Quebec. Soil and Tillage Research 104 (1), 134–139.

Alumira, J., Rusike, J., 2005. The green revolution in Zimbabwe. Journal of Agricultural Development Economics 2 (1), 50–66.

Andersson, J.A., Giller, K., 2012. On heretics and God's blanket salesmen: contested claims for Conservation Agriculture and the politics of its promotion in African smallholder farming. In: Contested Agronomy: Agricultural Research in a Changing World. Earthscan, London.

AU/NEPAD/NPCA, 2006. The Abuja Declaration on Fertilizers for an African Green Revolution: Status of Implementation at Regional and National Levels (Retrieved from Addis Ababa and Midrand).

Aune, J.B., Bationo, A., 2008. Agricultural intensification in the Sahel—the ladder approach. Agricultural Systems 98 (2), 119—125.

Aune, J.B., Coulibaly, A., 2015. Microdosing of mineral fertilizer and conservation agriculture for sustainable agricultural intensification in Sub-Saharan Africa. In: Lal, R., Singh, B.R., Mwaseba, D.L., Kraybill, D., Hansen, D.O., Eik, O.L. (Eds.), Sustainable Intensification to Advance Food Security and Enhance Climate Resilience in Africa. Springer International Publishing, Switzerland, pp. 223—234.

Bayer, C., Gomes, J., Zanatta, J.A., Vieira, F.C.B., Piccolo, M.d.C., Dieckow, J., Six, J., 2015. Soil nitrous oxide emissions as affected by long-term tillage, cropping systems and nitrogen fertilization in Southern Brazil. Soil and Tillage Research 146, 213—222. https://doi.org/10.1016/j.still.2014.10.011.

Bhan, S., Behera, U.K., 2014. Conservation agriculture in India – problems, prospects and policy issues. International Soil and Water Conservation Research 2 (4), 1—12. https://doi.org/10.1016/S2095-6339(15)30053-8.

Blessing, O.C., Ibrahim, A., Safo, E.Y., Yeboah, E., Abaidoo, R.C., Logah, V., Monica, U.I., 2017. Fertilizer micro-dosing in West African low-input cereals cropping: benefits, challenges and improvement strategies. African Journal of Agricultural Research 12 (14), 1169—1176.

Calegari, A., Hargrove, W., Rheinheimer, D.D.S., Ralisch, R., Tessier, D., de Tourdonnet, S., de Fatima Guimarães, M., 2008. Impact of long-term no-tillage and cropping system management on soil organic carbon in an Oxisol: a model for sustainability. Agronomy Journal 100 (4), 1013—1019.

Chianu, J.N., Chianu, J.N., Mairura, F., 2012. Mineral fertilizers in the farming systems of sub-Saharan Africa. A review. Agronomy for Sustainable Development 32 (2), 545—566. https://doi.org/10.1007/s13593-011-0050-0.

Chinsinga, B., 2011. Seeds and subsidies: the political economy of input programmes in Malawi. IDS Bulletin 42 (4), 59—68.

Daryanto, S., Wang, L., Jacinthe, P.-A., 2017. Impacts of no-tillage management on nitrate loss from corn, soybean and wheat cultivation: a meta-analysis. Scientific Reports 7 (1), 12117.

Denning, G., Kabambe, P., Sanchez, P., Malik, A., Flor, R., Harawa, R., Magombo, C., 2009. Input subsidies to improve smallholder maize productivity in Malawi: toward an African Green Revolution. PLoS Biology 7 (1), e1000023.

Derpsch, R., 2008. No-tillage and conservation agriculture: a progress report. No-Till Farming Systems 7—39. Special Publication(3).

Derpsch, R., Franzluebbers, A., Duiker, S., Reicosky, D., Koeller, K., Friedrich, T., Weiss, K., 2014. Why do we need to standardize no-tillage research? Soil and Tillage Research 137, 16—22.

Derpsch, R., Lange, D., Birbaumer, G., Moriya, K., 2016. Why do medium-and large-scale farmers succeed practicing CA and small-scale farmers often do not?—experiences from Paraguay. International Journal of Agricultural Sustainability 14 (3), 269—281.

Dorward, A., Kydd, J., 2004. The Malawi 2002 food crisis: the rural development challenge. The Journal of Modern African Studies 42 (3), 343—361.

Du Preez, C.C., Van Huyssteen, C.W., Mnkeni, P.N., 2011. Land use and soil organic matter in South Africa 2: A review on the influence of arable crop production. South African Journal of Science 107 (5—6), 35—42.

Eicher, C.K., 1995. Zimbabwe's maize-based green revolution: preconditions for replication. World Development 23 (5), 805–818.

Erenstein, O., Sayre, K., Wall, P., Hellin, J., Dixon, J., 2012. Conservation agriculture in maize-and wheat-based systems in the (sub) tropics: lessons from adaptation initiatives in South Asia, Mexico, and Southern Africa. Journal of Sustainable Agriculture 36 (2), 180–206.

FAO, 2014. What Is Conservation Agriculture. Retrieved from. http://www.fao.org/ag/ca/1a.htm.

FAO, 2015. World Fertilizer Trends and Outlook to 2018. ISBN 978-92-5-108692-6, Annual Report 14. (Retrieved from FAO, Rome).

Farooq, M., Siddique, K.H.M., 2015. Conservation agriculture: concepts, brief history, and impacts on agricultural systems. In: Farooq, M., Siddique, K.H.M. (Eds.), Conservation Agriculture. Springer International Publishing, pp. 3–17.

Fisher, B., Turner, K., Zylstra, M., Brouwer, R., De Groot, R., Farber, S., Harlow, J.J. E.a., 2008. Ecosystem services and economic theory. Integration for Policy-relevant Research 18 (8), 2050–2067.

Giller, K.E., Witter, E., Corbeels, M., Tittonell, P., 2009. Conservation agriculture and smallholder farming in Africa: the heretics' view. Field Crops Research 114 (1), 23–34.

Godfray, H.C., Garnett, T., 2014. Food security and sustainable intensification. Philosophical Transactions of the Royal Society of London B Biological Sciences 369.

González-Sánchez, E.J., Ordóñez-Fernández, R., Carbonell-Bojollo, R., Veroz-González, O., Gil-Ribes, J.A., 2012. Meta-analysis on atmospheric carbon capture in Spain through the use of conservation agriculture. Soil and Tillage Research 122, 52–60. https://doi.org/10.1016/j.still.2012.03.001.

Govaerts, B., Sayre, K.D., Deckers, J., 2005. Stable high yields with zero tillage and permanent bed planting? Field Crops Research 94 (1), 33–42. https://doi.org/10.1016/j.fcr.2004.11.003.

Hayashi, K., Abdoulaye, T., Gerard, B., Bationo, A., 2008. Evaluation of application timing in fertilizer micro-dosing technology on millet production in Niger, West Africa. Nutrient Cycling in Agroecosystems 80 (3), 257–265.

Henao, J., Baanante, C., 2006. Agricultural production and soil nutrient mining in Africa: implications for resource conservation and policy development (In).

Hobbs, P.R., 2007. Conservation agriculture: what is it and why is it important for future sustainable food production? The Journal of Agricultural Science - Cambridge 145 (2), 127.

Hobbs, R.J., Arico, S., Aronson, J., Baron, J.S., Bridgewater, P., Cramer, V.A., Zobel, M., 2006. Novel ecosystems: theoretical and management aspects of the new ecological world order. Global Ecology and Biogeography 15 (1), 1–7. https://doi.org/10.1111/j.1466-822X.2006.00212.x.

Holland, J.M., 2004. The environmental consequences of adopting conservation tillage in Europe: reviewing the evidence. Agriculture, Ecosystems & Environment 103 (1), 1–25. https://doi.org/10.1016/j.agee.2003.12.018.

Homann-Kee Tui, S., Valbuena, D., Masikati, P., Descheemaeker, K., Nyamangara, J., Claessens, L., Erenstein, O., van Rooyen, A., Nkomboni, D., 2015. Economic trade-offs of biomass use in crop-livestock systems: Exploring more sustainable options in semi-arid. Zimbabwe. Agricultural Systems 134, 48–60. https://doi.org/10.1016/j.agsy.2014.06.009.

Hove, L., Mashingaidze, N., Twomlow, S., Nyathi, P., Moyo, M., Mupangwa, W., Masvaya, E., 2008. Micro Doses, Mega Benefits Promoting Fertilizer Use in Semi-arid Zimbabwe (Vol. Manual). International Crops Research Institute for the Semi-Arid Tropics.

Ibrahim, A., Abaidoo, R.C., Fatondji, D., Opoku, A., 2016. Fertilizer micro-dosing increases crop yield in the Sahelian low-input cropping system: a success with a shadow. Soil Science & Plant Nutrition 62 (3), 277–288.

IFDC, 2012. Fertilizer Deep Placement (FDP) Brochure. In: International Fertilizer Development Center. IFDC. http://issuu.com/ifdcinfo/docs/fdp_8pg_final_web?e=1773260/1756718.

Jat, M.L., Saharawat, Y.S., Gupta, R., 2011. Conservation agriculture in cereal systems of South Asia: nutrient management perspectives. Karnataka Journal of Agricultural Sciences 24 (1), 100–105.

Joshi, P., 2011. Conservation agriculture: an overview. Indian Journal of Agricultural Economics 66 (1), 53.

Kaizzi, K.C., Mohammed, M.B., Nouri, M., 2017. Fertilizer use optimization: principles and approach. In: Wortmann, C.S., Sones, K. (Eds.), Fertilizer Use Optimization in Sub-saharan Africa, vol. 17. CAB International, Nairobi, Kenya.

Kassam, A., Friedrich, T., Derpsch, R., 2010. Conservation agriculture in the 21st century: a paradigm of sustainable agriculture. In: Paper Presented at the European Congress on Conservation Agriculture.

Kassam, A., Friedrich, T., Derpsch, R., 2017. Global spread of conservation agriculture: interim update 2015/16. In: Paper Presented at the 7th World Congress on Conservation Agriculture, 1–4 August 2017, Rosario, Argentina.

Kleinman, P.J., Sharpley, A.N., Saporito, L.S., Buda, A.R., Bryant, R.B., 2009. Application of manure to no-till soils: phosphorus losses by sub-surface and surface pathways. Nutrient Cycling in Agroecosystems 84 (3), 215–227.

La Notte, A., D'Amato, D., Mäkinen, H., Paracchini, M.L., Liquete, C., Egoh, B., Crossman, N.D., 2017. Ecosystem services classification: a systems ecology perspective of the cascade framework. Ecological Indicators 74, 392–402.

Lal, R., 2015. A system approach to conservation agriculture. Journal of Soil and Water Conservation 70 (4), 82A–88A.

Legoupil, J.-C., Lienhard, P., Khamhoung, A., 2015. Conservation agriculture in southeast Asia. In: Conservation Agriculture. Springer, pp. 285–310.

Liverpool-Tasie, S., Omonona, B., Awa, S., Ogunleye, W., Padilla, S., Reardon, T., 2016. Growth and Transformation of Chicken & Eggs Value Chains in Nigeria (Retrieved from).

Marenya, P.P., Barrett, C.B., 2009. Soil quality and fertilizer use rates among smallholder farmers in western Kenya. Agricultural Economics 40 (5), 561–572.

Masvaya, E.N., Nyamangara, J., Descheemaeker, K., Giller, K.E., 2017. Tillage, mulch and fertiliser impacts on soil nitrogen availability and maize production in semi-arid Zimbabwe. Soil and Tillage Research 168, 125–132. https://doi.org/10.1016/j.still.2016.12.007.

Moyo, M., 2014. Effectiveness of a Contract Farming Arrangement: A Case Study of Tobacco Farmers in Mazowe District in Zimbabwe. Stellenbosch University, Stellenbosch.

Mugwira, L., Nyamangara, 1998. Organic carbon and plant nutrients in soils under maize in Chinamhora Communal Area, Zimbabwe. In: Bergstrom, L.F., Kirchmann, H. (Eds.), Carbon Nutrient Dynamics in Natural Agricultural Tropical Ecosystems. CAB International, UK, pp. 15–21.

Müller, F., Burkhard, B., 2012. The indicator side of ecosystem services. Ecosystem Services 1 (1), 26–30.

Murendo, C., Wollni, M., 2015. Ex-post impact assessment of fertilizer microdosing as a climate-smart technology in sub-saharan Africa. ICRISAT.

Musodza, C., 2015. Zimbabwe's Fast Track Land Reform Programme and the Decline in National Food Production: Problems of Implementation, Policy and Farming Practices.

Mutegi, J.K., Munkholm, L.J., Petersen, B.M., Hansen, E.M., Petersen, S.O., 2010. Nitrous oxide emissions and controls as influenced by tillage and crop residue management strategy. Soil Biology and Biochemistry 42 (10), 1701–1711.

Muzangwa, L., Mnkeni, P.N.S., Chiduza, C., 2017. Assessment of conservation agriculture practices by smallholder farmers in the eastern Cape province of South Africa. Agronomy 7 (3), 46.

Mwangi, W.M., 1996. Low use of fertilizers and low productivity in sub-Saharan Africa. Nutrient Cycling in Agroecosystems 47 (2), 135–147.

Natcher, D., Bachmann, E., Pittman, J., Kulshreshtha, S., Baco, M.N., Akponikpe, P., Peak, D., 2016. Knowledge diffusion and the adoption of fertilizer microdosing in Northwest Benin. Sustainable Agriculture Research 5 (3), 1.

Ncube, B., Dimes, J., Twomlow, S., Mupangwa, W., Giller, K., 2007. Raising the productivity of smallholder farms under semi-arid conditions by use of small doses of manure and nitrogen: a case of participatory research. Nutrient Cycling in Agroecosystems 77 (1), 53–67.

Negin, J., Remans, R., Karuti, S., Fanzo, J.C., 2009. Integrating a broader notion of food security and gender empowerment into the African Green Revolution. Food Security 1 (3), 351–360.

Nyamangara, J., Masvaya, E.N., Tirivavi, R., Nyengerai, K., 2013. Effect of hand-hoe based conservation agriculture on soil fertility and maize yield in selected smallholder areas in Zimbabwe. Soil and Tillage Research 126, 19–25.

Omonode, R.A., Vyn, T.J., 2013. Nitrification kinetics and nitrous oxide emissions when nitrapyrin is coapplied with urea–ammonium nitrate. Agronomy Journal 105 (6), 1475–1486.

Orr, A., Orr, S., 2002. Agriculture and micro enterprise in Malawi's rural south: Overseas development institute (ODI). Agricultural research & extension network (AgREN), Network Paper No. 119.

Osmond, D., Riha, S., 1996. Nitrogen fertilizer requirements for maize produced in the tropics: a comparison of three computer-based recommendation systems. Agricultural Systems 50 (1), 37–50.

Palé, S., Mason, S., Taonda, S., 2009. Water and fertilizer influence on yield of grain sorghum varieties produced in Burkina Faso. South African Journal of Plant and Soil 26 (2), 91–97.

Palm, C.A., Gachengo, C.N., Delve, R.J., Cadisch, G., Giller, K.E., 2001. Organic inputs for soil fertility management in tropical agroecosystems: application of an organic resource database. Agriculture, Ecosystems & Environment 83 (1), 27–42.

Reynolds, T.W., Waddington, S.R., Anderson, C.L., Chew, A., True, Z., Cullen, A., 2015. Environmental impacts and constraints associated with the production of major food crops in Sub-Saharan Africa and South Asia. Food Security 7 (4), 795–822. https://doi.org/10.1007/s12571-015-0478-1.

Rockström, J., Kaumbutho, P., Mwalley, J., Nzabi, A., Temesgen, M., Mawenya, L., Damgaard-Larsen, S., 2009. Conservation farming strategies in East and Southern Africa: yields and rain water productivity from on-farm action research. Soil and Tillage Research 103 (1), 23–32.

Rukuni, M., Tawonezvi, P., Eicher, C., Munyuki-Hungwe, M., Matondi, P., Harare, 2006. Zimbabwe's Agricultural Revolution Revisited, pp. 119–140.

Rurinda, J., Wijk, M.T., Mapfumo, P., Descheemaeker, K., Supit, I., Giller, K.E., 2015. Climate change and maize yield in southern Africa: what can farm management do? Global Change Biology 21 (12), 4588–4601.

Rusinamhodzi, L., 2015. Tinkering on the periphery: labour burden not crop productivity increased under no-till planting basins on smallholder farms in Murehwa district, Zimbabwe. Field Crops Research 170 (0), 66–75. https://doi.org/10.1016/j.fcr.2014.10.006.

Rusinamhodzi, L., Corbeels, M., Nyamangara, J., Giller, K.E., 2012. Maize–grain legume intercropping is an attractive option for ecological intensification that reduces climatic risk for smallholder farmers in central Mozambique. Field Crops Research 136, 12–22.

Sanchez, P.A., 2002. Soil fertility and hunger in Africa. Science 295 (5562), 2019–2020. https://doi.org/10.1126/science.1065256.

Sime, G., Aune, J.B., 2014. Maize response to fertilizer dosing at three sites in the Central Rift Valley of Ethiopia. Agronomy 4 (3), 436–451.

Singh, B., Ajeigbe, H., 2007. Improved cowpea-cereals-based cropping systems for household food security and poverty reduction in West Africa. Journal of Crop Improvement 19 (1–2), 157–172.

Sserunkuuma, D., 2005. The adoption and impact of improved maize and land management technologies in Uganda. Journal of Agricultural Development Economics 2 (1), 67–84.

Stevenson, J.R., Serraj, R., Cassman, K.G., 2014. Evaluating conservation agriculture for small-scale farmers in sub-saharan Africa and South Asia. Agriculture, Ecosystems & Environment 187 (0), 1–10. https://doi.org/10.1016/j.agee.2014.01.018.

Steward, P.R., Dougill, A.J., Thierfelder, C., Pittelkow, C.M., Stringer, L.C., Kudzala, M., Shackelford, G.E., 2018. The adaptive capacity of maize-based conservation agriculture systems to climate stress in tropical and subtropical environments: a meta-regression of yields. Agriculture, Ecosystems & Environment 251, 194–202.

Tabo, R., Bationo, A., Amadou, B., Marchal, D., Lompo, F., Gandah, M., Fatondji, D., 2011. Fertilizer microdosing and "warrantage" or inventory credit system to improve food security and farmers' income in West Africa. In: Innovations as Key to the Green Revolution in Africa. Springer, pp. 113–121.

Tabo, R., Bationo, A., Hassane, O., Amadou, B., Fosu, M., Sawadogo-Kabore, S., Koala, S., 2008. Fertilizer microdosing for the prosperity of resource poor farmers: a success story. In: Paper Presented at the Increasing the Productivity and Sustainability of Rainfed Cropping Systems of Poor Smallholder Farmers: Proceedings of the International Workshop on Rainfed Cropping Systems, Tamale, Ghana.

Tadele, Z., 2017. Raising crop productivity in Africa through intensification. Agronomy 7 (1), 22.

Twomlow, S., Rohrbach, D., Dimes, J., Rusike, J., Mupangwa, W., Ncube, B., Maphosa, P., 2011. Micro-dosing as a pathway to Africa's Green Revolution: evidence from broad-scale on-farm trials. In: Innovations as Key to the Green Revolution in Africa. Springer, pp. 1101–1113.

van Ittersum, M.K., van Bussel, L.G., Wolf, J., Grassini, P., van Wart, J., Guilpart, N., Mason-D'Croz, D., 2016. Can sub-Saharan Africa feed itself? Proceedings of the National Academy of Sciences 113 (52), 14964–14969.

Vandamme, E., Ahouanton, K., Mwakasege, L., Mujuni, S., Mujawamariya, G., Kamanda, J., Saito, K., 2018. Phosphorus micro-dosing as an entry point to sustainable intensification of rice systems in sub-Saharan Africa. Field Crops Research 222, 39–49.

Ecosystem services

Ecosystem services are the conditions and processes through which natural ecosystems and the species that them sustain and fulfill human life in other words, ecosystem services are benefits people obtain from ecosystems (Millenium Ecosystem Assessment, 2005). Ecosystem services are categorized into provisioning, regulating, supporting, and cultural services. The material benefits obtained from ecosystems such as water, food, wood, and other goods are called provisioning services. The maintenance services that ecosystems provide such as maintaining the quality of air and soil, providing flood and disease control, or pollinating crops are referred to as regulating services. Supporting services of ecosystems involve providing living spaces and maintaining diversity of the biota. The nonmaterial benefits obtained from ecosystems are called cultural services. They include inspiration, cultural identity, sense of home, and spiritual experience related to the natural environment. Through their activities, humans have been able to reduce and reinforce these ecosystem system services. For instance, in their quest to produce more food, humans engage in agricultural production. Agriculture influences and is influenced by all the categories of ecosystem services. The effects of agriculture on ecosystem services is largely influenced by the type of agriculture that is practiced. For instance, intensive agriculture, which entails high use of chemical inputs, is often accompanied by negative effects on the natural resource base jeopardizing its future productive potential and ability to produce ecosystem services. Thus approaches which minimize negative effects have increasingly been advocated in the past several decades. These approaches are variously known as sustainable agriculture intensification, ecosystem approaches to agriculture, or simply sustainable agriculture. Their basic premise is the adoption of intensive but sustainable practices, critical to realizing the benefits of ecosystem services and sustainable food systems. CA is one such system promoted under the rubric of sustainable agriculture or ecosystem enhancing systems.

Sustainable food systems

A sustainable food system is a collaborative network that integrates sustainable food production, processing, distribution, consumption, and waste management to enhance the environmental, economic, and social health of a particular place (Drewnowski and Ecosystem Inception Team, 2018). It is thus understood as a system that ensures food security and nutrition for all without compromising the economic, social, and environmental bases of such systems for future generations (Bilal et al., 2017). The definition demonstrates the importance of seeking sustainability in three dimensions: environmental, economic, and social, at every stage of a food system, from agricultural production, processing, retailing, consumption to waste management. For example, agriculture not only suffers from the impacts of climate change but also together with land use changes account for about a quarter of greenhouse gas (GHG) emissions. Agriculture has the potential to be an important part of the solution through mitigation (reducing and/or removing a significant amount of global emissions). Hence a sustainable food system is climate smart and simultaneously increases agricultural productivity, enhances climate resilience, and reduces GHGs for agriculture and related land use change.

Sustainable food system is a dynamic concept the conditions that ensure sustainability in food systems can vary widely across countries and regions, as well as across different stakeholders. Furthermore, sustainable food systems need to increase agricultural productivity, improve climate resilience, and reduce greenhouse gas emissions for agriculture and related land use change. Promoters of CA have used these as the pillars upon which the farming system is based (The World Bank, 2012).

Conservation agriculture and ecosystem services nexus

When a full range of agronomic practices for CA (i.e., reduced or no tillage, permanent organic soil cover by retaining crop residuals, and crop rotations, including cover crops) (Palm et al., 2013) are optimally applied, they provide substantial ecosystem services that play a key role in sustaining the livelihoods of smallholder farmers, particularly in the rural communities. The ultimate goal of optimally applied CA agronomic practices is to increase crop yields and enhance environmental sustainability by leveraging several ecosystem service parameters specified by the Millennium Ecosystem Assessment (2005), particularly supporting (soil formation, nutrient cycling, and primary production), regulating (climate and water regulation), and provisioning (food security) ecosystem services. For example, Sanchez (1995) argues that one of the primary goals of CA is to reverse environmental degradation and improve sustainability. This goal can be achieved through the soil enhancement properties of CA which fall under the supporting ecosystem services. Similarly, Garrity (2004) argues that CA can potentially help over one billion smallholder farmers around the world to reverse land degradation, improve the environment, and enhance their livelihoods by replenishing soils, protecting water catchments, restoring water catchments, and conserving biodiversity. Thus, natural ecosystems and agroecosystems are closely intertwined. The range of ecosystem services that agroecosystems can provide entirely depends on the ecosystem services received from natural, unmanaged ecosystems. These include supporting services (e.g., genetic biodiversity for use in breeding crops and livestock, soil formation and structure, soil fertility, nutrient cycling, and the provision of water) and regulating services (e.g., pollinators and natural enemies that move into agroecosystems from natural vegetation). In turn, agroecosystems provide a range of provisioning services like food, forage, fiber, bioenergy, and pharmaceuticals (Power, 2010).

The broad range of ecosystem services that CA can offer is perhaps better assessed through the climate smart landscape framework (FAO, 2013; Scherr et al., 2012). In this framework, climate smart agronomic practices are applied at three levels: (1) field and farm scale; (2) diversity of land use across the landscape;

and (3) management of land use interactions at landscape scale (FAO, 2013; Scherr et al., 2012). These are explained below.

Field and farm scale level

At the field and farm level, the suite of agronomic practices through which CA directly provides ecosystem services include soil, water, and nutrient management, along with agroforestry, livestock, husbandry, and forest and grassland management techniques (FAO, 2013; Scherr et al., 2012). The express goal of such agronomic practices at the field and farm level is to increase provisioning services by increasing the yields. However, these management practices also enhance other ecosystem services, such as pollination, biological pest control, soil fertility and structure, water regulation, and support for biodiversity because the manner in which habitats are managed with agroecosystems can determine the availability of the required resources for pollinators or natural enemies (Power, 2010).

Diversity of land use across landscapes level

At the diversity of land use across the landscapes level, CA principally addresses supporting, regulating, and provisioning ecosystem services. These ecosystem services are achieved through a high level of diversity in terms of land cover, land use, and species and varietal diversity of plants and animals. This is important because diversity provides important climate mitigation and adaptation functions which in turn immensely enhance the provisioning services in terms of food security. Scherr et al., (2012, p. 4) supported by FAO (2013), list some of the regulation and provisioning services in this category as: (1) reducing risks of production and livelihood losses, particularly the risks related to pests and diseases and vulnerability that usually ravage homogenous crop cover during unexpected weather conditions; (2) utilizing areas of the landscape strategically as emergency reserves for food, feed, fuel, and incomes, for example, as sources of "famine foods", such as wild greens, tree fruits, and roots; and (3) sustaining minimally disturbed habitats within the landscape mosaic that also serve as carbon sinks, for example, by maintaining and conserving other types of land cover within the landscape, such as perennial grasslands, woodlands, forests, and wetlands.

Land use interactions at landscape level

In terms of land use interactions at landscape level, the structure of the landscape in which the agroecosystem is embedded will determine the range of ecosystem services that the system can provide. Complex mosaics of diverse cropping systems that are embedded in a natural habitat matrix will provide a much richer suite of ecosystem services compared with simple landscapes that are dominated by one or two cropping systems (Power, 2010). Thus, according to Scherr et al. (2012, p. 14), integrated agricultural landscape management practices are characterized by

(1) landscape interventions that are designed to achieve multiple objectives, including human well-being, food and fiber production, climate change mitigation, and conservation of biodiversity and ecosystem services; (2) ecological, social, and economic interactions among different parts of the landscape are managed to seek positive synergies among interests and actors or reduce negative trade-offs; (3) the key role of local communities and households as both producers and land stewards is acknowledged; (4) a long-term perspective is taken for sustainable development, adapting strategies as needed to address dynamic social and economic changes; and (5) participatory processes of social learning and multistakeholder negotiation are institutionalized, including efforts to involve all parts of the community and ensure that the livelihoods of the most vulnerable people and groups are protected or enhanced.

The climate smart landscape framework is in agreement with sustainable food systems in that it subsumes the three important dimensions of sustainability: environmental, social, and economic. Many observers have called for the promoters of CA among smallholder farmers in sub-Saharan Africa to pay particular attention to sociocultural, institutional, and economic issues arguing that these factors impinge on its success among such an important group of farmers and ultimately its ability to contribute to the provision of ecosystem services.

Social cultural issues in CA relate to the division of labor and responsibilities that have to do with ownership and access to production assets and resources, decision-making regarding the type of crops, and hectarage to plant for a particular season, providing labor for weeding, harvesting, and preparation of produce for market. Although the common perception is that men disproportionately dominate decision making in CA, Umar (2017) found that joint decision-making was more dominant in their case study from the Eastern Province of Zambia. Conversely, Beuchelt and Badstue (2013) note that, in many instances the market niches, resources, and products, including the women's traditional crops, are taken over by men once they become lucrative. Generally, CA is a gendered activity with clear roles for men and women, and even children in some instances, as was established by Mwonjera et al. (2017) in Uganda and Tanzania. The "gendering" is context-specific and is usually based on the cultural norms that obtain in the local area.

A myriad of institutions plays one role or another in the implementation of CA, ranging from community service organizations, nongovernmental, government, media, universities, financial institutions, and international agencies (Mwonjera et al., 2017; Dougill et al., 2017). CA is mostly promoted by international agricultural development organizations (e.g., FAO, 2013; CGIAR, 2013) using persuasive "multiple wins" messages from different case studies, the so called success stories (Campbell et al., 2014). However, despite substantial donor budgets and enthusiasm for CA, the dominant narrative of CA being successful across varied agroecological regions and diverse livelihood strategies has been questioned by some analysts (Giller et al., 2009). Still, CA provides perhaps unmatched climate change mitigation and adaptation benefits in different ecological settings and livelihood contexts.

The key economic parameters that determine the success or failure of CA in many African countries, as argued by Mwonjera et al. (2017), relate to poor road networks, markets, and pot-harvest infrastructure. They list a myriad key marketing constraints affecting CA in Uganda and Tanzania, including (1) lack of access to market information, (2) inadequate market infrastructure, (3) nonstandardized weights and measures, (4) exploitation by farm-level brokers, and (4) high transportation costs (Mwonjera et al., 2017, p. 2000). Thus the extent to which CA succeeds in providing ecosystem services depends on the invariable synergistic interaction of several factors. In the next section, we examine the development and practice of CA in sub-Saharan Africa by using the country of Zambia as a case study. The choice of Zambia is deliberate. The country has been hailed as having the highest adoption of CA practices among smallholder farmers in the region (FAO, 2011b).

Conservation agriculture development and practice in Africa—case of Zambia

Since independence in 1964, the Zambian government wanted to change the status quo of being economically highly dependent on mining by expanding and intensifying land use and food production in the country through the introduction of high-input agriculture and animal traction subsidies to smallholders (van Donge, 1984). Therefore immediately after independence, an agricultural policy that primarily focused on the promotion of maize was introduced. Maize production received large-scale support in form of marketing and extensive fertilizer and input subsidies. This led farmers to devote ever-larger pieces of land to maize production (Haggblade and Tembo, 2003; IESR, 1999; Zulu et al., 2000). The support for maize production and marketing by the Zambian government was further expanded in the 1980s, when the country's declining copper industry prompted government to develop rural alternatives for an increasingly unemployed urban population (Ferguson, 1999; Gould, 1997). Farmers received tractor and plow credit and subsidized rental schemes which further encouraged expansion of cropped area via ploughing. The guaranteed maize market provided further incentive for farmer adoption of the high-input maize packages. For about two and half decades, the Zambian agricultural soils were heavily laden with mineral fertilizers and herbicides and were ploughed. As a result of the heavy application of chemical fertilizers, herbicides, and continued extensive ploughing, Zambian agriculture entered the 1990s with significantly declining land quality and productivity. In addition to soil degradation the Zambian agriculture of the 1990s suffered another blow as the government subsidies for smallholder farmers were abruptly discontinued under structural adjustment in 1991 (Baudron et al., 2007). At the same time, drought and livestock diseases drastically reduced cattle numbers. Accordingly, the incidence of late planting increased, which impacted negatively on yields in plough-based smallholder farming. This occurrence resulted in an increased interest to extend CA to smallholder farmers in Zambia.

CA is promoted as a solution to agricultural problems such as soil degradation that result from agricultural practices that deplete the organic matter and nutrient content of the soil and to address the problem of intensive labor requirements in smallholder agriculture (Hebblethwaite et al., 1996; Steiner et al., 1998; Fowler and Rockstrom, 2001; Derpsch, 2003; Hobbs, 2007). In Zambia, as much as CA is recognized and embraced by the government, its implementation is highly dependent on aid from donor agencies that support the promotion of CA practice among farmers (Amelia et al., 2014). The development and promotion activities are championed by many NGOs such as the Conservation Farming Unit (CFU) of the Zambia National Farmers Union, the Institute of Agricultural and Environmental Engineering Project, the Golden Valley Agricultural Research (Haggblade and Tembo, 2003), and the Community Markets for Conservation (COMACO). The focus and emphasis of these organizations in their promotion of CA varies. For instance, COMACO emphasizes organic CA where only manure should be used while CFU encourages the use of mineral fertilizers and herbicides.

Development trajectory of conservation agriculture in Zambia

The current development path of CA in Zambia is that it is now embraced by the Ministry of Agriculture as an official policy of the Zambian government. Accordingly, the government has included CA in the National Agricultural Policy (MAL, 2013). Ironically, the Zambian government has not been in the forefront of CA research and extension (Andersson and D'Souza, 2014). There is a suggestion that government may actually obstruct the spread of CA practices such as crop rotation (Umar et al., 2012) because a hefty share of its agricultural budget goes to incentives for maize production through the Farmer Input Support Program (The World Bank, 2010) and maize marketing through the Food Reserve Agency, a public agency which purchases maize from smallholder farmers at government- determined pan territorial and pan seasonal floor prices (Umar et al., 2012). Rather than government, CA promotion in Zambia was driven to scale by the CFU (Aagaard, 2012). It was first promoted in areas where there was cotton and commercial maize production thus rendering it firmly in a narrative of commercial smallholder production. Currently, CA uptake in Zambia has become increasingly incentivized by means of direct support to farmers. In large CA projects funded by the FAO and the Norwegian government, such support has ranged from providing seedlings for the leguminous tree, winter thorn (*Faidherbia albida*), and CA farm implements (Aune et al., 2012) to mineral fertilizer and hybrid seeds (Ndiyoi et al., 2012; FAO, 2011a).

Over the years, CA in Zambia has evolved to include various agricultural development interventions which have been implemented in different regions of the country. In Eastern province, programs such as the Soil Conservation and Agroforestry Extension, Agricultural Credit Management Program, and the Agricultural

Sector Investment Program launched in the 1980 and 1990s were the forerunners. More recently, the FAO- funded Conservation Agricultural Scaling up (CASU) program was introduced. CASU was a national CA program implemented in 31 districts across the country. A closer look at these programs shows that they are all donor- led and donor- funded. This gives a clear indication of the donor dependent nature of the CA development trajectory in Zambia. One would argue that if donors pulled out today, smallholder farmers practice of CA would reduce significantly.

Challenges and successes of conservation agriculture practice in Zambia

To measure the successes and challenges of CA practice, this chapter will consider the adoption and nonadoption of the three principles of CA: minimum soil disturbance; permanent organic soil cover with crop residues or cover crops; and diversified crop rotations (FAO, 2010a,b). To realize the full benefits of CA, the CA cropping system must be adopted as a full package and not adopting only one of the principles (Twomlow et al., 2008). However, adoption of CA in Zambia is not fully registered. The Conservation Farming Unit reported about 60, 000—180, 000 farmers practicing CA in various regions of the country (Thierfelder and Wall, 2010). Nevertheless, the current situation is that CA in Zambia is only conducted as part of farmers' farming practice. Farmers still use other practices or mix some practices from CA with conventional agricultural practices. Practicing CA in the true sense (adoption of all the three principles) among Zambian farmers is much lower than the maximum number of farmers who experiment with only one or two components of CA (Thierfelder and Wall, 2010).

It is believed that during the CA project implementation phases when there is active promotion of technologies by NGOs, adoption of CA among farmers seems high. The adoption levels however significantly drop after the project ends. This is because of the temporary influence of the project rather than a sustained change in agricultural practice (Haggblade and Tembo, 2003; Habanyati et al, 2018). For instance, Haggblade and Tembo (2003) reported that 75,000 Zambian smallholder farmers practiced conservation farming in 2002/2003, from about 20,000 in the 2001/2002 season because of the 60,000 starter packs issued as a drought-relief measure by a consortium of donors. They estimated that some 15,000 were spontaneous adopters, while the remaining 60,000 practiced conservation farming as a condition for receiving their inputs. In addition to false adoption of CA because of the incentives provided by the promoters during the project implementation phases, most farmers practice partial adoption of CA.

With regard to crop residue retention, this is especially challenging in the Southern province of the country where farmers engage in both crop production and keeping livestock. In such a place, crop residues which should be retained in fields are eaten up by livestock (Umar et al, 2011). This makes this principle of CA difficult to practice for these farmers. In areas where the soils are heavily infested with pests, farmers believe that burning crop residue abates the infestation: thus they burn the residue after a farming season.

Despite the above challenges, CA has recorded some successes in terms of increased food security especially during dry years, reduced labor because of mechanization (ripping); increased crop diversity due to crop rotation, improved soil health, and planting of trees for various purposes. These positive effects are elaborated below in the implications of the nexus between CA and ecosystem services on sustainable food systems.

Implications of the conservation agriculture-ecosystem services nexus on sustainable food systems

In the quest to rejuvenate degraded soils, increase profits, and ensure household food security, CA has been used as a sustainable food production system by conserving and enhancing the natural resource base and ecosystem services (Palm et al., 2013). Despite its rapid adoption in Australia, in the United States and in South America in the past three decades, the adoption of CA has been slow in developing countries of sub-Saharan Africa. Yet it is among African smallholder farmers that CA could enhance ecosystem services, improve farm productivity, and tackle food insecurity. Crop productivity among smallholder African farmers is currently the lowest in the world as traditional agricultural practices have diminished soil productivity to the extent that many agricultural soils are depleted of nutrients and are unable to naturally sustain crop productivity (Naab et al., 2017). The urgent need for approaches that would sustainably address these challenges cannot be emphasized. CA has a number of benefits to sustainable food systems, including enhancing ecosystem services through various ecosystem components and processes. However, whether CA increases or decreases ecosystem delivery depends on how it is practiced in different locations in respect to its principles. There are implications of the CA-ecosystem services nexus on sustainable food systems discussed below.

Reduction in crop yield variability

Owing to its capacity to retain soil moisture for longer, it is argued that CA leads to reduced crop yield variability especially during extreme climate events such as droughts and floods (Umar et al., 2012). In general CA often reduces water and wind erosion by protecting the soil surface with crop residues and increases water infiltration, as well as decreased run-off (Verhulst et al., 2010). Thus in areas where CA is practiced fully, households are able to attain higher crop yields even during droughts because the soil cover protects the soil from moisture loss and the undisturbed soil holds much more water than ploughed soil. This implies that CA households could be more food secure and could sustain themselves during extreme climatic events that would otherwise wipe out, or significantly reduce their crop production.

Premised on the principle of crop rotation, CA households could reduce total crop production variability because of the diversity of crops that are planted and alternated in the same agricultural field i.e., cereals such as maize and wheat then followed by legumes (FAO, 2012). Thus, CA households often have a variety of crops because of the nature of the agricultural system as compared with households engaged in conventional farming practices only.

Improved household food security

CA can lead to improved household food security through the application of its three principles and associated sound agronomic practices. For example, Palm et al., (2013) argued that soil moisture retention can be higher with CA, resulting in higher and more stable yields during dry seasons. CA has a positive impact on household food security not only in terms of improvements in maize security (Nyambose and Jumbe, 2013) but also on the production and consumption of legumes such as groundnuts (*Arachis hypogeae*), soya beans (*Glycine max*), and cowpeas (*Vigna unguiculata*) (Nyanga, 2012). Nyambose and Jumbe (2013) found that household food security among smallholder farmers in Malawi was higher for CA adopters than for the nonadopters. In Zambia, CA has the potential to improve food security among smallholder farmers, more specifically in dry areas of Southern Zambia such as Choma and Livingstone because of high water holding capacity in CA. The contribution of CA to household food security is evident through increased per capital crop consumption, intraseasonal production, and increased number of meals per day and availability of food throughout the year among adopters (Ngwira et al., 2014). CA adopters ensure household food stability through "green harvest" which shortens their hunger period. Because they plant earlier, CA households start to harvest their maize earlier, which they consume before it starts to dry. Hence the reduction in the hunger period.

Improved soil health

When its three principles are adhered to, CA is reported to improve soil quality (fertility and health), improve crop yields, and reduce input costs (Palm et al., 2013). The principle of retaining crop residues on the soil surface in association with the two other CA principles is intended to increase carbon inputs and enhance ecosystems benefits such as soil fertility, improved soil water retention, and biological properties (Palm et al., 2013). Consequently, adherence to and application of the CA principles can maintain higher soil quality through recycling of nutrients and increased crop yield compared with conventional practices (Ngwira et al., 2013).

When households consistently apply the principles of minimum tillage and permanent soil cover with crop residues or live mulches, there is a possibility for reduced soil erosion and increased soil fertility in their agricultural fields which would entail higher crop yields (Naab et al., 2017). Improved soil health as a result of nutrient cycling from CA is also attained through crop rotation or intercropping

(Naab et al., 2017). Besides, there is higher biological diversity in CA fields contributing to improved soil health. However, it must also be noted that there are trade-offs in instances where CA is fully practiced as compared with when CA is partially practiced. In this regard the benefits of CA may not be fully realized in cases where farming households partially adopt CA.

Higher dietary diversity among households practicing conservation agriculture

CA could lead to dietary and nutritional improvements among households due to increased own production and consumption of cereals and legumes (Herforth and Harris, 2014). For example, Kumar et al. (2015) argued that greater agricultural diversity is associated with greater dietary diversification. In the case of Zambia, it was found that households with greater agricultural diversity had fewer stunted children (Kumar et al., 2015). Thus addition of a variety of foods resulting from crop rotation, intercropping in CA, and integration with fruit trees (agroforestry) may help to meet nutrition gaps prevalent in most rural households and eventually improve household dietary diversity.

References

Aagaard, P.J., 2012. History of the Conservation Farming Unit—Zambia.

Amelia, D.F., Kopainsky, B., Nyanga, P.H., 2014. Exploratory model of conservation agriculture adoption and diffusion in Zambia: a dynamic perspective. In: Proceedings of the 32nd International Conference of the System Dynamics Society, Delft, The Netherlands, pp. 20–24.

Andersson, J.A., D'Souza, S., 2014. From adoption claims to understanding farmers and contexts: a literature review of Conservation Agriculture (CA) adoption among smallholder farmers in Southern Africa. Agriculture, Ecosystems & Environment 187, 116–132.

Aune, J.B., Nyanga, P., Johnsen, F.H., 2012. *A Monitoring and Evaluation Report of the Conservation Agriculture Project 1 (CAP1) in Zambia*. Department of International Environment and Development Studies, Noragric, Norwegian University of Life Sciences.

Baudron, F., Mwanza, H., Triomphe, B., Bwalya, M., 2007. Conservation Agriculture in Zambia: A Case Study of Southern Province. African Conservation Tillage Network (ACT), Nairobi, Kenya.

Beuchelt, T.D., Badstue, L., 2013. Gender, nutrition- and climate-smart food production: Opportunities and trade-offs. Food Security 5 (5), 709–721.

Bilal, S., Torres, C., Wermuth, E., 2017. *Towards* more sustainable food systems - Editorial. GREAT Insights Magazine 6 (Issue 4) (September/October).

Campbell, B.M., Thornton, P., Zougmoré, R., van Asten, P., Lipper, L., 2014. Sustainable intensification: What is its role in climate smart agriculture? Current Opinion in Environmental Sustainability 8, 39–43.

CFU, 2007. Conservation farming and conservation agriculture: handbook for Ox farmers in agro-ecological regions I and IIa. In: Conservation Farming & Conservation Agriculture. Conservation Farming Unit, Lusaka, Zambia.

CGIAR, 2013. Annual Report 2013: Featuring Climate-Smart Agriculture. CGIAR, Montpellier, France, p. 64.

Chappell, A., Agnew, C.T., 2004. Modelling climate change in West African Sahel rainfall (1931–90) as an artifact of changing station locations. International Journal of Climatology 24 (5), 54.

Derpsch, R., 2003. Conservation tillage, no-tillage and related technologies. Conservation Agriculture 181–190.

Dougill, A.J., Whitefield, S., Stringer, L.C., Vincent, K., Wood, B.T., Chinseu, E.L., Stewart, P., Mkwambisi, D.D., 2017. Mainstreaming conservation agriculture in Malawi: knowledge gaps and institutional barriers. Journal of Environmental Management 195, 25–34.

Drewnowski, A., The Ecosystem Inception Team, 2018. The Chicago consensus on sustainable food systems science. Front. Nutr. 4, 74. https://doi.org/10.3389/fnut.2017.00074.

FAO, 2009. Food Security and Agricultural Mitigation in Developing Countries: Option for Capturing Synergies. Food and Agricultural Organization of the United Nation, Rome.

FAO, 2010a. Climate Smart Agriculture: Policies, Practices, and Financing for Food Security Adaptation and Mitigation. Food and Agricultural Organization of the United Nations, Rome, Italy.

FAO, 2010b. What is Conservation Agriculture? [Online] Available:. accessed on 04/02/2019. http://www.fao.org/conservation-agriculture/en/last.

FAO, 2011a. Terminal Report Conservation Agriculture Scaling up for Increased Productivity and Production (CASPP) Project. FAO/Norwegian Embassy/ministry of Agriculture and Cooperatives, Lusaka.

FAO, 2011b. Socio-economic analysis of conservation agriculture in Southern Africa. REOSA Network Paper 02. Johannesburg, South Africa.

FAO, 2012. Towards the Future We Want: End the Hunger and Make the Transition to Sustainable Agriculture and Food Systems. Report on Rio+20 Rome: FAO., Rome.

FAO, 2013. Climate-smart agriculture sourcebook. FAO, Rome.

Ferguson, J., 1999. Expectations of Modernity: Myths and Meanings of Urban Life on the Zambian Copperbelt. University of California Press, Berkeley.

Fowler, R., Rockstrom, J., 2001. Conservation tillage for sustainable agriculture—an agrarian revolution gathers momentum in Africa. Soil and Tillage Research 61, 93–107.

Friedrich, T., Derpsch, R., Kassam, A., 2012. Overview of the global spread of conservation agriculture. Field Actions Science Reports, Special Issue 6. http://factsreports.revues.org/1941.

Garrity, D.P., 2004. Agroforestry and the achievement of the Millennium development goals. Agroforestry Systems 61, 5–17.

Giller, K.E., Witter, E., Corbeels, M., Tittonell, P., 2009. Conservation agriculture and smallholder farming in Africa: the heretics' view. Field Crops Research 114, 23–34.

Gould, J., 1997. Localising modernity: Action, interests and association in rural Zambia. Finish Anthropological Society, Helsinki.

Habanyati, E.J., Nyanga, P.H., Umar, B.B., 2018. Factors contributing to disadoption of conservation agriculture among smallholder farmers in Petauke, Zambia. Kasetsart Journal of Social Sciences. In press. https://www.sciencedirect.com/science/article/pii/S2452315117304459.

Haggblade, S., Tembo, G., 2003. Conservation Farming in Zambia. International Food Research Policy Research Institute: Washington. D.C., USA. Discussion Paper No. 108.

Hebblethwaite, J., Soza, R., Faye, A., Hutchinson, N., 1996. No-till and reduced tillage for improved crop production in sub-Saharan Africa. Achieving greater impact from research

investments in Africa. In: Proceedings of the Workshop Developing African Agriculture: Achieving Greater Impact from Research Investments, Addis Ababa, Ethiopia, pp. 195–199. September 26–30, 1995.
Herforth, A., Harris, J., 2014. Understanding and Applying Primary Pathways and Principles *Brief #1. Improving Nutrition through Agriculture Technical Brief* Series. Arlington VA: USAID/Strengthening Partnerships, Results, and Innovations in Nutrition Globally (SPRING) Project.
Hobbs, P.R., 2007. Conservation agriculture: what is it and why is it important for future sustainable food production? The Journal of Agricultural Science 145 (02), 127–137.
Hobbs, P.R., Sayre, K., Gupta, R., 2008. The role of conservation agriculture in sustainable agriculture. Philosophical Transactions of the Royal Society B 363, 543–555.
Institute of Economic and Social Research (IESR), 1999. Agricultural Sector Performance Analysis, 1997-99. Ministry of Agriculture, Food and Fisheries, Lusaka.
Johansen, C., Haque, M.E., Bell, R.W., Thierfelder, C., Esdaile, R.J., 2012. Conservation agriculture for small holder rain fed farming: opportunities and constraints of new mechanized seeding systems. Field Crops Research 132, 18–32.
Kassam, A., Friedrich, T., Shaxson, F., Pretty, J., 2009. The spread of conservation agriculture: justification, sustainability and uptake. International Journal of Agricultural Sustainability 7 (4), 292–320.
Kumar, N., Rawatt, R., Harris, J., 2015. If they grow it, will they eat and grow? Evidence from Zambia on agricultural diversity and Child undernutrition. Journal of Development Studies 51 (8), 1060–1077. https://doi.org/10.1080/00220388.2015.1018901.
MAL, 2013. Draft of National Agricultural Policy. Policy objectives and Measures, Vol. Objective 10. Ministry of Agriculture and Livestock, Lusaka Zambia.
Mazvimavi, K., 2011. Socio-economic Analysis of Conservation Agriculture in Southern Africa. Food and Agriculture Organization, Rome, Italy. Network Paper No. 2.
Millennium Ecosystem Assessment, 2005. Ecosystems and Human Well Being: Synthesis. Island Press, Washington DC.
Mwonjera, C., Shikuku, K.M., Twyman, J., Laderach, P., Ampaire, E., Astern, P.V., Twomlow, S., Winowiecki, L.A., 2017. Climate smart agriculture rapid apraisal (CSA-RA): a tool for prioritising context-specific climate smart agriculture technologies. Agricultural Systems 151, 192–203.
Naab, J.B., Mahama, G.Y., Yahaya, I., Prasad, P.V.V., 2017. Conservation agriculture improves soil quality, crop yield, and incomes of smallholder farmers in North Western Ghana. Frontiers of Plant Science 8, 996. http://doi.org/10.3389/fpls.2017.00996.
Ndiyoi, M., Mwala, M., Mupindu, S., Marongwe, S., 2012. Beneficiary assessment report on capacity building for CA: expansion of the Farmer Input Support Response Initiative (FISRI) to rising prices of agricultural commodities in Zambia. In: Draft Final Report: Final Evaluation of the Farmer Input Support ResponseInitiative (FISRI), Zambia.
Ngwira, A.R., Thierfelder, C., Lambert, D.M., 2013. Conservation agriculture systems for Malawian smallholder farmers: long-term effects on crop productivity, profitability and soil quality. Renewable Agriculture and Food Systems 28, 350–363.
Ngwira, A., Johnsen, F.H., Aune, J.B., Mekuria, M., Thierfelder, C., 2014. Adoption and extent of conservation agriculture practices among smallholder farmers in Malawi. Journal of Soil and Water Conservation 69 (2), 107–119.
Nyamangara, J., Masvaya, E.N., Tirivavi, R., Nyengerai, K., 2013. Effect of hand-hoe based conservation agriculture on soil fertility and maize yield in selected smallholder areas in Zimbabwe. Soil and Tillage Research 126, 19–25.

Nyambose, W., Jumbe, C., 2013. Does Conservation Agriculture enhance household food security: evidence from smallholder farmers in Nkhotakota in Malawi. In: 4th International Conference of the African. Association of Agricultural Economists, Hammemet, Tunisia.
Nyanga, P.H., 2012. Food security conservation agriculture and pulses; evidence from smallholder farmers in Zambia. Journal of Food Research 1 (2), 120–138.
Palm, C., Blanco-Canqui, H., DeClerck, F., Gatere, L., Grace, P., 2013. Conservation agriculture and ecosystem services: an overview. Agriculture, Ecosystems & Environment 187, 87–105.
Power, A.G., 2010. Ecosystem services and agriculture: tradeoffs and synergies. Philosophical Transactions of the Royal Society, B: Biological Sciences 365, 2959–2971.
Sanchez, P.A., 1995. Science in agroforestry. Agroforestry Systems 30, 5–*55*.
Scherr, S.J., Shames, S., Friedman, R., 2012. From climate-smart agriculture to climate-smart landscapes. Agriculture & Food Security 1–*12. 2012*.
Steiner, K., Derpsch, R., Koller, K., 1998. Sustainable management of soil resources through zero tillage. Agricultural and Rural Development 5, 64–66.
The World Bank, 2010. Zambia: Impact Assessment of the Fertilizer Support Program, Analysis of Effectiveness and Efficiency, Report No. 54864-ZM, Sustainable Development Department Agriculture and Rural Development Africa Region. World Bank.
The World Bank, 2012. The World Bank Annual Report 2012 Volume 1. *Main Report*. World Bank: Washington D.C., USA.
Thierfelder, C., Wall, P.C., 2009. Effects of conservation agriculture techniques on infiltration and soil water content in Zambia and Zimbabwe. Soil and Tillage Research 105 (2), 217–227. https://doi.org/10.1016/j.still.2009.07.007. https://doi.org/10.1016/j.still.2009.07.007.
Thierfelder, C., Wall, P.C., 2010. Rotation in conservation agriculture systems of Zambia: effects on soil quality and water relations. Experimental Agriculture 46 (3), 309–325.
Twomlow, S., Urolov, J., Jenrich, M., Oldrieve, B., 2008. Lessons from the field–Zimbabwe's conservation agriculture task force. Journal of SAT Agricultural Research 6 (1), 1–11.
Umar, B.B., Aune, J.B., Johnsen, F.H., Lungu, O., 2011. Options for Improving Smallholder Conservation Agriculture in Zambia. Journal of Agricultural Science 3 (3), 50–62.
Umar, B.B., Aune, J.B., Johnsen, F.H., Lungu, I.O., 2012. Are smallholder Zambian farmers economists? A dual-analysis of farmers' expenditure in Conservation and Conventional Agriculture Systems. Journal of Sustainable Agriculture 36, 908–929.
Umar, B.B., 2012. Reversing agro-based land degradation through conservation agriculture: Emerging experiences from Zambia's smallholder farming sector. Sustainable Agriculture Research 1 (2), 72–83.
Umar, B.B., 2017. Conservation agriculture promotion and uptake in Mufulira, Zambia- A political agronomy approach. Journal of Sustainable Development 1 (1), 156–169.
van Donge, J.K., 1984. Rural-urban migration and the rural alternative in mwase Lundazi, eastern province, Zambia. African Studies Review 27, 83–96.
Verhulst, N., François, I., Govaert, B., 2010. Conservation agriculture, improving soil quality for sustainable production systems? CIMMYT, Mexico.
Zulu, B., Nijhoff, J.J., Jayne, T.S., Negassa, A., 2000. Is the glass half empty or half full? An analysis of agricultural production trends in Zambia. Working Paper No.3. Lusaka: Food Security Research Project.

CHAPTER

Ecosystem services from different livestock management systems

7

Irvin Mpofu, BSc Hons, MSc, MBA, PhD
Professor, Animal Production and Technology, Chinhoyi University of Technology, Chinhoyi, Zimbabwe

Introduction

This chapter deals with the basic question "How does livestock production systems benefit humans and the environment?" Livestock farming involves the rearing of animals for food products, nonfood products (leather, wool, and even organic fertilizer), and other human uses, such as draft power and social protection. Livestock are assets that have life and hence the term, and this normally applies to domesticated animals such as beef cattle, dairy cows, chickens, goats, pigs, horses, donkeys, mules, rabbits, insects, and sheep. The chapter will take the view that undomesticated animals such as wildlife which can be an asset in one way or the other are also livestock.

Livestock provisioning services

Livestock ecosystems provide direct food rich in protein, e.g., beef, pork, milk, and eggs. These can be value added and processed into mince, pies, sausages, cheese, yoghurt, and egg powder.

Livestock also provide terrestrial land cover by agriculture (72% of total) because grassland, shrubs, herbs, and sparse vegetation constitute 44%, tree covers 39%, and cropland covers 17%. The majority of milk production is based on forage especially in intensive operations. As a result, expansive pastures provide carbon sink for greenhouse gases such as carbon dioxide. Large livestock numbers are normally high in those areas where local breeds are adaptive advantage on grass and roughage.

Livestock ecosystems also support services in different grassland types, e.g., habitat provisioning 34% and nutrient cycling 28%.

Cultural services in different grassland types are (1) cultural, historical, and natural heritage 22%, (2) knowledge systems, (3) education 20%, and (4) landscape values 22%.

The Role of Ecosystem Services in Sustainable Food Systems. https://doi.org/10.1016/B978-0-12-816436-5.00007-X

Ecosystem services from different livestock management systems

Intensive livestock ecosystems

Intensive livestock farming means the production of animals where their environment in total is provided for them to maximize profit out of their rearing. The environment here includes housing, water, nutrition, veterinary protection, temperature, humidity, and management. While in the developed world, intensive livestock farming has graduated from the generic commercial production into industrial and factory farming, intensification in the developing countries has remained largely rudimentary except for few cases. This has allowed the goods of livestock farming to be more broadly accessible and inexpensive to buy in some cases, but per-capita consumption of meat for example has remained low because of the high retail costs. While animal products have benefited human kind more and more, some practices of intensive livestock farming have resulted in undesirable consequences such as food safety, animal welfare, and environmental impacts (Greentumble Editorial Team, 2016).

Advantages of intensive livestock farming

Intensive livestock farming has the largest potential in contribution to gross domestic product (GDP) compared with semi-intensive or extensive. Global estimates are that intensive livestock farming combined have a direct contribution amounting to USD883 billion (Greentumble Editorial Team, 2016). This excludes services either downstream or upstream of the livestock value chain, e.g., slaughter houses, feed manufacturing, transport companies, and equipment manufacturers. From a food and nutrition security point of intensive livestock systems support at least 1.3 billion people (FAO, 2018). This has been made possible because intensification is hinged on higher yield obtained with high efficiency that compresses cost through judicious feeding, disease control, and rearing animals under housing with climate control systems. In a way intensification helps mankind with land use because more and more animals can be kept on a small land area through use of vertical space especially in poultry production. Animals kept in a small area are easier to manage, including controlling disease outbreaks.

Disadvantages of intensive livestock farming

The disturbing feature of intensive livestock farming hinges mainly on animal welfare and environmental and health concerns despite impressive productivity and return per animal. Intensive animal production impact positively on GDP of nations, but the cost-saving techniques often used have potential negative effect on the health and well-being of the animals. Countries in the European Union and New Zealand for example have instituted strict legislation that compels livestock farmers and policymakers to recognize animals as "sentient beings" and not merely as commodities. The legislation makes it clear that farmers and any animal handler appreciate the freedoms the animals inherently have. These are freedom from undue emotions like pain and distress and freedom from diseases, hunger, thirst, and any manner of stress. Transportation of animals over long distances without a protocol that gives them a semblance of comfort is prohibited much as the inhuman painful slaughtering.

Every effort needs to be taken such that when benefiting from intensive livestock ecosystems cases of animal cruelty are to a bare minimum, but it would be more desirable if it is nil. The biggest problem with animal welfare is over crowding because of maximization of profit per unit area. Disease outbreak and spreading can be explosive to both the crowded animals and also to humans if the health challenges are zoonotic.

The intensively reared animals also pose serious environmental damages contributing over 14% of human-induced greenhouse-gas emissions. Records show that this contribution is higher than that of the transport sector. The pollution occurs because of the huge volumes of waste and the velocity and variety of the affluent. In many cases farmers have been found to pollute water bodies and in some cases decimate the ecosystem's biodiversity.

The entry points to maximize ecosystem service in intensive livestock and the development trajectory

Responsible food security and nutrition security policies and strategies are imperative. These should be continuously improved so that sustainable livestock intensification can be achieved. Sustainable intensification of livestock operations integrates the compelling need to produce enough food and nonfood animal products with the desire to creating eco-friendly approaches. Some agro-industrialists have come up with ways of helping farmers produce livestock while protecting the environment from deteriorating and also ensuring animal welfare. Through Milk Matters, there is an initiative to keep customers up to date on current developments in the dairy production industry and provide advice to dairy farmers who want to improve their operation.

Semi-intensive/extensive ecosystems

Semiintensive livestock ecosystems are based on animals to some extent fending for themselves through grazing or feed scavenging. The same challenges faced with the intensive livestock ecosystems subsist but on a lower scale. The disadvantages exist but on a reduced scale. Animal welfare challenges exist though on a lower rate. The impact on the environment is equally measured with the reduced level of intensification.

The entry points to maximize ecosystem service in semi-intensive livestock and the development trajectory

There is always tendency to upgrade semiintensive operations into intensive ones overtime because of the profit pull factor. For cattle, sheep and goats for example, semiintensive livestock ecosystems offers opportunities for sustainable utilization of rangelands. This is possible if a measured approach is instituted where a balance matrix of goats, sheep, and cattle followed. For example, cattle are heavy grazers and can graze really low, while sheep graze on middle parts of grass, and goats are predominantly browsers. This means a farmer can utilize the rangelands resources sustainably with a livestock mix that utilizes different levels of the rangeland resources without completion and without deteriorating the ecosystem resources.

Extensive livestock ecosystems

In this chapter livestock refers to any living animal that has value and for the purpose and objective of this chapter this shall include wildlife.

Domesticated livestock like cattle reared under extensive livestock ecosystems provide the least stress on animals and the environment. Humankind therefore benefits from such a system in a sustainable manner provided the stocking density and the human population is manageable. Good returns on investment can be obtained; however, this requires use of vast pieces of land. Land is finite and has become a scarce resource especially in the developed countries. In developing countries, human population is manageable, but productivity is very low because of low use of technology. The latter is because of lack of technical know-how and also because of poverty. Farmers can reap more from extensive livestock systems if they have buying power to procure basic technologies such as forage cutters, fodder banking, and disease control.

Wildlife production peripheral areas/environments

There is need to deploy and improve science and technology research for developing countries' poverty reduction and social development efforts through the priority area of Agriculture ecosystems through livestock. Peripheral areas are remote wildlife production peripheral areas and normally in Transfrontier Conservation Areas (TFCAs). TFCAs provide integrated development and conservation opportunities for remote and neglected transboundary areas through wildlife production. They are therefore normally referred to as wildlife peripheral areas. They offer exceptional farming systems because they are not in competition with crop production. While the soils are generally unproductive and precipitation patterns hamper crop and intensive livestock production, ecotourism is a viable option. However, these opportunities are not fully utilized in many developing countries because of meagre technical knowhow and low resources for improvement.

The entry points to maximize ecosystem service in extensive livestock and the development trajectory

Interesting enough some developing countries have policies that discourage intensive and even semiintensive crop production, and the local communities nevertheless always try to produce their own food to avert starvation. In situations where the local communities are close to these areas do not benefit from the wildlife systems in a manner that mitigate poor crop production.

In some countries notably Kenya, South Africa, Zimbabwe, and Namibia these areas are conserved with the sole objective to promote the conservation of wildlife populations and supporting the development of a wildlife economy with a positive bearing on local development and enhancement of household farm incomes.

This could be significant to African countries and Southern Africa in particular because a significant proportion of populations have food security and nutrition challenges.

Incentivizing efforts by farmers who improve management of livestock ecosystems.

There is scope in coming up with incentives for improving management of livestock ecosystems so that sustainable production ensues. First of all, it is necessary to identify evidence from case studies on the likelihood of uptake of incentives in a particular context. Then if positive, a policy may be crafted which can be applied through guaranteed participation by farmers through a social contract with set targets to be achieved in terms of livestock productivity and protection of the environment in a sustainable manner. After that, it is necessary to apply the policy and check for policy response. Once positive uptake is achieved, wider implementation is effected accompanied by an instrument to monitor, evaluate, and learn. A report is produced, and refinements are done for sustainable utilization of livestock ecosystems. The exploitation of the livestock ecosystems should include both food and nonfood items.

Constraints to sustainable exploitation of livestock ecosystems

Like with any systems, a plethora of constraints affects the sustainable exploitation of livestock ecosystems. These are

- **(i)** Environmental changes which are beyond the control of the farmer. Climate change is real and well documented, and its effects on livestock numbers and productivity are now clear.
- **(ii)** Societal factors which do not encourage farmers to move with the trends being promoted for sustainable utilization of heritage.
- **(iii)** Lack of research and also research results dissemination; scientist in the developing countries value more the number of publications instead of coming up with research agenda that addresses Modern societies, and research results inform decision-making, and if it is not availed, then humankind is doomed.

(iv) Linked to research is lack of appropriate knowledge especially technical knowledge.
(v) Lack of policies hampers any prospects of progressing.
(vi) Lack of adequate income especially farm income does not create impetus to invest in the livestock ecosystems for the benefit of the farmers.
(vi) Insecurity predisposes livestock farmers to poor production potential hence low income.
(vii) Lack of recognition plays negatively to the efficient production potential.

Conclusions

All these challenges or constraints can easily be transformed into opportunities that may allow livestock farmers to exploit their heritage profitably and sustainably. It is also imperative especially in the developing countries to raise awareness about the important ecosystem services provided by livestock species. This should be buttressed by the use of modern assessment methods and tools to develop results-based incentive systems. This may promote gene banks that can be used in restocking of livestock through assisted reproduction.

References

FAO, 2018. FAO's Role in Animal Production. FAO's 2014–2017 Strategic Framework. http://www.fao.org/animal-production/en/?%DC%FB%BF%07.
Greentumble Editorial Team, 2016. Advantages and Disadvantages of Intensive Livestock Farming. http://www.ad-nett.org/livestock_farming.html.

Further reading

https://greentumble.com/the-main-causes-of-land-pollution. Retrieved 12th February, 2019.
Hoffmann, I., 2015. Livestock as Providers of Ecosystem Services. Seminar• Wageningen University. June 2015.

CHAPTER

Crop-livestock integration to enhance ecosystem services in sustainable food systems

8

Sabine Homann-Kee Tui[1], Roberto O. Valdivia[2], Katrien Descheemaeker[3], Trinity Senda[4], Patricia Masikati[5], Milton T. Makumbe[6], Andre F. van Rooyen[6]

[1]Senior social scientist, International Crops Research Institute for the Semi-Arid Tropics (ICRISAT), Matopos Research Institute, Bulawayo, Zimbabwe; [2]Assistant Professor, Senior Researcher, Department of Applied Economics, Oregon State University, Corvallis, OR, United States; [3]Assistant Professor, Plant Production Systems, Wageningen University, Wageningen, Gelderland, The Netherlands; [4]International Livestock Research Institute, (ILRI), Nairobi, Kenya; [5]Scientist, World Agroforestry Centre (ICRAF), Elm Road Woodlands, Lusaka, Zambia; [6]Principal Scientist, International Crops Research Institute for the Semi-Arid Tropics (ICRISAT), Matopos Research Institute, Bulawayo, Zimbabwe

Introduction

Agricultural systems in developing countries are challenged to produce more food from the same area of land and from scarce water resources. Integrated crop-livestock systems hold intensification options that can improve the production of nutritious food and environmental sustainability, while reducing people' vulnerability to climate-related hazards (Tarawali et al. 2011; Lemaire et al., 2013; Garrett et al., 2017). Through an integrated relationship between crop and livestock components, farmers can increase agricultural productivity per unit of land and water, beyond the productivity of the individual components (Bonaudo et al., 2014; Peyraud et al., 2014). More diversified crop-livestock systems provide more opportunities for integration compared with systems that are composed of single farm components (Kremen et al., 2012; Valbuena et al., 2015). The multiple benefits deliver important ecosystem functions and services which help to use increasingly scarce resources more efficiently and conserve the natural environment (Hendrickson et al., 2008; Thornton and Herrero, 2015). In a market-oriented production context, returns on integration and diversification increase as farmers produce more biomass of better nutritional quality. They can do this using available technologies, while becoming less dependent on external inputs; this, in turn, makes them less sensitive to price fluctuations (Ryschawy et al., 2012; Homann-Kee Tui et al., 2015).

In many cases, however, crop-livestock systems still fall short of accomplishing the goals of high productivity and production, lifting people out of poverty or

The Role of Ecosystem Services in Sustainable Food Systems. https://doi.org/10.1016/B978-0-12-816436-5.00008-1

reducing greenhouse gas emissions, which depends on the level of "integration", often a function of other factors in the system's (Herrero et al., 2010; Nin-Pratt et al., 2011). Persistent low agricultural production, yield gaps, and limited access to land, combined with increasing human populations and degradation of environmental resources, policy and institutional issues, climate variability, and change have increased pressure on these systems (Anderson, 1992; Tiffen, 2003). Much is said about the need for these systems to intensify; quantitative information on the impact on multiple criteria, not only agricultural productivity but also economic, social, and environmental, is, however, scarce. In addition, the contribution of crop-livestock interactions is often hidden and unaccounted for in technical, economic, and policy decision-making, which means that there is insufficient investment in their protection and management (Lemaire et al., 2013; Lipper et al., 2014). Research can play an important role in better understanding potential trade-offs and synergies in these different domains, particularly where limited access to biomass, nutrients, water, and labor increases the trade-offs (Giller et al., 2009; Baudron et al., 2012). Research, therefore, plays a crucial role in guiding policy decisions and recommending transformational processes needed to improve the crop-livestock interactions, address trade-offs, and take advantage of possible synergies that lead to improved ecosystem services.

Zimbabwe, for instance, is a country with high urgency to transform its agricultural systems (Rippke et al., 2016). For a long time the unanimous mandate had been to ensure food security through provision of maize seed and fertilizer, backed up by a maize focused support system (Grant et al., 2012; Homann-Kee Tui et al., 2013). During the 1990s, when maize production was intensively promoted, yields were commonly around 1500 kg ha^{-1}; currently, with low input levels, yield levels fluctuate around 750 kg ha^{-1} (Government of Zimbabwe, 2002; Mudimu, 2003). Efforts to sustainably raise yields using conservation agriculture have come a long way with high fertilizer application rates to prevent N immobilization when mulching. However, the positive effects of mulching take time to be realized. This, combined with competition between crops and livestock feed, creates a strong trade-off with biomass (Rusinamhodzi et al., 2011; Valbuena et al., 2015). Without strong linkages to the other systems components, small grains, legumes, livestock, and markets which all add value to farm outputs, vulnerable agricultural systems record very low returns per unit land (Homann-Kee Tui et al., 2015; Masikati et al., 2015).

This chapter addresses two research questions:

1. How can improved integration of crop-livestock systems enhance ecosystem services in smallholder farming systems in drylands such as those in Southern Africa?
2. What agricultural transformation can reduce trade-offs and/or create synergies in farm resource allocations while improving ecosystem services within the specific context of Southern Africa?

We first link challenges and opportunities for crop-livestock systems to reduce trade-offs and take advantage of possible synergies and then demonstrate how agricultural transformation can lead to improved ecosystem services in sustainable food systems. We are using the case study of Nkayi district in Central Zimbabwe, where the Agricultural Model Intercomparison and Improvement Project (AgMIP) approach was implemented to assess impacts of climate change. We extended the analyses to demonstrate how transformation of crop livestock systems can enhance ecosystem services. A high potential for crop-livestock integration supporting food systems was anticipated, but the land remains underutilized (Masikati et al., 2015; Antle et al., 2017; Homann-Kee Tui et al., in preparation).

Crop-livestock systems and ecosystem services

This section synthesizes fundamental crop-livestock linkages, ecosystem services they can provide, and how they can enhance the sustainability of food systems.

Key characteristics of crop-livestock systems

Crop residues for dry season feeding of livestock

Crop residues are a vital dry season feed resource, providing livestock with feed when other resources are scarce (Valbuena et al., 2015). Combined with crude protein carriers and dual purpose and forage legumes, affordable quality feed can be provided (FAO, 2011; Masikati et al., 2013). Feed markets that attribute higher value to dual purpose residues than to grains are evolving (Berhanu et al., 2012; Blummel et al., 2012). Supporting livestock during the most critical times of the year, maintaining body condition, and reducing mortalities translates directly to more and better-quality animal products and services.

Livestock manure for soil fertility amendment and increased crop yields

Animal manure, along with green manure (crop rotation, intercropping, agroforestry, Mapfumo and Giller, 2001; Rufino et al., 2006) is the major available on-farm resource for soil fertility improvement. Manure application enhances the efficacy of inorganic fertilizer and is important for sustaining soil health and improving soil structure, slow release of nutrients, and carbon sequestration, and with intensification, manure is becoming a private resource of its own (Abdoulaye and Sanders, 2005).

Use of animals for draft power for timely preparation of crop fields

Access to draft animals at critical times of the year is a strategy for increasing the levels of crop production and in turn results in more biomass for livestock feed (Tittonell et al., 2010; Homann-Kee Tui et al., 2013). Farmers feeding crop residues to their livestock in enclosures recorded improved feeding and manure production efficiency, thereby increasing overall farm productivity. Draft power addresses labor constraints while allowing farm production levels beyond food self-sufficiency.

Ecosystem services provided by crop-livestock systems

Ecosystem services, defined as the multitude of benefits (well-being, health, and economy) that nature provides to society, are delivered by and to agriculture as agriculture both benefits from and supports ecosystem services (http://www.fao.org/ecosystem-services-biodiversity/en/). Improved crop-livestock integration provides ecosystem services in many ways (Table 8.1). For instance, if farms can provide enough fodder resources, rangeland pressure can be reduced, allowing for improved carbon recycling and habitat protection functions (Milton et al., 2003; Swinton et al., 2007).

Ecosystem services can be classified in four categories, presented here in relation to the benefits from crop-livestock integration:

Provisioning

Food, feed, and water are the resources obtained from ecosystems; rural households often depend directly on these services for their livelihoods. They can also be traded in markets. Integrated and diversified crop and livestock production results in a more stable food supply and improved nutrition, together with increased income through more rapid nutrient and energy cycling and better water use efficiency (Sumberg, 1998; Descheemaeker et al., 2011). Poor families in particular often depend on low cost, plant-based carbohydrate-dominated diets, whereas human diets require a more diversified and nutritious diet (Grace et al., 2018). Access to plant and animal-based foods improves physical and cognitive development of children in areas with high levels of malnutrition.

Supporting

Ecosystem services are necessary for the production of other ecosystem services. For instance, maintaining a diversity of crops and livestock, the integration of crops and livestock benefits nutrient cycling, and the more efficient use of natural resources. Integration also improves synergies between improved feed and nutrition, soil health, and environment (Thornton et al., 2009). Increasing water holding capacity is important in drylands where climatic fluctuations are high. Local access to nutrients and the reduction in cost of inorganic fertilizer is crucial, particularly in remote areas (Descheemaeker et al., 2011; Gil et al., 2015).

Regulating

Maintaining the benefits obtained from regulation of the ecosystem processes is important, particularly where poverty and exploitative natural resource use induce degradation. For instance, biodiversity, diversification of croplands, and expansion of legumes prevents pests (Pelley, 2009). In addition, confining livestock in designated areas helps to halt the expansion of agriculture into the natural environments, controlled rangeland and manure management help to increase carbon sequestration on croplands and rangelands (Mbow et al., 2014; Thornton and Herrero, 2015), improving breed and feed quality and increasing productivity and offtake reduce greenhouse gas emissions (FAO, 2013; Herrero et al., 2015).

Table 8.1 Vision on how crop-livestock integration can benefit ecosystem service delivery and the sustainability of food systems.

Crop-livestock components	Functions of ecosystem services	Contributions to sustainable food systems
- Multiple purpose crops provide food, feed, soil fertility, and fuel. - With higher crop yields, there is more feed biomass and improved feed quality for livestock. - Better fed livestock produce more manure and draft power, increased grain and biomass production. The energy for household use can replace charcoal and wood.	**Provisioning**, material benefits people get from ecosystem, such as the production of food, biomass and water; draft power, energy and raw material, and genetic resources	- Increased food and nutrition security and profits - Income converted from livestock sustains prompt food and nutritional needs, especially in dry areas where crop production is risky - Diversified income sources buffer against trade, price, and climate fluctuations
- Livestock manure is an important source of organic nutrients, maintenance of soil productivity, soil structure, water, and nutrient retention, in crop lands and rangelands. - Crop rotation and intercropping increase soil fertility and moisture, and therewith crop and livestock productivity	**Supporting,** are necessary for the production of all other ecosystem services, such as soil fertility, soil formation, primary production, and habitat	- Resource efficient intensification of land use and increased agricultural productivity - Reduced production costs, e.g., for inorganic fertilizers and stock feed - Mitigation of drought effects, through more stable production at reduced risk
- Investment in organic material from crops and livestock helps to reduce emissions and build soil organic carbon - Diversified systems, with substantial shares of cereals, legumes, and livestock, have less pest and disease infestations - Diversification of crops and livestock also hedges against risks, e.g., climatic and market price fluctuations	**Regulating**, such as the control of ecosystem processes, climate, carbon sequestration, pest, disease and weed control, as well as market risk	- Climate change mitigation, reduced GHG emission, increased carbon sequestration - Negative environmental impacts of food production reduced, less pesticide use, and better soil erosion control

Continued

Table 8.1 Vision on how crop-livestock integration can benefit ecosystem service delivery and the sustainability of food systems.—*cont'd*

Crop-livestock components	Functions of ecosystem services	Contributions to sustainable food systems
- Improved access to knowledge, networks, resilience, and profitability, reducing vulnerability and poverty - Poor farmers in particular find more opportunities, more stable livelihood options, self-esteem - Better off farmers face less risks under climate variability and change	**Cultural**, nonmaterial benefits people gain from ecosystem, such as identify, spiritual health, landscape esthetics, and welfare	- Capacity and self-esteem, as farmer recognized for competence - Learning and knowledge sharing enhanced where there are more benefits from farming - Creation of welfare in rural areas

Cultural

The nonmaterial benefits include recreation, knowledge systems, and social relations. With reference to crop-livestock linkages and markets as a way to underpin effective crop-livestock integration, we recognize the cultural benefits from engaging in diversified production and market processes. Income from crops and livestock is already spent mostly on education, food, and human health (Powell et al., 2004; van Rooyen and Homann-Kee Tui, 2009). Cultural benefits include the ability of farmers to improve their livelihoods and nutrition, along with their economic welfare and well-being (Ryschawy et al., 2012; Lemaire et al., 2013).

Challenges and trade-offs for crop-livestock integration enhancing ecosystem services

In sustainable food systems, identifying and reducing trade-offs of agriculture and ecosystem services is a major challenge, while at the same time grabbing opportunities and helping reduce people's vulnerability to climate change and other setbacks. The influence of crops and livestock on ecosystem services can be positive or negative, a result of the dynamics, and interactions of complex systems. More croplands and growing livestock populations risk aggravating negative consequences for the environment, unless the systems diversify, integrate, and intensify in a sustainable way (Franzluebbers et al., 2014).

Low productivity of both crops and livestock

Constraints and threats in smallholder farming systems limit the integration of crops and livestock. Poor soil fertility and management, where soils response to fertilizer is limited or risky, results in low crop production and livestock feed. Manure management is often poor: large volumes are lost to open grazing, and the little that is collected is poorly stored, leading to loss of nutrients and increase in greenhouse gas emissions (McIntire et al., 1992; Schlecht et al., 2004). In some instances large herds are kept on farms for draft power services during growing periods, insurance, and as assets. Feed gaps during the dry season result in seriously undernourished animals, often because of the high numbers. Rainfall variability regularly depletes the feed resources. Feed shortages force farmers to sell off livestock below market rate to sustain food security during periods of low crop yields. Meanwhile, crops are bred for the sole purpose of food security; the genetic potential for increasing feed digestibility and nutritional value, without undermining crop yield performance, is not utilized (Blummel et al., 2012). With increasing demand and poor availability of alternative biomass, farm resources are increasingly contested and dwindle (Valbuena et al., 2015).

Complexity and interactions

Farming systems typically include multiple subcomponents with different management requirements, all competing for limited biomass, labor, and cash investments. As most impact studies do not recognize component interactions, they look at isolated system components to analyze potential effects and not at the entire farming system. Interventions are, therefore, not designed to engineer the integration but, rather, support single technical components. An often-cited example is the promotion of conservation agriculture in drylands, postulating retention of crop residues for soil fertility amendment, but without organizing alternative biomass for livestock feed (Giller et al., 2009; Valbuena et al., 2012). Policy directives can reinforce these challenges. For instance, cereal crops, typically grown for subsistence, receive greater government and extension support but are mostly of lower nutritional value. Legumes provide higher nutrient density for food and feed, as well as offer soil fertility benefits. They are traditionally women's crop for their multiple benefits to the household but receive less support, and input and output markets are often underdeveloped, hence they are grown on lower shares of agricultural land (Homann-Kee Tui et al., 2013). Farmers with livestock are usually wealthier and have more land as they invest in and use livestock to build up their farms and livelihoods (Harris and Orr, 2014; Homann-Kee Tui et al., 2015). Managing livestock assets more productively is hampered by a lack of support for improved participation in markets, lack of incentive mechanisms and missing support services (Moll et al., 2005). Increasing marketable livestock offtake, which enables reinvestments in farming, also requires banking systems to store the cash for when it is needed. In addition, there is an interaction with off-farm opportunities, embedded in larger income strategies, which would allow farmers to make decisions more strategically, rather than being needs driven (Barrett and Constas, 2014). While most farmers try to diversify their income portfolio with off-farm activities, many lack the opportunities.

Spatial heterogeneity

Heterogeneity, often high in farm household populations, is a challenge for analysis and recommendations. Farming systems differ in farm size and composition; there are also behavioral differences in agroecologies and intensity of crop-livestock systems because of the farm decision makers' knowledge and experience (Valbuena et al., 2012; Antle et al., 2017; Valdivia et al., 2017). At community level, natural resource endowments, biomass availability, and access to off-farm resources influence the type and degree of crop-livestock integration. Large heterogeneity within farming communities results in a large variety of management and adaptive capacity which influences the ability to provide food over the annual cycle, the capacity of a system to withstand a shock or disruption, and long-term responses to climate change. Blanket recommendations that do not address spatial heterogeneity frequently do not work, and adoption rates remain low.

Climate change effects

Although magnitude and direction might differ by region, through higher temperatures, rainfall variability, changes in crop and variety suitability, reduced yield potential, biomass availability and nutritional quality, grain, and residues are expected to increase biomass and water trade-offs (Jones and Thornton, 2009; Masikati et al., 2015). Livestock changes in rangeland composition and herbage quality are projected to further affect feed availability and quality (Herrero et al., 2015; Descheemaeker et al., 2016). This directly augments the negative effects on livestock production, the impact of heat and limited drinking water, and animal physiology and health, thereby reducing livestock performance (Thornton et al., 2009; Descheemaeker et al., 2011).

Socioeconomic drivers

Effects of socioeconomic institutional and demographic changes are probably the strongest drivers shaping pressures and trade-offs in farming systems (Steiner and Franzluebbers, 2009; Valbuena et al., 2015). Expanding human populations compete for food and water supplies, while poorly developed infrastructure, markets, and services restrict agricultural potential and aggravate risk, limiting the responses to threats such as climatic fluctuations, price distortions, pests, and diseases (Anderson, 1992; Jayne et al., 2006; Valdivia et al., 2012). In the absence of economic development and labor markets, employment outside of agriculture has remained limited

(Rauch et al., 2016; Cousins and Hall, 2013). Governments under pressure to produce and supply more affordable food are geared toward provision of cheap, nutrient-poor, and carbohydrate-dense food, perpetuating malnutrition and food safety issues and health disparities (Ecker and Quaim, 2011; Gómez et al., 2013). With poor diets and limited food sufficiency, people depending on these agricultural systems are highly vulnerable, and the physical and cognitive development of current and future generations is being seriously compromised.

Agricultural transformation for improving ecosystem services

Enabling agricultural transformation into sustainable food systems requires an approach that accepts the complexity of the behavior of these systems while seeking to create appropriate synergies between farm subcomponents, cereals, legumes, and livestock. Effectively creating synergies means action and interventions across the food systems, with a greater consciousness of where to intervene (Meadows, 1999; Malhi et al., 2009). The seemingly dichotomous goals for achieving high quality food production while improving environmental quality can then be reconciled.

Paradigm and goals

There is need to align paradigms and pathways toward achieving sustainability goals. A shift toward holistic solutions can take a long time but can be more effective than linear solutions that have been trying to push single technologies and largely failing (Malhi et al., 2009). Agricultural diversity, integration, and associated ecosystem services need to be enhanced at all levels of organization, all agro-ecological regions, and within all local farming contexts (Lemaire et al., 2013; Moraine et al., 2016). At farm level, investment in integration means strengthening the positive links between crops and livestock through mechanisms that reward them. This requires the consideration of the impact of agricultural policies on nutrition and food safety, diversified production synchronized to demand improved conditions for farmers to participate in input and output markets, and food pricing that makes healthy feeds and foods more affordable. At regional level there is need to reverse the decoupling of crop and livestock production and to encourage livestock use of landscapes that supports crop production and good management of rangelands to help increase carbon storage in the soil.

Systems structure and feedback

Structural change in food systems requires consideration of sustainability issues across agricultural as well as related sectors, notably health and education. Awareness of sustainable food systems requires broad engagement of actors—policy, private sector, development, and research—on policy and decision-making toward enhancing the role of ecosystem services through integrated agricultural practices (Franzluebbers et al., 2014; Firbank et al., 2018). An understanding of farming systems must guide management decisions, investments, and adaptations, and there must be a closer link between research and policy processes for informed decision-making. New information and extension structures are required, enabling sound diagnosis and prototyping in context, communication, and capacity development for upscaling, particularly since diversified and integrated systems require knowledge and skills beyond investing in a few farm activities (Garrett et al., 2017).

Structural elements

While efforts are usually put into improving structural subsystem-specific elements, addressing these without the above change in paradigm and structures will be incoherent and ineffective in achieving sustainable food systems (Malhi et al., 2009). Here we list examples of possible structural elements that can enhance synergies toward sustainable food systems, encompassing economic, social, environmental, and regulatory contexts.

- Policies: Promotion of dual purpose crop varieties and seed systems, notably legumes; food safety control, labeling, and implementation; nutrition standards in public organizations such as schools and hospitals; agriculture, nutrition, and health introduced as components of the educational curricula.
- Market incentives: The increase in demand for food presents opportunities for farmers who are able to improve agricultural production, gain access to markets, and benefit from its income. Investment in local value addition, e.g., fodder and manure transactions and value chain and enterprise developments, that address bottlenecks in replacing input subsidies.
- Technologies: Suites of technologies that increase productivity and nutrient density, integrating suitable cereal and legume varieties and livestock breeds with a combination of locally sourced and supplementary inputs, increasing overall farm profitability.

How can we test and analyze priorities and options for this transformation?

The search for more sustainable food systems has motivated multidimensional impact assessment (economic, social, and environmental), simulation models that help to evaluate the performance of existing farming systems, possible modifications of those systems, and trade-offs among the various dimensions (Antle, 2011; Antle et al., 2014). Thereby the usefulness of prospective changes in agricultural systems can be assessed in areas where they are not yet in widespread use, as well as the use of existing or new technologies under future climate and socioeconomic conditions.

The goal is to influence decision-making for transforming and reorienting agricultural systems to support food security and nutrition, under policy and technology changes as well as other changes in the context. The AgMIP has developed a framework for integrated assessments that can be used to evaluate the impact of intensifying farming systems in a way that enhances ecosystem services and involving the following three elements:

Regional integrated assessment

Interventions and comprehensive solutions need to be developed, taking into account the contextual conditions (i.e., technological, biophysical, and socioeconomic) instead of trying to fix isolated technical problems that arise. Novel integrated modeling approaches of farming systems provide methods and tools to simulate uptake and assess the impact of interventions and their multiple benefits in mixed farming systems (Antle et al., 2015; Shikuku et al., 2017). Farming systems are evaluated as an integrated unit, rather than individual production activities, and linked to other environmental and social outcomes that they may impact. Multidisciplinary research teams use the characterization of current systems to identify key system components, and the corresponding data and models assess them alongside alternative and improved systems. The whole-farm and household modeling approach uses inputs from climate, crop, livestock, and economic simulation modeling components and provides a range of economic outcome indicators and their distributional effects for heterogeneous communities.

Codesigning improved management strategies and future development pathways

The assessment process involves scientists, experts, and stakeholders representing knowledge and experience from climate, crops, livestock, and economic disciplines in testing (1) improved management strategies that are realistic and feasible for farming systems within 5 years and (2) representative agricultural pathways (RAPs) for mid-century conditions of farming, at the same time considering the impact of climate change (Valdivia et al., 2015). Parameters that represent farming systems and production methods (i.e., technologies) are combined with the physical environment (i.e., the climate) and the economic, policy, and social environment in which the systems operate (i.e., the socioeconomic setting). The quantitative information is used to define inputs for the crop, livestock, economic models such as farm, herd, and household size, production and market participation, market prices and costs of production, off-farm income, etc. The narratives include information that helps characterize some of the key interrelations in the agricultural system but cannot be modeled, like gender and nutrition, such as pests and diseases, water use, institutional constraints, labor availability, and attitude toward investment.

Stakeholder engagement to prepare conditions for uptake

Scientists and decision makers require exchange of data and information on how agricultural systems might change and adjust within the local context and how capacity can be built to support and implement changes in these systems (Valdivia et al., 2015). RAP's iterative development process between scientists, experts, and stakeholders, across disciplines and institutions, helps to develop and set priorities in research and policy making for adjustment and change at multiple scales. The understanding of possible direction, magnitude, and interplay of biophysical and socioeconomic drivers, associated capabilities, challenges, and opportunities supports the development of technologies, institutions, and policies to achieve a positive impact. RAPs incorporate feedback from interventions for building an informed dialog addressing future challenges and opportunities and thus provide a process to effectively engage stakeholders in the development and evaluation of technological options.

Application: an example from smallholder crop-livestock systems in semiarid Zimbabwe

The AgMIP approach was applied within a case study in Nkayi district, Central Zimbabwe, which illustrated typical features of the multiple challenges that restrict the full contribution of crop-livestock integration to ecosystem services (Homann-Kee Tui et al., 2013; Homann-Kee Tui et al., 2015; Antle et al., 2017).

The case: farming in Nkayi district

Nkayi district in Central Zimbabwe was selected as a case study site where farmers, in a cross-country comparison with Malawi and Mozambique, already make greater use of crop-livestock integration.

Crop production is rain-fed, and average annual rainfall ranges from 450 to 650 mm, making the system vulnerable to erratic rainfall with a drought year frequency of two in every 5 years. Long-term average maximum and minimum temperatures are 26.9°C and 13.4°C, respectively. The soils vary from inherently infertile deep Kalahari sands, which are mainly nitrogen and phosphorus deficient, to equally nutrient deficient clay and clay loams that have been continuously cropped without soil replenishment. Farmers use mainly a monocereal cropping system with addition of low amounts of inorganic and organic soil amendments. Natural pasture provides the main feed for livestock, and biomass availability is seasonal. During the wet season feed quantity and quality is appreciable, while during the dry season there is low biomass of poor quality.

As in many parts of Africa, rural populations live from mixed crop—livestock farming. A farm household has less than 2 ha land, mainly under maize, with smaller portions of sorghum, groundnuts, and cowpeas as staple crops, combined with the use of communal rangelands, fallow land, and crop residues for livestock production. About two-thirds of the households' own cattle and herd sizes vary from a few to up to 40 head per household. Farmers also keep some donkeys and goats. Livestock offer opportunities for farm diversification and intensification, spreading risk and providing significant livelihood benefits. Animals are kept to complement cropping activities through the provision of manure for soil fertility maintenance, draft power for cultivation, transport, cash, and food, while crop residues are used as supplementary dry-season feed.

The increased demand for crop and livestock products could benefit small scale farmers as they gain access to markets, if they are able to diversify and intensify production in a sustainable way. This could reduce risk and increase resilience by providing farmers with diverse sources of income. Low productivity, however, is the critical challenge because of a combination of factors that include unfavorable climatic conditions, poor and depleted soils, environmental degradation, and low level of capital endowment, leading to limited uptake of improved technologies as well as poor market access and adverse policies. Climate variability and change stressors, superimposed on the structural problems, chip further away at food security and increase vulnerability.

Enhancing ecosystem services through crop-livestock integration

The transformative management options were carefully selected by scientists, together with district and regional experts from crops and livestock extension and rural communities in a co-design process. Simulations of business-as-usual interventions, such as improved soil fertility management and new crop varieties, had shown that this would improve food security and feed supply to some extent but not fulfill the ultimate objective of lifting farmers out of poverty (Masikati et al., 2015).

To achieve near-term options, scientists and experts worked on a vision to remove barriers to farming within the next 5 years. Stakeholders chose to combine improved cereal and legume management (higher planting densities, improved varieties, and soil fertility management) with a policy to convert land that was not needed to achieve food self-sufficiency from cereals to food and feed legumes; price incentives for better aggregation of quality legumes sold drove the expansion of legumes. Fig. 8.1 illustrates the positive feedback around more diversified and better integrated farming systems, providing more and higher value outputs which can be reinvested to support the system, while resources are being used more prudently, leading to overall increased efficiency.

Fig. 8.2 illustrates the synergetic effects of these interventions on ecosystem services in entire farms and rural communities.

- **Provisioning**: Strong increases in crop and livestock production due to combined effects of better genetic potential, and (in) organic fertilizer application and increased planting density, together with rotational benefits for greater shares of maize with legumes, suggest that positive impacts of sustainable intensification options would be high.

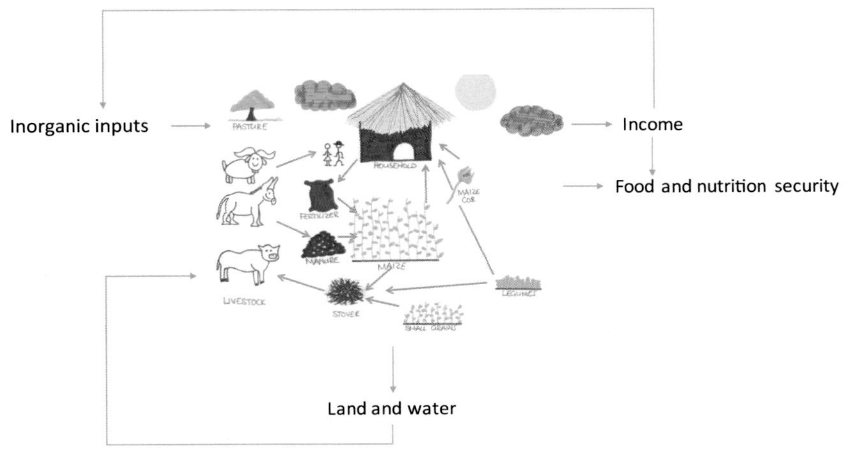

FIGURE 8.1

Simplified influence diagram on crop-livestock integration enhancing ecosystem services and contributing to sustainable food systems.

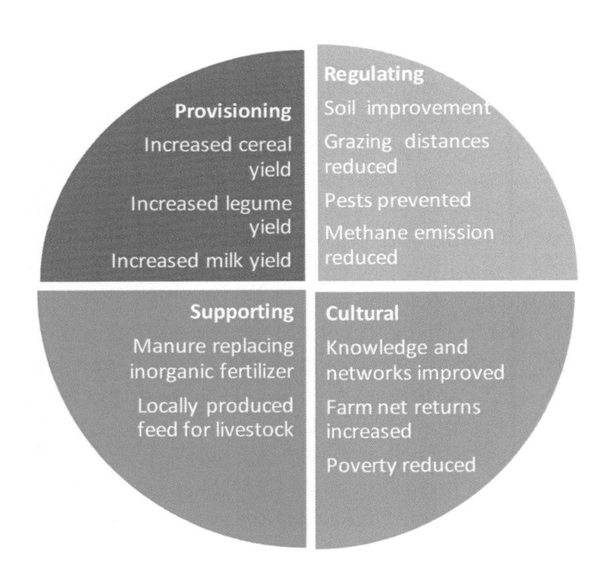

FIGURE 8.2

Enhancing ecosystem services through crop-livestock integration.

- **Supporting**: Soil fertility responses would increase through application of manure (1.1 t ha^{-1}), combined with microdose mineral fertilizer (20 kg N ha^{-1}). More biomass of higher feed quality (4.3-5 t ha^{-1} Mucuna) would add value to crop residues and help to alleviate dry season feed gaps through locally processed fodder.

- **Regulating**: Farmer behavior change would direct operations toward greater agrobiodiversity, which would in turn enhance the regulation functions in ecosystem services, contributing to the reduction of soil and rangeland degradation, organic manure building soil organic carbon, and improved feed, leading to reduced methane emission.
- **Cultural**: Livelihood, welfare, and social benefits would also be high. Diversification would support opportunities for the extremely poor, e.g., cultivating more profitable legumes would result in fodder biomass transfers. Poverty rates would be reduced substantially.

Ecosystem services in sustainable versus fast economic growth pathways

For transformative change options by mid-century and taking into account the effects of climate change, scientists and experts contrasted two plausible future scenarios that would characterize the conditions of countries like Zimbabwe if they were to invest in a sustainable development pathway (green) versus a fast economic growth pathway (gray) with less consideration for agricultural sustainability. Tables 8.2 and 8.3, and Fig. 8.3 present trends and narratives of the pathways and their possible impacts on ecosystem services.

Under the sustainable development pathway, emphasis was placed on efficient resource use, inclusive development and concern for the environment. Diversification and integration of crop and livestock functions drove agricultural intensification. The interventions created synergies between the different categories of ecosystem services, with no perceived trade-offs.

The fast economic growth pathway was driven by competitiveness and inequality, with the view of bringing fast, short-term gains with food security, and income. Support for the agricultural sector was heavy on external, inorganic inputs. For this pathway there were indeed trade-offs: faster economic growth led to fewer ecosystem services.

Provisioning

Both paths provided more food but were widely different in magnitude, composition, and sensitivity to climate change. The sustainable development pathway supported more diversified and healthy quality food, which was less sensitive to climate change and offered more equitable distribution. The fast economic growth path, incorporating more inorganic fertilizer and stock-feed for livestock, was far more sensitive to climate change.

Supporting

The sustainable development pathway enhanced positive systems dynamics through incentives for diversification and integration. Prudent resource use and low-risk

Table 8.2 Ecosystem services under mid-century conditions, sustainable development (green), and fast economic growth (gray) pathways.

Sustainable development (green)	Fast economic growth (gray)
Provisioning	
Crop yields in future systems were better than in current systems but also more sensitive to climate change with nutrient limitations partly alleviated through improved soil fertility management. Crop production was diversified into greater shares of food and feed legumes. Combining crop genetic potential with microdosing, manure application, crop rotation, and increased planting densities increased maize and sorghum yields by more than 100% and groundnut yields by more than 200%.	Higher inorganic fertilizer rates led to higher crop yields in future systems, but also to more sensitive responses to climate change when compared to current systems and under the sustainability pathway.
In contrast to crops, livestock production in future systems was less sensitive to climate change compared with current systems. In future systems, crop residues, fodder from Mucuna, and concentrates were fed to cattle to alleviate feed gaps. Livestock benefited from improved supply in the form of more and better-quality feed. As crude protein availability was the main limiting factor, the supplementary feeding alleviated feed gaps more profoundly, resulting in further improvements in animal productivity.	Livestock production under this pathway was less sensitive to climate change compared with current systems. Here higher rates of concentrates than crop residues and Mucuna were fed. As a result, variation in rangeland and on-farm fodder production played a minor role. Only better-off farms with bigger herds of cattle were noticeably sensitive.
Supporting	
Incentives for diversification and integration of crops and livestock work on positive agroecology dynamics within the farming systems, with prudent resource use and low risk investments. Multipurpose, robust, and adaptive crops and livestock as well as integrated farm management strengthen the link between soil fertility and livestock nutrition. Information services and mechanization support resilience and profitability.	The gray pathway would support increased production of staple and cash crops through high costs and energy intensive investments, genotypes with quick return, commercial input supply, inorganic soil fertility management, pest and disease campaigns, and industrial mechanization.
Regulating	
Positive agroecology dynamics between farms and the environment, agrobiodiversity at farm and landscape level, together with integrated land use planning, enhance water use efficiency, soil and rangeland quality improvement, and plant and animal health.	Under the gray pathways the regulating function will be through resource intensive investments and overutilization of resources.

Table 8.2 Ecosystem services under mid-century conditions, sustainable development (green), and fast economic growth (gray) pathways.—*cont'd*

Sustainable development (green)	Fast economic growth (gray)
Cultural	
There would be more equitable development at farm and community level. Women would be key actors in decision processes, and gender balance would result in positive impacts on farming for nutrition. Poverty rates would be curbed to 36% of the farm living below poverty line. More farms would benefit from adaptation to climate change, including the extremely poor.	Poverty rates would be higher as compared with the sustainable development path, with about 47% living below poverty line. This path compromises on social and human health, excludes women and vulnerable groups from development, and there is a trade-off between economic growth, environment, women's engagement, and nutrition.

investments made use of multipurpose, robust, and adaptive crops and livestock and integrated farm management, which strengthened the link between soil fertility and livestock nutrition, together with mechanization support. The fast economic growth pathway supported genotypes with quick returns, commercial input supply, inorganic soil fertility management, pest and disease campaigns, and industrial mechanization.

Regulating

The sustainable development pathway worked on positive agroecology systems dynamics with the environment at farm and landscape level, effects of agrobiodiversity, integrated land use planning, enhanced water use efficiency, soil and rangeland improvement, and plant and animal health. Under the fast economic growth pathway, the control function was through resource intensive investments to reduce risks.

Cultural

Under the sustainable development pathway there was more equitable development at farm and landscape levels, vulnerability was lowered, and adaptation benefited more people, especially the poor, and it improved gender balance and positive impacts on nutrition. The fast economic growth pathway compromised on social and human health, excluded women and vulnerable groups from development and created trade-offs between economic growth, environment, women's engagement, and nutrition.

Both pathways had a third or more farmers under extreme poverty. Research and policy will be challenged to come up with safety net interventions that lift these

Table 8.3 Trends and narratives for ecosystem services under sustainable development (green) and fast economic growth (gray) pathways.

	Sustainable development (green)		**Fast economic growth (gray)**	
Provisioning				
Parameter	Trend	Implications for ecosystem services	Trend	Implications for ecosystem services
Cereal production	↗ (dotted)	Improved production through market incentives for better ecosystem services, less sensitive to climate change	↗	Intensive production of food crops and for industrial uses, following market demand, more sensitive to climate change
Legume production	↗ (dotted)	Improved production through market incentives for better ecosystem services, less sensitive to climate change	↗ (dotted)	Intensive production of food crops and for industrial uses, following market demand, more sensitive to climate change
Livestock production	↗ (dotted)	Improved production through market incentives for better ecosystem services, less sensitive to climate change	↗	Commercialized production, owners of large herds, more sensitive to climate change
Supporting				
Fertilizer application	↗ (dotted)	Use of efficient technologies will lead to prudent uses of inorganic fertilizer	↗	Market incentives and subsidies encourage uptake of inorganic fertilizer
Improved seed use	↗	Use of agro-ecologically suitable high yielding varieties	↗	Market incentives and subsidies encourage uptake of improved seed
Livestock feed	↗	Investment in dry season technologies, local feed processing and fodder markets, complemented by stock feed	↗	Market incentives will encourage uptake of improved feed technologies, new large-scale feeding systems
Mechanization	↗ (dotted)	Targeted mechanization for removing labor constraints at strategic entry points	↗	Industrialized mechanization for land preparation, production, postharvesting, and processing

Table 8.3 Trends and narratives for ecosystem services under sustainable development (green) and fast economic growth (gray) pathways.—*cont'd*

	Sustainable development (green)		Fast economic growth (gray)	
Regulating				
Rangeland health	↗ (dotted)	Controlled land use, pasture management, fodder reserves, and reversing degradation	→	Inequitable and extractive use of rangelands, resulting in islands of productive and overused rangelands
Pest control	↗ (dotted)	Pest prevention through diversification and integrated measures	↗	Monocultural production will accelerate the risk of pest infestation
Groundwater availability	→	Sustainable water management and rehabilitation, water saving technologies, and reversing degradation	→	Inequitable access to water for commercial uses, with irrigation for specialized production and disadvantages for downstream users
Energy use efficiency	→	Switch to renewable energies, curbing demand for high energy use technologies	↘ (dotted)	Energy waste, e.g., through water pumping, electricity, inorganic fertilizer use, counteracting efforts to reduce input costs
Cultural				
Food security and nutrition	↗	Food security sustained also in lean periods, and for agro-ecological regions with high climate variability. Livestock, especially small ruminants, and other sectors help sustaining food security and overcoming food deficits. Greater food diversity of diets and micronutrients, malnutrition reduced through promotion of food programs, supported by education	↗ (dotted)	Carbohydrate-based, industrially produced food of limited nutritional value will be available at relatively low prices. Agriculture will be geared to fast increases in production, with trade-offs to food diversity and nutrition.

Continued

Table 8.3 Trends and narratives for ecosystem services under sustainable development (green) and fast economic growth (gray) pathways.—*cont'd*

	Sustainable development (green)		Fast economic growth (gray)	
		and gender empowerment.		
Gender equality and equity		Effects of empowerment, social support mechanisms and positive dynamics support gender equality and equity. Gender will be mainstreamed in economic, environment, and social development.		Gender not mainstreamed in agricultural policies, fast economic growth, and aggressive investment culture will exclude women from opportunities
Women's asset ownership and influence on decisions		Women will have strong influence on decisions, ownership, control, and security over productive assets. Improved access to market opportunities and support services will enhance their capabilities.		Women's influence on decisions, ownership, access to and control of key productive assets will be over-shadowed by the drive for profiteering

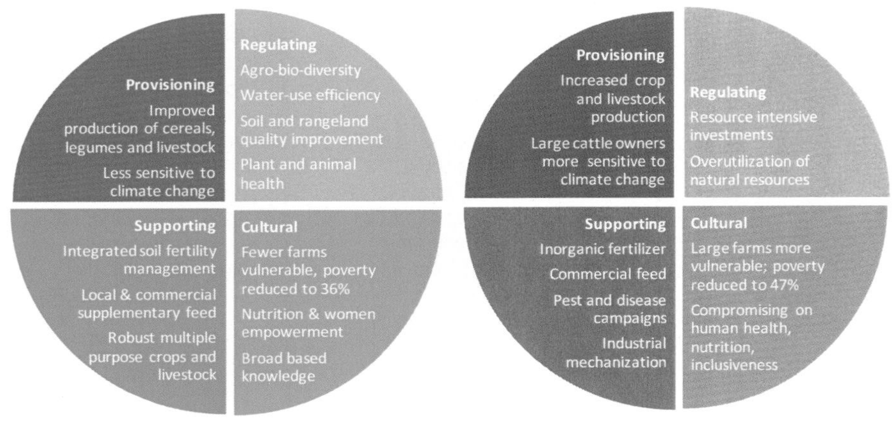

FIGURE 8.3

Enhancing ecosystem services under sustainable development (green) and fast economic growth (gray) pathways.

households to a state of well-being, e.g., creating links to off-farm income. The other two-thirds of farm households were likely to step up to better livelihoods. Here it is important to note that even though positive outcomes are likely, a high proportion of households are still vulnerable to losses. Policies need to focus not only on achieving the positive economic impacts but also on reducing vulnerability.

Conclusions

The paradigm of sustainable agriculture and food systems is to minimize negative impacts on the environment while optimizing production by protecting, conserving, and enhancing natural resources and using them efficiently; ecosystem services play a critical yet often underestimated role in sustainable agriculture (http://www.fao.org/ecosystem-services-biodiversity/en/).

Well integrated crop-livestock systems create synergies on a farm, recycle resources efficiently, by-products of one component serve as a resource for the other, or through exchange of residues, manure, and draft power across farms. Integration thereby enhances crop and livestock production and safeguards the environment through prudent and efficient resource use. Shifting toward diversification and improving integration of crop-livestock systems increases opportunities for synergies, across provisioning, supporting, regulating, and cultural services.

In this section we share concrete lessons from the case study in Nkayi district, Central Zimbabwe, on what can be done to support crop-livestock integration enhancing ecosystem services.

Importance of crop livestock integration

Our case from Zimbabwe illustrates very well that better integration of crops and livestock can support delivery of ecosystem services for food security and nutrition, income, and environmental benefits, by smallholder famers that are vulnerable to climate change and other types of stresses. These results are not hypothetical; related studies suggest that smallholder farmers are prepared to invest in crop-livestock linkages and expand impacts of ecosystem services, if a conducive environment is given (Komarek et al., 2015; Homann-Kee Tui et al., 2015). It is clear that institutional constraints hinder farmers' investments.

The study substantiates that, beyond biophysical processes, economic mechanisms can be reinforcing and accelerating the functional linkages within and across farms: The profits from crops and livestock increase with diversification and reallocation of land and improved market access (Homann-Kee Tui et al., in preparation). Under sustainable development pathways social cooperation and engagement of women supported higher production levels and nutritional benefits. This is in line with other studies that suggest that incentives for diversification and integration enhance overall farm net returns, e.g., increased demand for quality biomass, for higher and more sustained flows of crops and livestock to markets and thereby initiate new forms of exchange as new source of income in rural communities (Moll, 2005; van Rooyen and Homann-Kee Tui, 2009).

The analyses of adoption and net returns also highlight that to transform and reorient agricultural systems to support food security through sustainable farming practices, farmers' have to gain enough value from the reduced or reversed negative environmental impacts. In line with others our case study suggests that unless there are adequate incentives, poorly integrated and linear extension and support services continue to push mainstream technologies and prevent uptake of options that enable ecosystem service delivery for sustainable food systems (Firbank et al., 2018). Facilitating knowledge, skills, and behavioral change by farmers and support services requires that this type of results informs institutional innovation (Röling et al., 2012; Schutt et al., 2016).

Assessment approaches

Clear quantification and validation of ecosystem services is required to ensure that ecosystem services are considered in decision-making. Crop-livestock linkages supporting ecosystem services however remain largely unaccounted for. What is needed are appropriate data structures that can assess key indicators, for example ecosystem services, to inform planning approaches for trajectories, investment options, and progress in achievement of sustainable development goals (SDGs).

Approaches like the AgMIP regional integrated assessment provide this type of information (Antle et al., 2014, 2015; Valdivia et al., 2015). Illustrating the trade-offs between economic growth, environment, social and human health, food, and nutrition security, also under the effects of climate change, helps to define policies and institutions that would support the achievement of goals like SDGs. The approach can help strengthening the contribution of crop-livestock integration for ecosystem services. Assessing ecosystem services was not the focus of the Nkayi study, but available data, simulation results, and expert consultations during scenario development allowed us to quantity the contributions of crops and livestock and possible impacts to the ecosystem services delivery.

Through the integrated assessments, priorities can be set with a level of context specificity that is useful to researchers and decision makers, who influence investments in agricultural research and development. The outcomes can be specified for particular landscapes and farm types, unlike aggregations and representative farms; it is important to understand what certain technologies benefit which type of farmers. They can also inform timing of interventions, helping to understand which type of farmer will be more vulnerable, when and what provisions need to be made in response to early warning systems. Pathways and scenarios can guide preparing for future requirements and roles, institutional, policy, capacities, and cross-cutting issues such as nutrition and the environment. Better clarity in priority setting can then help reengineering research and development processes, guiding governments and investors decisions.

Significance for policy and scaling

This case study illustrates how sustainability pathways for agriculture and the environment can help prepare more conducive context for ecosystem services, where the options cannot be tested in real life situations (Valdivia et al., 2017). There was high potential for reconfiguration of agricultural systems through integrated interventions (technologies, institutions, and policies) to diversified, integrated, and profitable agriculture, which will improve delivery of ecosystem services (Homann-Kee Tui et al., in preparation). For upscaling there is need to capture more data at landscape scale, le, impacts of sustainable crop and livestock production on water and nutrient cycling. When improved, the AgMIP approach can be useful for upscaling, planning, and monitoring of countries agricultural investment plans.

The codesign process as part of cross-scale dialog informs and motivates change in the larger set of socioinstitutional conditions. Scenarios for addressing institutional barriers, comparing priorities and trade-offs under different pathways, support stakeholder dialog and policy planning for high-intensity and large-scale impacts (Jochnick, 2012). The process influences how decision makers see ecosystem services and the implications for the natural environment, thereby improving proactiveness of policymakers rather than being reactive.

Future priorities for sustainable food systems must engage more on processes for learning and guidance on how to reach outcomes more effectively (Malhi et al., 2009). Science plays a new role to fasten decision and implementation processes (technical and institutional) along feasible pathways. Legitimacy for policy and practice is instilled through the codesign process of scenarios across scales. Testing out-of-the-box transformative interventions for diverse and integrated agricultural systems using participatory foresight and scenarios building could be a strategy to propose pathways for change.

Acknowledgments

The data for the case in this chapter were generated through the AgMIP Crop Livestock Intensification Project (CLIP), with support from the UK Department for International Development's UKaid. This work was implemented as part of the CGIAR Research Program on Climate Change, Agriculture and Food Security (CCAFS) and the CGIAR Research Program on Water Land and Ecosystems (WLE), which is carried out with support from CGIAR Fund Donors and through bilateral funding agreements. For details please visit https://ccafs.cgiar.org/donors. The views expressed in this document cannot be taken to reflect the official opinions of these organizations. The authors thank Violette Kee Tui for editing.

References

Abdoulaye, T., Sanders, J.H., 2005. Stages and determinants of fertilizer use in semiarid African agriculture: the Niger experience. Agricultural Economics 32, 167–179.

Anderson, J.R., 1992. Difficulties in African agricultural systems enhancement? Ten hypotheses. Agricultural Systems 38, 387–409.

Antle, J., 2011. Parsimonious multi-dimensional impact assessment. American Journal of Agricultural Economics 93 (5), 1292–1311.

Antle, J.M., Stoorvogel, J.J., Valdivia, R.O., 2014. New parsimonious simulation methods and tools to assess future food and environmental security of farm populations. Philosophical Transactions of the Royal Society B 369, 20120280.

Antle, J.M., Valdivia, R.O., Boote, K.J., Janssen, S., Jones, J.W., Porter, C.H., Rosenzweig, C., Ruane, A.C., Thorburn, P.J., 2015. AgMIP's trans-disciplinary agricultural systems approach to regional integrated assessment of climate impact, vulnerability and adaptation. In: Rosenzweig, C., Hillel, D. (Eds.), Handbook of Climate Change and Agroecosystems: The Agricultural Model Intercomparison and Improvement Project Integrated Crop and Economic Assessments, Part 1. Imperial College Press, London.

Antle, J., Homann-Kee Tui, S., Descheemaeker, K., Masikati, P., 2017. Using AgMIP regional integrated assessment methods to evaluate climate impact, adaptation, vulnerability and resilience in agricultural systems. In: Zilberman, D., Lipper, L., McCarthy, N., Asfaw, S., Branca, G. (Eds.), Climate Smart Agriculture – Building Resilience to Climate Change. Elsevier Book (Chapter).

Barrett, C.B., Constas, M.A., 2014. Toward a theory of resilience for international development applications. Proceedings of the National Academy of Sciences 111, 14625–14630.

Baudron, F., Andersson, J.A., Corbeels, M., Giller, K.E., 2012. Failing to yield? Ploughs, conservation agriculture and the problem of agricultural intensification: an example from the Zambezi Valley, Zimbabwe. The Journal of Development Studies Routledge 48, 393–412.

Berhanu, T., Habtamu, Z.S., Twumasi-Afriyie, S., Blummel, M., Friesen, D., Mosisa, W., 2012. Breeding maize for food-feed traits in Ethiopia. In: Mosisa Worku, S., Twumasi-Afriyie, S., Legesse, W., Berhanu, T., Girma, D., Gezehagn, B., et al. (Eds.), Ethiopia Meeting the Challenges of Global Climate Change and Food Security through Innovative Maize Research. Proceedings of the Third National Maize Workshop of Ethiopia. 18–20 April, 2011. Addis Ababa (Ethiopia). EIAR/CIMMYT.

Blummel, M., Lukuyu, B., Zaidi, P.H., Duncan, A.J., Tarawali, S.A., 2012. Dual-purpose crop development, fodder trading and processing options for improved feed value chains. In: Mosisa Worku, S., Twumasi-Afriyie, S., Legesse, W., Berhanu, T., Girma, D., Gezehagn, B., et al. (Eds.), Ethiopia Meeting the Challenges of Global Climate Change and Food Security through Innovative Maize Research. Proceedings of the Third National Maize Work- Shop of Ethiopia. 18–20 April, 2011. Addis Ababa (Ethiopia). EIAR/CIMMYT.

Bonaudo, T., Bendahan, A.B., Sabatier, R., Ryschawy, J., Bellon, S., Leger, F.L., Magda, D., Tichit, M., 2014. Agroecological principles for the redesign of integrated crop–livestock systems. European Journal of Agronomy 57, 43–51.

Cousins, B., Hall, R., 2013. Rights without illusions: the potential and limits of rights-based approaches to securing land tenure in rural South Africa. In: Langford, M., Cousins, B., Dugard, J., Madlingozi, T. (Eds.), Symbols or Substance? The Role and Impact of Socio-Economic Rights Strategies in South Africa. Cambridge University Press, Cambridge.

Descheemaeker, K., Amede, T., Haileslassie, A., Bossio, D., 2011. Analysis of gaps and possible interventions for improving water productivity in crop–livestock systems of

Ethiopia. Experimental Agriculture 47 (S1), 21–38. https://doi.org/10.1017/S0014479710000797.
Descheemaeker, K., Oosting, S.J., Homann Kee-Tui, S., Masikati, P., Falconnier, G.N., Giller, K.E., 2016. Climate change adaptation and mitigation in smallholder crop–livestock systems in sub-Saharan Africa: a call for integrated impact assessments. Regional Environmental Change 16, 2331–2343.
Ecker, O., Qaim, M., 2011. Analyzing nutritional impacts of policies: an empirical study for Malawi. World Development 39 (3), 412–428. https://doi.org/10.1016/j.worlddev.2010.08.002.
Food and Agriculture Organization of the United Nations (FAO), 2011. Save and Grow: A Policymaker's Guide to the Sustainable Intensification of Smallholder Crop Production. FAO, Rome.
Food and Agriculture Organization of the United Nations (FAO), 2013. Climate-smart Agriculture Sourcebook. FAO, Rome.
Firbank, L.G., Eory, V., Gadanakis, Y., Lynch, Y.M., Sonnino, R., Takahashi, T., 2018. Grand challenges in sustainable intensification and ecosystem services. Frontiers in Sustainable Food Systems 28. https://doi.org/10.3389/fsufs.2018.00007.
Franzluebbers, A.J., Lemaire, G., Carvalho, P.C.F., Sulc, R.M., Dedieu, B., 2014. Towards agricultural sustainability through integrated crop-livestock systems: environmental outcomes. Introduction. Agriculture, Ecosystems & Environment 190, 1–3.
Garrett, R.D., Niles, M.T., Gil, J.D.B., Gaudin, A., Chaplin-Kramer, R., Assmann, A., Assman, T.S., Brewer, K., de Faccio Carvalho, P.C., Cortner, O., Dynes, R., Garbach, K., Kebreab, E., Mueller, N., Peterson, C., Reis, J.C., Snow, V., Valentim, J., 2017.
Social and ecological analysis of commercial integrated crop livestock systems: Current knowledge and remaining uncertainty. Agricultural Systems 155, 136–146.
Gil, J., Siebold, M., Berger, T., 2015. Adoption and development of integrated crop–livestock–forestry systems in Mato Grosso, Brazil. Agriculture, Ecosystems and Environment 199, 394–406.
Giller, K.E., Witter, E., Corbeels, M., Tittonell, P., 2009. Conservation agriculture and smallholder farming in Africa: the heretics' view. Field Crops Research 114, 23–34.
Gómez, M.I., Barrett, C.B., Raney, T., Pinstrup-Andersen, P., Meerman, J., Croppenstedt, A., Thompson, B., 2013. Post-green revolution food systems and the triple burden of malnutrition. Food Policy 42, 129–138. https://doi.org/10.1016/j.foodpol.2013.06.009.
Government of Zimbabwe, 2002. Central Statistical Office – Crops Sector Report. Harare.
Grace, D., Dominguez-Salas, P., Alonso, S., Lannerstad, M., Muunda, E., Ngwili, N., Omar, A., Khan, K., Otobo, E., 2018. The Influence of Livestock-Derived Foods on Nutrition during the First 1,000 Days of Life. ILRI Research Report 44. ILRI, Nairobi, Kenya.
Grant, W., Wolfaardt, A., Louw, A., 2012. Maize value chain in the SADC region. In: Southern Africa Trade Hup. AECOM International Development, USAID Southern Africa. Gabarone, Technical Report.
Harris, D., Orr, A., 2014. Is rainfed agriculture really a pathway from poverty? Agricultural Systems 123, 84–96. https://doi.org/10.1016/j.agsy.2013.09.005.
Hendrickson, J.R., Hanson, J., Tanaka, D.L., Sassenrath, G., 2008. Principles of integrated agricultural systems: introduction to processes and definition. Renewable Agriculture and Food Systems 23, 265–271.

Herrero, M., Thornton, P.K., Notenbaert, A.M., Wood, S., Msangi, S., Freeman, H.A., Bossio, D., Dixon, J., Peters, M., van de Steeg, J., Lynam, J., Rao, P.P., Macmillan, S., Gerard, B., McDermott, J., Seré, C., Rosegrant, M., 2010. Smart investments in sustainable food production: revisiting mixed crop-livestock systems. Science 327, 822–825.

Herrero, M., Wirsenius, S., Henderson, B., Rigolot, C., Thornton, P., Havlik, P., De Boer, I., Gerber, P.J., 2015. Livestock and the environment: what have we learned in the past decade? Annual Review of Environment and Resources 40, 177–202.

Homann-Kee Tui, S., Blummel, M., Valbuena, D., Chirima, A., Masikate, P., van Rooyen, A.F., Kassie, G.T., 2013. Assessing the potential of dual-purpose maize in southern Africa: a multi-level approach. Field Crops Research 153, 37–52. https://doi.org/10.1016/j.fcr.2013.07.002.

Homann-Kee Tui, S., Valbuena, V., Masikati, P., Descheemaeker, K., Nyamangara, J., Claessens, L., Erenstein, O., van Rooyen, A., Nkomboni, D., 2015. Economic trade-offs of biomass use in crop-livestock systems: exploring more sustainable options in semi-arid Zimbabwe. Agricultural Systems 134, 48–60.

Homann Kee Tui S., Descheemaeker, K., Masikati, P., Sisito, G., Valdivia, R., Crespo, O., Claessens L., in preparation. Re-designing Smallholder Farming Futures for Reduced Vulnerability to Climate Change in Semi-arid Southern Africa.

Jayne, T.S., Zulu, B., Nijhoff, J.J., 2006. Stabilizing food markets in eastern and southern Africa. Food Policy 31 (4), 328–341.

Jochnick, C., 2012. Systems, power, and agency in market-based approaches to poverty. In: Oxfam America Research Backgrounder Series. www.oxfamamerica.org/market-based-approaches-to-poverty.

Jones, P.G., Thornton, P.K., 2009. Croppers to livestock keepers: livelihood transitions to 2050 in Africa due to climate change. Environmental Science and Policy 12, 427–437. https://doi.org/10.1016/j.envsci.2008.08. 006.

Komarek, A.M., Bell, L.W., Whish, J.P., Robertson, M.J., Bellotti, W.D., 2015. Whole-farm economic, risk and resource- use trade-offs associated with integrating forages into crop–livestock systems in western China. Agricultural Systems 133, 63–72.

Kremen, C., Iles, A., Bacon, C., 2012. Diversified farming systems: an agroecological, systems-based alternative to modern industrial agriculture. Ecology and Society 17, 44.

Lemaire, G., Franzluebbers, A., de Faccio Carvalho, P.C., Dedieu, B., 2013. Integrated crop–livestock systems: strategies to achieve synergy between agricultural production and environmental quality. Agriculture, Ecosystems and Environment. https://doi.org/10.1016/j.agee.2013.08.009.

Lipper, L., Thornton, P., Campbell, B.M., Baedeker, T., Braimoh, A., Bwalya, M., Caron, P., Cattaneo, A., Garrity, D., Henry, K., Hottle, R., Jackson, L., Jarvis, A., Kossam, F., Mann, W., McCarthy, N., Meybeck, A., Neufeldt, H., Remington, T., Sen, P.T., Sessa, R., Shula, R., Tibu, A., Torquebiau, E.F., 2014. Climate-smart agriculture for food security. Nature Climate Change 4, 1068–1072.

Malhi, L., Karanfil, O., Merth, T., Acheson, M., Palmer, A., Finegood, D.T., 2009. Places to intervene to make complex food systems more healthy, green, fair, and affordable. Journal of Hunger and Environmental Nutrition 4 (3–4), 466–476. https://doi.org/10.1080/19320240903346448.

Mapfumo, P., Giller, K.E., 2001. Soil fertility management strategies and practices by smallholder farmers in semi-arid areas of Zimbabwe. ICRISAT and FAO, Bulawayo.

Masikati, P., Manschadi, A., van Rooyen, A., Hargreaves, J., 2013. Maize–mucuna rotation: a technology to improve water productivity in smallholder farming systems. Agricultural Systems 123, 62–70.

Masikati, P., Homann-Kee Tui, S., Descheemaeker, K., Crespo, O., Walker, S., Lennard, C.J., Claessens, L., Gama, A.C., Famba, S., van Rooyen, A.F., Valdivia, R.O., 2015. Crop–livestock intensification in the face of climate change: exploring opportunities to reduce risk and increase resilience in southern Africa by using an integrated multi- modeling approach. In: Rosenzweig, C., Hillel, D. (Eds.), Handbook of Climate Change and Agroecosystems: The Agricultural Model Intercomparison and Improvement Project Integrated Crop and Economic Assessments, Part 2. Imperial College Press, London, pp. 159–198.

Mbow, C., Van Noordwijk, M., Luedeling, E., Neufeldt, H., Minang, P.A., Kowero, G., 2014. Agroforestry solutions to address food security and climate change challenges in Africa. Current Opinion in Environmental Sustainability 6, 61–67. https://doi.org/10.1016/j.cosust.2013.10.014.

McIntire, J., Bourzat, D., Pingali, P., 1992. Crop-Livestock Interactions in Sub-Saharan Africa. World Bank, Washington, D.C.

Meadows, D.H., 1999. Leverage Points. Places to Intervene in a System. Sustainability Institute.

Milton, S.J., Dean, W.R.J., Richardson, D.M., 2003. Economic incentives for restoring natural capital in southern African rangelands. Frontiers in Ecology and the Environment 1, 247–254.

Moll, H.A.J., 2005. Costs and benefits of livestock systems and the role of market and nonmarket relationships. Agricultural Economics 32, 181–193. https://doi.org/10.1111/j.0169-5150.2005.00210.x.

Moraine, M., Duru, M., Therond, O., 2016. A social-ecological framework for analyzing and designing integrated crop–livestock systems from farm to territory levels. Renewable agriculture and food systems 1 (1), 1–14. https://doi.org/10.1017/S1742170515000526.

Mudimu, G., 2003. Zimbabwe Food Security Issues Paper. Working and Discussion Papers. ODI, UK.

Nin-Pratt, A., Johnson, M., Magalhaes, E., You, L., Diao, X., Chamberlin, J., 2011. Yield Gaps and Potential Agricultural Growth in West and Central Africa. Washington DC.

Pelley, J., 2009. Biodiversity is good for your health. Frontiers in Ecology and the Environment 7, 347.

Peyraud, J.L., Taboada, M., Delaby, L., 2014. Integrated crop and livestock systems in Western Europe and South America: a review. European Journal of Agronomy 57, 41–42.

Powell, J.M., Pearson, R.A., Hiernaux, P.H., 2004. Crop–livestock interactions in the West African drylands. Agronomy Journal 96 (2), 469–483.

Rauch, T., Beckmann, G., Neubert, S., Rettberg, S., 2016. Rural Transformation in Sub-Saharan Africa. SLE Strategic Paper, Berlin.

Rippke, U., Ramirez-Villegas, J., Jarvis, A., Vermeulen, S.J., Parker, L., Mer, F., Diekkrüger, B., Challinor, A.J., Howden, M., 2016. Timescales of transformational climate change adaptation in sub-Saharan African agriculture. Nature Climate Change. https://doi.org/10.1038/NCLIMATE2947.

Röling, N., Hounkonnou, D., Kossou, D., Kuyper, T., Nederlof, S., Sakyi-Dawson, O., Traoré, M., van Huis, A., 2012. Diagnosing the scope for innovation: linking smallholder practices and institutional context: introduction to the special issue. NJAS – Wageningen Journal of Life Sciences 60, 1–6.

Rufino, M.C., Rowe, E.C., Delve, R.J., Giller, K.E., 2006. Nitrogen cycling efficiencies through resource-poor African crop-livestock systems. Agriculture, Ecosystems and Environment 112, 261–282.

Rusinamhodzi, L., Corbeels, M., van Wijk, M.T., Rufino, M.C., Nyamangara, J., Giller, K.E., 2011. A meta-analysis of long term effects of conservation agriculture on maize grain yield under rain-fed conditions. Agronomy for Sustainable Development 31 (4), 657–673.

Ryschawy, J., Choisis, N., Choisis, J.P., Joannon, A., Gibon, A., 2012. Mixed crop-livestock systems: an economic and environmental-friendly way of farming? Animal 6, 1722–1730.

Schlecht, E., Hiernaux, P., Achard, F., Turner, M.D., 2004. Livestock related nutrient budgets within village territories in western Niger. Nutrient Cycling in Agroecosystems 68, 199–211.

Schut, M., van Asten, P., Okafor, C., Hicintuka, C., Mapatano, S., Nabahungu, N.L., 2016. Sustainable intensification of agricultural systems in the central African highlands: the need for institutional innovation. Agricultural Systems 145, 165–176. https://doi.org/10.1016/j.agsy.2016.03.005.

Shikuku, K.M., Valdivia, R.O., Paul, B.K., Mwongera, C., Winowieck, L., Laederach, P., Herrero, M., Silvestri, S., 2017. Prioritizing climate-smart livestock technologies in rural Tanzania: a minimum data approach. Agricultural Systems 151, 204–216.

Steiner, J.L., Franzluebbers, A.J., 2009. Farming with grass—for people, for profit, for production, for protection. Journal of Soil and Water Conservation 64, 75–80.

Sumberg, J., 1998. Mixed farming in Africa: the search for order, the search for sustainability. Land Use Policy 15 (4), 293–317.

Swinton, S.M., Lupi, F., Robertson, G.P., Hamilton, S.K., 2007. Ecosystem services and agriculture: cultivating agricultural ecosystems for diverse benefits. Ecological Economics 64, 245–252.

Tarawali, S., Herrero, M., Descheemaeker, K., Grings, E., Blummel, M., 2011. Pathways for sustainable development of mixed crop livestock systems: taking a livestock and pro-poor approach. Livestock Science 139, 11–21.

Thornton, P.K., Herrero, M., 2015. Adapting to climate change in the mixed crop and livestock farming systems in sub-Saharan Africa. Nature Climate Change 5, 830–836.

Thornton, P.K., van de Steeg, J., Notenbaert, A., Herrero, M., 2009. The impacts of climate change on livestock and livestock systems in developing countries: a review of what we know and what we need to know. Agricultural Systems 101, 113–127.

Tiffen, M., 2003. Transition in sub-Saharan Africa: agriculture, urbanization and income growth. World Development 31, 1343–1366.

Tittonell, P., Muriuki, A., Shepherd, K.D., Mugendi, D., Kaizzi, K.C., Okeyo, J., Verchot, L., Coe, R., Vanlauwe, B., 2010. The diversity of rural livelihoods and their influence on soil fertility in agricultural systems of East Africa – a typology of smallholder farms. Agricultural Systems 103, 83–97.

Valbuena, D., Erenstein, O., Homann-Kee Tui, S., Abdoulaye, T., Claessens, L., Duncan, A.J., Gérard, B., Rufino, M., Teufel, N., van Rooyen, A., van Wijk, M.T., 2012. Conservation agriculture in mixed crop-livestock systems: scoping crop residue trade-offs in Sub-Saharan Africa and South Asia. Field Crops Research 132, 175–184.

Valbuena, D., Homann-Kee Tui, S., Erenstein, O., Teufel, N., Duncan, A., Abdoulaye, T., Swain, B., Mekonnen, K., Germaine, I., Gérard, B., 2015. Identifying determinants, pressures and trade-offs of crop residue use in mixed smallholder farms in Sub-Saharan Africa and South Asia. Agricultural Systems 134, 107–118. https://doi.org/10.1016/j.agsy.2014.05.013.

Valdivia, R.O., Antle, J.M., Stoorvogel, J.J., 2012. Coupling the Tradeoff Analysis Model with a market equilibrium model to analyze economic and environmental outcomes of agricultural production systems. Agricultural Systems 110, 17–29.

Valdivia, R.O., Antle, J.M., Rosenzweig, C., Ruane, A.C., Vervoort, J., Ashfaq, M., Hathie, I., Homann-Kee Tui, S., Mulwa, R., Nhemachena, C., Ponnusamy, P., Rasnayaka, H., Singh, H., 2015. Representative agricultural pathways and scenarios for regional integrated assessment of climate change impact, vulnerability and adaptation. In: Rosenzweig, C., Hillel, D. (Eds.), Handbook of Climate Change and Agroecosystems: The Agricultural Model Intercomparison and Improvement Project Integrated Crop and Economic Assessments, Part 1. Imperial College Press, London.

Valdivia, R.O., Antle, J.M., Stoorvogel, J.J., 2017. Designing and evaluating sustainable development pathways for semi-subsistence crop–livestock systems: lessons from Kenya. Agricultural Economics 48, 11–26.

Van Rooyen, A., Homann-Kee Tui, S., 2009. Promoting goat markets and technology development in semi-arid Zimbabwe for food security and income growth. Tropical and Subtropical Agroecosystems 11, 1–5.

CHAPTER

Ecosystem services in doubled-up legume systems

9

Regis Chikowo, MPhil, PhD[1,2], Vimbayi Chimonyo, MSc, PhD[1], Chiwimbo Gwenambira, MSc[1], Sieg Snapp, MSc, PhD[1]

[1]*Plant Soil and Microbial Sciences Department, Michigan State University, East Lansing, MI, United States;* [2]*Crop Science Department, University of Zimbabwe, Harare, Zimbabwe*

Introduction

Legume-cereal intercropping systems have been practiced on smallholder farms in Africa as part of crop diversification, risk reduction, and harnessing the biological N_2-fixation benefits of more legumes on increasingly smaller farms. The most common goal of intercropping is to produce a greater yield on a given piece of land by making use of resources or ecological processes that would otherwise not be utilized by a single crop (Trenbath, 1976; Keating and Carberry, 1993). It is particularly important not to have crops competing with each other for physical space, nutrients, water, or solar radiation. Examples of intercropping strategies are planting a deep-rooted crop with a shallow-rooted crop or planting a tall crop with a shorter crop that requires partial shade. When staking materials are scarce, intercropping of maize and common bean varieties that need structural support makes production of such common beans varieties feasible. Often, this crop mutualism is associated with increased biodiversity as the habitat created is conductive to a variety of insects and soil organisms that would not be available in a single crop environment.

Soil nitrogen (N) and phosphorus (P) are often the most limiting nutrients to crop production in tropical soils. Without adequate external nutrients, intercropping systems will perform poorly, more so for the cereal component of the cereal/legume intercrop. When P is acutely deficient, as is widely the case (Snapp, 1998; Snapp et al., 1998) the legume component will perform poorly as well, at best, only managing to maintain system performance at the status quo. Thus the expected stimulus associated with more legumes in cropping systems remains elusive as legume N_2-fixation and primary productivity are constrained in poor soils. The ecosystem benefits of intercropping systems as widely practiced are insufficient to trigger crop production improvement on an intensification trajectory. Based on the important principle that legumes provide a pathway to amplify the benefits of the little fertilizer that is generally accessed by smallholder farmers in Africa, this chapter presents a novel doubled-up intercropping system that generates comparatively

The Role of Ecosystem Services in Sustainable Food Systems. https://doi.org/10.1016/B978-0-12-816436-5.00009-3

larger and high-quality crop residues, a necessary initial step for stimulating subsequent larger crop yields for cereal crops grown in rotation.

What are doubled-up legume systems?

The doubled-up legume technology entails intercropping two grain legumes, exploiting the opportunity presented by complementary growth habits and plant architecture. The most successful doubled-up legume intercropping system involves pigeonpea intercropped with groundnut in an additive design, with little intraspecific competition (Fig. 9.1). In this doubled-up system, groundnut and pigeonpea are planted at the same time. Pigeonpea grows very slowly during the first 3 months and only starts rapid growth when the groundnut component would be approaching maturity. The groundnut component is harvested at about 4 months after planting, making room for the full development of pigeonpea as a sole crop until maturity, a further 2–3 months later. This is a typical example of temporal intercropping.

For small farms, crop diversification is strongly constrained by limited land, as farmers allocate a large proportion of the farm to the stable crop, which is usually a cereal. The doubled-up technology provides an entry point that fulfills multiple objectives, including (1) enabling integration of more grain legumes when land is limiting, (2) providing a pathway for rehabilitating fields with poor soil fertility, and (3) facilitating extended ground cover in cropped lands as pigeonpea can be in the field for 6–8 months, depending on variety used. The doubled-up legume technology has been piloted with Malawian smallholder farmers who have small landholdings of between 0.5 and 2 ha.

FIGURE 9.1

Groundnut/pigeonpea doubled-up system that provides “double” grain and soil fertility benefits.

Light interception and water use drive the doubled-up system

Successful intercropping crop combinations are premised on capitalization on both spatial and temporal complementarities that result in an overall increase in resource use and use efficiency during the entire growing season (Fukai and Trenbath, 1993; Trenbath 1986). Complimentarily occurs when component crops use resources differently, especially greater light interception over time because of temporal differences in component canopies. Additionally, more benefits accrue with both temporal and spatial differences in root growth, which ensure efficient use of water and soil nutrients over the growing season.

The compatibility of the pigeonpea-groundnut doubled-up system is illustrated well by each crop's peak leaf area index (LAI) that occur at different times during the cropping season. With an earlier LAI peak the groundnut crop utilizes radiation with little partitioning required between the crops (Fig. 9.2). From emergence up to maximum canopy cover groundnut has adequate access to radiation since the overstorey pigeonpea canopy is developing at a very slow pace. At this stage extinction coefficient for pigeonpea allows radiation to reach the understorey groundnut. By the time pigeonpea canopy begins to expand and LAI increases, groundnut, which would be at full canopy by then, would not require large amounts of radiation as it would have started to senesce. It is important to note that because of the rigid wide row spacing adopted by smallholder farmers in much of Africa, the LAI of groundnut rarely exceeds 3, the LAI at which light interception has been reported to be at 95% complete (Duncan et al., 1978). Thus at optimum plant populations

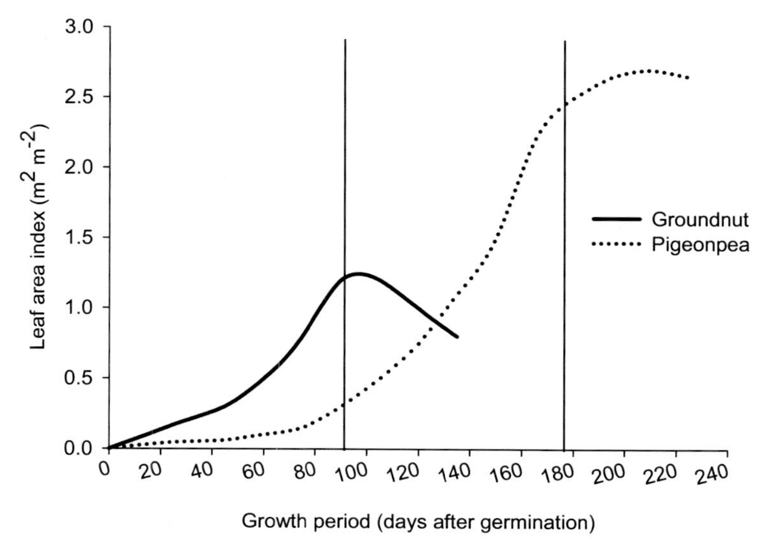

FIGURE 9.2

Leaf area index (LAI) development for pigeonpea and groundnut planted in a doubled-up intercropping system.

the slow growing pigeonpea plants are able to access adequate light for normal growth as groundnut LAI is well below the light extinction threshold.

Crop water use is associated with the interaction of roots and their ability to scavenge water in the soil plus the capacity of the corresponding canopy to transpire the captured water efficiently (Morris and Garrity, 1993). Water uptake is a function of rooting density distribution, soil-root system conductivities, and soil available water (Ogindo and Walker, 2003). For pigeonpea and groundnut, increased root density (temporal and spatially) and differences between rooting patterns (depth, width and length) in crop mixtures ensures that a larger volume of soil water can be exploited and thus improve water use efficiency. However, difficulties in studying root systems and root water extraction dynamics in multicrop systems have led to few studies actually quantifying water uptake in intercropping systems.

Canopy dynamics that influence crop water use are related to crop species, plant canopy features and evapotranspiration (ET) (Allen et al., 1998). The shape, size, and duration of a canopy influence ET and its partitioning into soil evaporation (E) and crop transpiration (T) (Ogindo and Walker, 2003; Morison et al., 2008). The larger the canopy presented by intercropping of groundnut and pigeonpea the greater the proportion of water "lost" through T in exchange for carbon dioxide than soil E (Seran and Brintha, 2010). Morris and Garrity (1993) also articulated enhanced water use from extending duration of maximum LAI. The duration of soil cover and attained maximum canopy cover is maintained for more than 6 months for pigeonpea-groundnut doubled-up system relative to 4 months for sole groundnut. Similar to radiation interception, intercropped pigeonpea and groundnut have critical stages of water demand that do not overlap.

The different growth habits for this unique legume-legume intercrop results in minimal intraspecific competition for nutrients, water, and radiation, while also adding high N leafy material to the soil. Thus the system "doubles" biological N_2-fixation contribution while also ensuring "double" grain legumes from the two legumes. The other promising doubled-up system is the pigeonpea-cowpea intercropping system that, however, requires more careful selection of compatible cowpea varieties. Creeping or late maturing cowpea varieties tend to be too competitive, suffocating the slow growing young pigeonpea plants. The pigeonpea plants become too weakened and ultimately fail to compensate adequately for the lost vigor (Fig. 9.3).

Implementing doubled-up systems: Malawi example

An array of groundnut/pigeonpea intercropping arrangements is possible, as dictated the agroecological environment and crop varieties. Component plant populations could range from substitutive designs to 100% additive designs, the chosen design often reflecting a combination of farmer production objectives and what the agroecological environment can support. For example, if groundnut is a major crop and the environment is conducive to groundnut production, the preferred option is to plant the groundnut component at its established optimal monoculture plant density and include the pigeon pea component at anything from 50% to 100% of its own optimal

(A) Optimal monoculture groundnut density

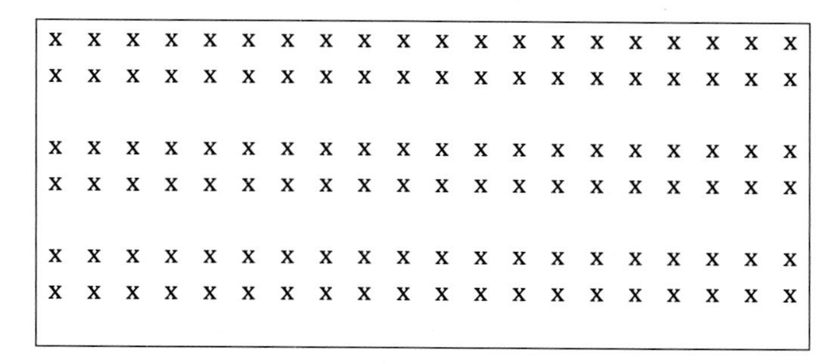

(B) Optimal monoculture pigeonpea density

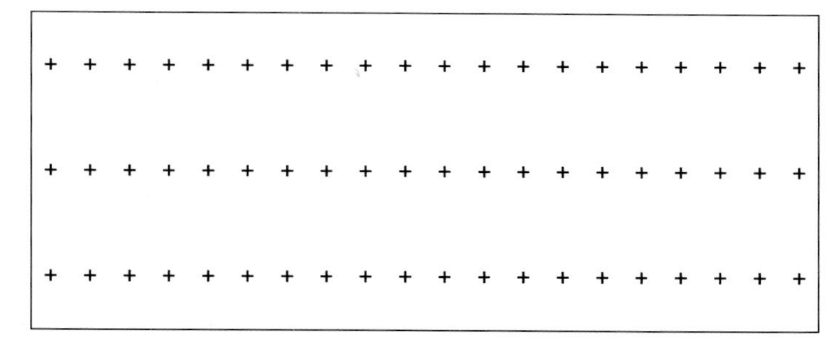

(C) Additive groundnut/pigeonpea doubled up system

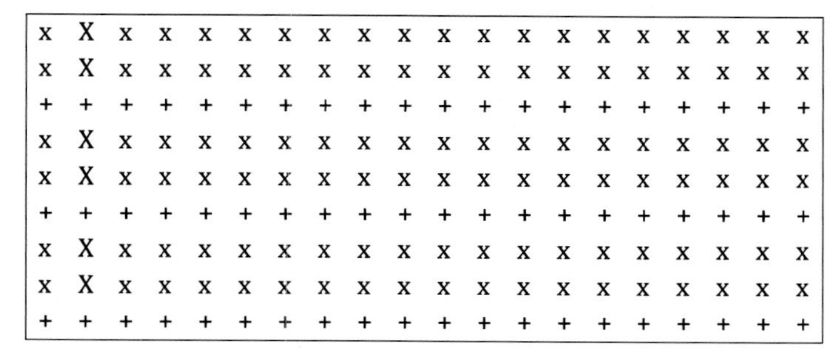

FIGURE 9.3

Diagrammatic illustration of additive design in groundnut/pigeonpea intercropping given the design of (A) groundnut monoculture, (B) pigeonpea monoculture, and (C) intercrop is established by planting both species exactly the way they are planted in the monoculture.

monoculture density. For runner type groundnut varieties, intraspecific competition that can be exerted on the pigeonpea component can be large; therefore some degree of substitute design must be considered. This also applies to more marginal areas, where the crop mixtures must be inclined toward the substitutive intercropping design. For practical purposes the optimal densities for the monocultures are adopted from what is locally advised by local extension agencies, which is often informed by empirical research as well as decades of experiential learning by local farmers.

While optimal monoculture densities in an environment can be easily established, a major hurdle in implementing intercropping for many crop species combinations is related to finding the optimal intercrop, where the degree of intraspecific competition for soil water varies from season to season as a function of rainfall season quality. Put differently, an intercropping design that is optimal for one season may not hold for another due to a different set of resources available to the intercrop, primarily soil water under rain-fed conditions. Fortunately the groundnut/pigeonpea intercrop is very robust because of its temporal nature as well as the vastly different plant architecture, resulting in much less challenging intercrop optimization.

Ecosystems services from doubled-up systems

Biological N_2-fixation and N cycling

The doubled-up cropping system in unique in that both intercrop components are capable of symbiotic biological N2-fixation using native rhizobia. Compatibility with native rhizobia that is widely present in many tropical soils enables both groundnut and pigeonpea to be grown across a range of agroecologies and by farmers in communities that hardly have access to legume inoculants. In experiments conducted in Malawi across several sites the groundnut-pigeonpea system fixed between 50 and 148 kg ha^{-1} N, performing better than other alternative legume cropping systems (Table 9.1). These data are drawn from experiments that were fertilized with no more than 10 kg ha^{-1} P.

Enhanced belowground biomass inputs and nutrient mobilization

Many tropical soils are severely depleted in soil organic matter, whose replenishment has remained intractable using mainstream cropping regimes (Smalling et al., 1997). Poverty in rural areas in sub-Saharan Africa is intricately linked to declining soil fertility (Vanlauwe et al., 2018). Traditionally, smallholder farmers burn crop residues as part of land preparation or remove the above ground residues from the field during crop harvesting. Therefore for crops with a small root biomass component, there is little opportunity for meaningful restoration of soil organic

Table 9.1 Nitrogen inputs (kg ha^{-1} N) through biological N_2-fixation for sole cropping or doubled-up systems.

Cropping system	N fixed (kg ha^{-1})	Source
Sole groundnut	21–102	Mhango et al. (2017)
Sole soyabean	36–74	Mzumara (2016), Njira et al. (2013)
Sole pigeonpea	45–120	Mhango et al. (2017), Njira et al. (2013)
Groundnut + pigeonpea	50–148	Mhango et al. (2017), Njira et al. (2013)

matter (SOM). An analysis by Jackson et al. (2017) suggests that root inputs are approximately five times more likely than an equivalent mass of aboveground litter to be stabilized as SOM. Microbes, particularly fungi and bacteria, and soil faunal food webs strongly influence SOM decomposition at shallower depths, whereas mineral associations drive stabilization at depths greater than ~30 cm.

The doubled-up technology accelerates soil N cycling resulting in improved productivity of crops grown in sequence. Directly the pigeonpea component adds substantial quantities of root biomass in the top soil layer and fine roots at deeper layers compared with other legume rotational systems that do not have a pigeonpea component. The activities of living roots and the decomposition of dead roots are central to belowground ecosystem cycling of carbon and nutrients. Fine roots exude compounds into the rhizosphere, particularly sugars that stimulate microbial communities that are essential for organic matter decomposition and the mobilization of soil nutrients (Dakora and Phillips, 2002; Bertin et al., 2003). Additionally the higher concentrations of N in fine roots coupled with the absence of recalcitrant C common in structural roots make fine roots highly decomposable (Gordon and Jackson, 2000). As such, fine root production and subsequent death and decomposition represent an important flux of labile carbon to soils, which is readily decomposed by microbes. Pigeonpea has a large coarse root component, confined mostly within the 0–40 cm depth (Fig. 9.4). These roots are far less dynamic than fine roots and so are comparatively a strong feature of the soil fabric over a longer time. When dead and decomposed, deep roots may result in deep capillary channels that enhance rainfall infiltration.

Increased nitrogen use efficiency for cereal crops grown in rotation

Under low input–low out production systems an ideal technology is one that is characterized by high yield and stability and high N use efficiency (NUE). In experiments conducted over several seasons and across several sites in central Malawi, maize in rotation with the doubled-up technology had higher yields, showed stability across a range of sites and had higher NUE relative to the other technologies (Fig. 9.5).

Maximizing ecosystems services

- High yielding improved pigeonpea varieties that are widely available and are highly susceptible to pest damage. This has been a major impediment for widespread adoption of pigeonpea, especially in farming communities that have not traditionally grown pigeonpea. In Malawi, farmers have produced an unimproved land race, which is however resistant to the common insect pests. This variety is enabling farmers to practice the doubled-up technology, albeit at a small scale, and acquire the knowledge that is essential for scaling the technology.
- Destructive wild fires are seasonal phenomena, especially in agroecologies that are characterized by long dry period after crop harvest. N-rich crop residues are often destroyed by fires, thus forfeiting the potential soil fertility benefits. Sadly,

FIGURE 9.4

Relative root biomass inputs by depth for a pigeonpea-groundnut intercropping systems. Fine roots at deeper layers are useful for nutrient cycling and increased water use efficiency.

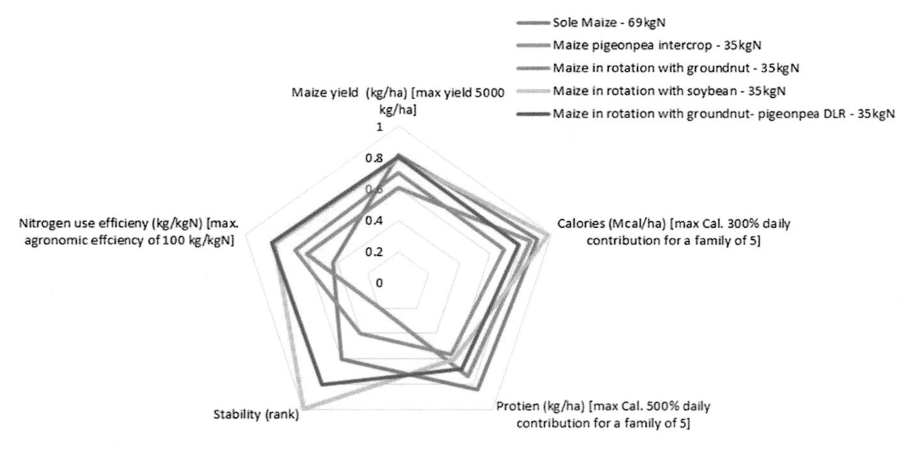

FIGURE 9.5

Radar chart showing overall yield, stability, and nitrogen use efficiency of maize in response to different legume technologies.

some of the fires are deliberately used by farmers as quick strategy to clear fields for the next cropping season. Local by-laws that criminalize use of fire in cropped lands and concerted efforts at educating farmers on the need for appropriate nutrient cycling through protecting crop residues could be viable options to ensure that the benefits of this the doubled-up technology are fully realized.
- Scaling—if farmers can increase the proportion of legumes on the farms from the current $< 15\%$ to $>30\%$, then a critical amount of N cycling could result in measurable impact at farm scale

Conclusions

The doubled-up cropping system is an intercropping technology that can be used as an entry point to support sustainable intensification on resource-constrained smallholder farms in sub-Saharan Africa. The utility of the system hinges on maximizing the ecological benefits that accrue from biological N2-fixation by two compatible grain legume crops. When poorly planned, intercropping systems often suffer from crowding, especially if nutrients and soil water are limiting. However, our experience with on-farm crop assessments confirmed that farmers achieve suboptimal plant densities because of a number of reasons, including use of nonviable seeds, too wide row spacing, planting under suboptimal moisture conditions, and plants dying early because of pest infestation or diseases. The doubled-up technology is an improvement on the contemporary cereal-legume intercropping and bring ecosystems benefits to the fore on resource-constrained smallholder farms.

Acknowledgments

The authors gratefully acknowledge funding from USAID through the Africa RISING program implemented by the International Institute of Tropical Agriculture (IITA) and its partners.

References

Allen, R.G., Pereira, L.S., Raes, D., Smith, M., 1998. FAO Irrigation and drainage paper No. 56. In: Rome Food Agric. Organ. United Nations, 56, p. e156.

Bertin, C., Yang, X., Weston, L.A., 2003. The role of root exudates and allelochemicals in the rhizosphere. Plant and Soil 256, 67–83.

Dakora, F.D., Phillips, D.A., 2002. Root exudates as mediators of mineral acquisition in low-nutrient environments. Plant and Soil 245, 35–47.

Duncan, W.G., Mc Cloud, D.E., Mc Graw, R.L., Boote, K.J., 1978. Physiological aspects of peanut yield improvement. Crop Science 18, 1015–1020.

Fukai, S, Trenbath, B.R., 1993. Processes determining intercrop productivity and yields of component crops. Field Crops Research 34, 247–271.

Gordon, S., Jackson, R.B., 2000. Nutrient concentrations in fine roots. Ecology 81, 275–280.

Jackson, R.B., Lajtha, K., Crow, S.E., Hugelius, G., Kramer, M.G., Pineiro, G., 2017. The ecology of soil carbon: pools, vulnerabilities, and biotic and abiotic controls. Annual Review of Ecology, Evolution and Systematics 48, 419–445.

Keating, B.A., Carberry, P.S., 1993. Resource capture and use in intercropping: solar radiation. Field Crops Research 34, 273–301.

Mhango, W.G., Snapp, S., Kanyama-Phiri, G.Y., 2017. Biological nitrogen fixation and yield of pigeonpea and groundnut: quantifying response on smallholder farms in northern Malawi. African Journal of Agricultural Research 12, 1385–1394.

Morison, J.I.L., Baker, N.R., Mullineaux, P.M., Davies, W.J., 2008. Improving water use in crop production. Philosophical Transactions of the Royal Society of London Series B Biological Sciences 363, 639–658. https://doi.org/10.1098/rstb.2007.2175.

Morris, R.A., Garrity, D.P., 1993. Resource capture and utilization in intercropping: water. Field Crops Research 34, 303–317.

Njira, K.O.W., Nalivata, P.C., Kanyama-Phiri, G.Y., Lowole, W.M., 2013. Effects of sole cropped, doubled up legume residues and inorganic nitrogen fertilizer on maize yields in Kasungu, Central Malawi. Agricultural Science Research Journal 3, 97–106.

Ogindo, H.O., Walker, S., 2003. The estimation of soil water extraction by maize-bean intercrop and its sole crop components in a semi-arid region. Crop Science 6, 430–435.

Seran, T., Brintha, I., 2010. Review on maize based intercropping. Journal of Agronomy 9, 135–145. https://doi.org/10.3923/ja.2010.135.145.

Smaling, E.M.A., Nandwa, S., Janssen, B.H., 1997. Soil fertility in Africa is at stake! In: Buresh, R.J., Sanchez, P.A., Calhoun, F. (Eds.), Recapitalization of Soil Nutrient Capital in Sub-saharan Africa. ASSA/CSSA/SSSA, Madison, Wisconsin, pp. 47–61.

Snapp, S.S., 1998. Soil nutrient status of smallholder farms in Malawi. Communications in Soil Science and Plant Analysis 29, 2571–2588. https://doi.org/10.1080/00103629809370135.

Snapp, S.S., Mafongoya, P.L., Waddington, S., 1998. Organic matter technologies for integrated nutrient management in smallholder cropping systems of Southern Africa. Agriculture, Ecosystems & Environment 71, 185–200. https://doi.org/10.1016/S0167-8809(98)00140-6.

Trenbath, B.R., 1976. Plant interactions in mixed cropping communities. In: Papendick, R.I., Sanchez, A., Triplett, G.B. (Eds.), Multiple Cropping. ASA Special Publication 27. American Society of Agronomy, Madison, WI, pp. 129–169.

Trenbath, B.R., 1986. Resource use by intercrops. In: Francis, C.A. (Ed.), Multiple Cropping. Macmillan, New York, pp. 57–81.

Vanlauwe, B., Six, J., Sanginga, N., Adesina, A.A., 2018. Soil fertility decline at the base of rural poverty in sub-Saharan Africa. Nature Plants 1. https://doi.org/10.1038/nplants.2015.101, 15101.

Further reading

Vandermeer, J.H., 1989. The Ecology of Intercropping. Cambridge University Press.

CHAPTER

Ecosystem services in paddy rice systems

10

P. Chivenge, PhD [1,2], O. Angeles, PhD [1], B. Hadi, PhD [1], C. Acuin, PhD [1], M. Connor, PhD [1], A. Stuart, PhD [1], R. Puskur, PhD [1], S. Johnson-Beebout, PhD [1]

[1]*International Rice Research Institute, Metro Manila, Manila, Philippines;* [2]*Senior Scientist, Sustainable Impact, International Rice Research Institute, Los Baños, Laguna, Philippines*

Introduction

Rice production in a global context

Rice (*Oryza sativa* L.) is the most widely consumed staple food globally, grown in 112 countries on more than 160 million hectares (FAO, 2016). Together with wheat (*Triticum aestivum aestivum*) and maize (*Zea mays*), they constitute the main cereal crops grown and consumed globally. However, rice is the major staple food in many low-income and middle-income countries and is thus important for food security and poverty alleviation. In 2016 global rice production was more than 740 million tonnes (FAO, 2016), 90% of which is produced and consumed in Asia (GRiSP, 2013). Rice and its production is a source of livelihood for more than 144 million rice farming households (GRiSP, 2013) and is important for economic development, particularly in Asia. With the increasing global population the demand for rice is increasing, especially in sub-Saharan Africa, a region that is also dominated by low-income countries.

Rice cultivation has been in existence since the beginning of organized agriculture, more than 3500 years ago (Sweeney and McCouch, 2007). Rice is an annual grass native to Asia and Africa but is now grown under varying climatic conditions from temperate to tropical and in a wide range of ecosystems across 112 countries, with China leading global rice production followed by India (GRiSP, 2013). It is produced in the humid tropics receiving more than 5000 mm annual precipitation, e.g., in Myanmar's Arakan Coast to dry deserts receiving less than 100 mm annually (GRiSP, 2013). The Jumla District in Nepal, at the foot of the Himalayas, is the highest elevation where rice is grown (>3000 m above sea level). However, rice productivity in the different ecosystems varies widely, depending on environmental factors and management. Research and development efforts have significantly increased global rice production, with global yields averaging 3.8 t ha^{-1}, but Africa lags behind at 1.8 t ha^{-1} (Norman and Kebe, 2006).

In Asia rice production significantly increased in the last half century, benefiting from the Green Revolution, which saw an increase in yields, because of the combined use of improved rice varieties that are mostly semidwarf, fertilizers, and pesticides and

The Role of Ecosystem Services in Sustainable Food Systems. https://doi.org/10.1016/B978-0-12-816436-5.00010-X

improved access to irrigation. The Green Revolution resulted in intensification of rice cropping systems, with continuous double-rice and triple-rice cropping systems being common in irrigated ecosystems. This resulted in increased rice production and reduced hunger and poverty among rice farmers and consumers. However, double burdens of malnutrition, with coexisting micronutrient deficiencies and excessive weight, are common among rice consumers, while monocropping systems have resulted in reduced biodiversity across the ecosystems. Nonetheless, lowland rice ecosystems provide various ecosystem services, which will be discussed in this chapter.

What are paddy rice systems and how do they differ from upland agrosystems?

The main rice growing ecosystems are classified as irrigated lowland, rainfed lowland, and rainfed upland (Fig. 10.1). Irrigated and rainfed lowland rice is grown on bunded fields creating semiaquatic environments where irrigation ensures production of one, two, or three crops per year, while rainfed lowland environments are flooded with rainwater kept within bunds. Of the 160 million hectares under rice production, 93 million hectares are irrigated and produce about 75% of the world's rice (GRiSP, 2013). In contrast rainfed lowland rice ecosystems occupy 53 million hectares and contribute 19% of the global rice produced, while rainfed upland ecosystems occupy 15 million hectares (GRiSP, 2013). Consequently about 94% of the rice is produced in irrigated and rainfed lowland environments; hence this chapter

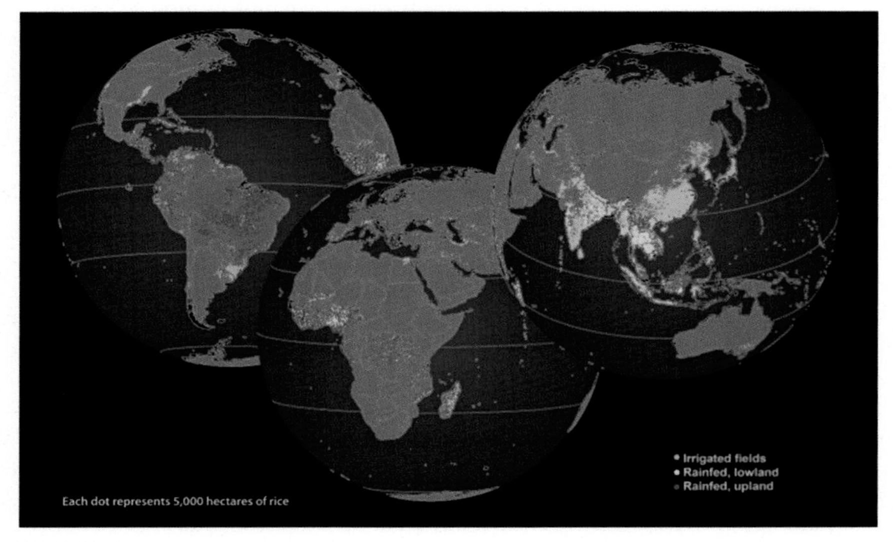

FIGURE 10.1

Major rice growing areas in the world and the distribution of the ecosystems where rice is grown (GRiSP, 2013).

largely focuses on these two ecosystems. These two ecosystems constitute what is generally regarded as paddy rice fields (Kögel-Knabner et al., 2010). Paddy rice is primarily puddled and transplanted, although wet-direct and dry-direct seeding is on the increase. Rainfed upland rice ecosystems are highly heterogeneous and prone to multiple abiotic stresses.

Most paddy rice ecosystems are wetlands, inundated by water part of the major growing season or continuously. The hydrological characteristics make paddy rice ecosystems vastly different from upland agroecosystems. Typical paddy rice soils are heavy textured, with a reduced soil horizon showing signs of permanent or intermittent waterlogging such as segregations of iron and manganese in the soil matrix (Kirk, 2004). Additionally a plow pan with high bulk density is often found at a depth between 15 and 25 cm, reducing water percolation and promoting surface flooding. Both these conditions limit root development for most upland crops. The manipulation of rice soils through puddling creates artificial submergence associated with reduced soil conditions, masking the original character of the soil (Kirk, 2004). Paddy soils are considered as manmade wetlands, and they constitute the second largest wetland area, after natural freshwater wetlands (Yoon, 2009; Kögel-Knabner et al., 2010). This has resulted in the creation of manmade soils, classified as Hydragic Anthrosols, that form from a wide range of parent material (IUSS Working Group, 2006). A typical soil profile of a Hydragic Anthrosols is shown (Fig. 10.2), with horizon designation done according to Jahn et al. (2006).

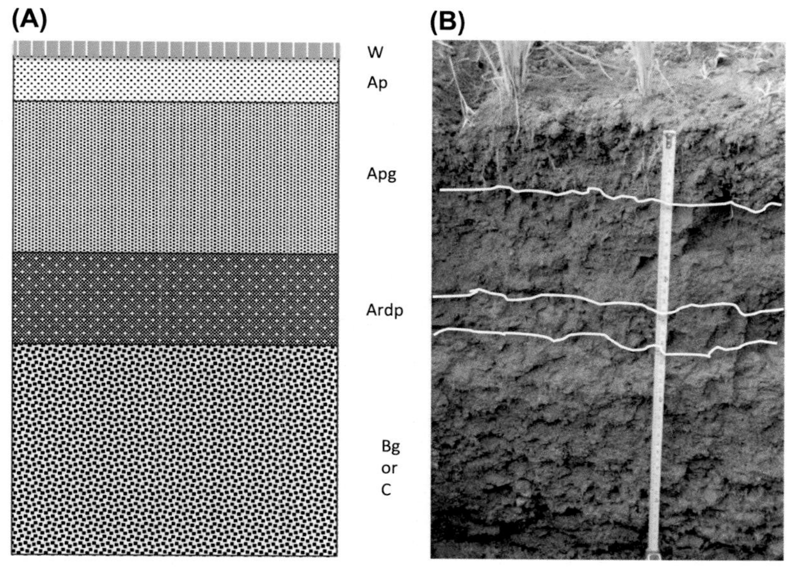

FIGURE 10.2

(A) A typical paddy soil profile, (B) a picture of a soil profile that has been recently converted to direct seeded rice with an aerobic topsoil.

W: a thin layer of standing water, which is a habitat of bacteria, phytoplankton, macrophytes, aquatic invertebrates, and vertebrates. Although there are diurnal fluctuations in the oxygen content with biological activity, this layer is generally oxic.
Ap: the interface of the standing water and soil, which is also generally oxic. The thickness of the layer ranges from a few millimeters to several centimeters.
Apg: the upper part of an anthraquic horizon that is puddled and flooded, showing a reduced matrix and some oxidized root channels. The layer has stagnic conditions with a distinct mottling pattern depicting oxidation and reduction occurring during alternate soil flooding and drainage.
Ardp: the lower part of the anthraquic horizon characterized with a plow pan. The layer has a high bulk density and highly compact with a platy or massive structure that restricts water infiltration. It is a reduced layer with stagnic conditions.
B or C: a hydragic horizon and usually depicts redoximorphic features, in some instances with iron and manganese concretions.

Rice has historically been grown in places with abundant water (e.g., monsoonal Asia) and consequently has a larger water footprint than most of the major cultivated crops, utilizing 2–3 times more water than upland crops (Bouman, 2007; Mekonnen and Hoekstra, 2011). Most of the rice is produced under continuous or intermittent flooding, e.g., alternate wetting and drying (Belder et al., 2004). Water is used for land preparation where land is puddled after flooding the soil for a few days. Puddling breaks down soil aggregates under the floodwater, removing air from soil pores, purposefully destroying soil structure to make a thick mud layer, which decreases water percolation and loss into deeper soil layers, thereby saturating the soil with water. In continuously flooded systems the ponded water is kept from about one to 10 centimeters above the soil surface until harvest. Under intermittent flooding soil is submerged during the first 3 weeks after transplanting and at flowering, and water is allowed to recede down to the root zone during the vegetative stage. The dry period oxidizes the soil and temporarily changes nutrient availability.

Rice is considered a semiaquatic annual crop and together with taro (*Colocasia esculenta)* is adapted to saturated soil conditions that are common in river deltas. Most of these areas are characterized by dry and wet seasons, with rice production occurring in both seasons when irrigated, ironically with greater yields during the dry season because of the increased solar radiation (and often temperature) than the wet season. In rainfed lowland ecosystems one rice crop is generally grown per year followed by a pulse crop or a fallow period. In irrigated ecosystems two or three rice crops are produced with the main systems being rice-rice, rice-rice-rice, rice-rice-pulse, rice-wheat, rice-maize, rice-rice-vegetable, or some other locally adapted combination. The majority of the irrigated ecosystems are intensive rice monocropping with high dependency on fertilizers, pesticides, and herbicides, whose use can cause environmental degradation and impact biodiversity and other ecosystem services derived from rice production.

Trends in intensification and water use

While traditional lowland rice-growing agri-food systems have provided ecosystem services sustainably for >3500 years, they have not been able to provide food security as global population has increased over the past century. Because the growing human population puts increasing pressure on land and water, it has been necessary for rice cropping systems to intensify, producing more grain on decreasing land area (for the past 50 years) and to become more water-efficient, producing more grain with decreasing irrigation water (for the past 15 years). These two trends, intensification of production and improvement of water-use efficiency, are key drivers of ecosystem pressure and key opportunities for ecosystem services in the coming decades.

Ecosystem services in lowland rice systems

Paddy rice farming offers a number of ecosystem services, including providing human livelihood and sustenance and enriching culture and provides home to a diverse assemblage of plant and animal life. These ecosystem services vary depending on the growth stage of rice, growing ecosystem, and whether rice is grown as a monocrop or in sequence with other crops.

According to the Millennium Ecosystem Assessment, ecosystems services are benefits derived by human beings from ecosystems, and these can be classified into four categories: (1) provisioning, (2) regulating, (3) cultural, and (4) supporting (MA, 2005). However, recently La Notte et al. (2017) underscored the challenges with current ecosystem services classification systems: (1) the inconsistency across concepts, terminology, and definitions and (2) the mix up of processes and end-state benefits or flows and assets. With the necessity to avoid ambiguity and improve comparability among ecosystem service–based approaches, a modified cascade framework was proposed to shift the concept of ecosystem services toward hierarchical ecological perspective. Hence in this chapter, ecosystem services are grouped into three categories: (1) provisioning, (2) regulating and maintenance, and (3) cultural.

Provisioning services

Food provisioning

The most obvious ecosystem services from paddy rice farming are the provisioning of food, energy, and livelihood. Paddy rice ecosystems provide staple food for most low-income and middle-income countries in South and Southeast Asia. Smallholder farmers produce about 80% of rice, and most of it is consumed locally.

The biomass harvested from rice crops is mainly divided into four portions: (1) milled grain, (2) rice bran, (3) rice husk, and (4) rice straw. The rice bran layer is retained with the milled grain (to make brown rice), removed with the husks (and polished to make white rice), or separated for production of bran oil or other products. Milled grain directly serves as food for human consumption, providing energy and nutrition. Most of the dietary fiber, micronutrients, lipids, and some of the proteins in the rice grain are found in the bran; hence brown rice is more nutritious than white rice. Additionally, white rice has higher glycemic index, a measure of the rate at which the sugar content of foods is metabolized by the human body, than brown rice. This has been associated with the rising overweight, obesity, and diabetes prevalence among rice consuming populations.

Rice is the staple food to more than half the global population, providing about 25% of global human energy consumption per capita and about 16% of protein (GRiSP, 2013). Globally, calories from rice have been steadily increasing from 391 in 1961 to 541 kcal person $^{-1}$ day^{-1} in 2013. In the past decade, much of this growth has been from Africa, but Asia continues to produce and consume the most calories from rice at 780 kcal person^{-1} day^{-1} compared with 238 in Africa (FAO, 2016). Rice has considerably higher carbohydrate and protein content than cassava, maize or sorghum, and millet and is mainly consumed in the form of whole grains (Norman and Kebe, 2006). Rice is typically boiled after washing, both processes resulting in loss of nutrients, including zinc, proteins, and lipids. Parboiling on the other hand results in the conservation of nutrient and water-soluble vitamins. However, populations highly dependent on rice tend to have diets that have limited diversity, and consequently rice becomes a key dietary contributor of proteins and micronutrients. This has motivated efforts to increase the nutrient density of rice through fortification during processing and biofortification through plant breeding. Biofortification is seen as a long-term solution because of its cost-effectiveness and its potential to reach marginalized rural communities that may have limited access to fortified products or other dietary sources of micronutrients (Bouis and Saltzman, 2017).

While rice is the main food product, there are also other significant food products that are produced as part of paddy rice systems. Rice-fish systems are common in East and South East Asia (Koohafkan and Furtado, 2004; Xie et al., 2011), providing fish as a source of protein in predominantly starch-based diets (Ahmed and Garnett, 2011). Similarly, in rice-prawn farming systems, prawns are a food source, and the integration can lead to greater income for farmers (Nair et al., 2014). Poultry farming is also integrated with rice farming, providing chicken and ducks and contributing to the nutrition of rice farming communities. While rats and mice are generally considered as pests to rice grain during the ripening phase of rice and affect food security (Singleton et al., 2010), they serve as a source of meat in some communities, as do molluscs, crustaceans, insects, reptiles, and amphibians (Halwart, 2006). Additionally aquatic plants growing in and around rice fields are used as vegetables in many Asian dishes. A survey in Northern Thailand found aquatic herbs such as *Nymphaea pubescens* (red water lily) and *Neptunia oleracea* (water mimosa) are harvested from rice fields as food source (Cruz-Garcia and Price, 2011).

Biomass provisioning for energy and other uses

Rice husk and straw, co-products when the grains are separated from the rest of the biomass, can be used for energy provisioning, through combustion or anaerobic digestion to produce biogas. Rice straw is produced when the plants are threshed during or just after harvest and therefore often has high moisture content, whereas rice husk is produced when paddy is milled after drying and therefore is often dryer than straw. A rice crop has an average harvest index (grain dry weight: total plant biomass dry weight) of 0.45−0.50 (Dobermann and Fairhurst, 2000). While there are no large-scale anaerobic digestion systems for rice straw in operation, there is much on-going research about how to improve and commercialize this process, with emphasis on keeping the digestor close to the source of the straw to minimize energy and cost in transporting the bulky wet material (Nguyen et al., 2016; Dai et al., 2018). For every ton of paddy harvested, up to 350 kg rice husk is produced. Rice husk can be formed into briquettes or pellets or burned directly and is used as a combustion energy source for small-scale steam engines, rice mills, and paddy drying (Nguyen, 2012). This is a cost-effective and efficient option because the husks are produced and consumed at the same location.

Rice biomass coproducts are also used for purposes other than energy. Carbonized rice husk is used as a soil amendment (Haefele et al., 2011) and as an absorbent for cleaning wastewater (Dias et al., 2018). Rice husk, which is high in silica, can be used in medium-density fiberboard, composite and organic plastics, low-roll-resistance tires, paint, and for nanosilica production (Kauldhar and Yadav, 2018). Rice straw is useful as mulch, livestock bedding material, substrate for mushroom culture and vermiculture, and livestock feed (Lim et al., 2012; Munder, 2013).

Livelihood provisioning

Rice, a staple food in Asia and parts of Africa and Latin America, provides food security and livelihood in these regions. While rice farmers and consumers constitute the majority of the poor, rice is a source of employment and income for more than 144 million farming families, most of whom are smallholders and live in rural areas. Consequently, rice is an important component of the economic and social order within rural communities. Rice is important for income generation, although it is a subsistence crop in some regions, e.g., West Africa. It is an accepted medium of exchange for coffee and cocoa, labor, and other farm goods in Africa (Norman and Kebe, 2006) or rent, wages, and debts in Asia (GRiSP, 2013). Rice is increasingly becoming important in Africa, particularly in East and Southern Africa where it is grown as a cash-crop by small-to medium scale farmers (Norman and Kebe, 2006). However, Africa's contribution to global rice production is still low (3.3%).

Postproduction operations such as parboiling offer additional income for societies (Akpokodje and Erenstein, 2001). In Northern Ghana parboiling is exclusively done by women with little or no formal education (Demuyakor et al., 2013). The income generated from rice production and postproduction is used for general household expenses, education, health, and social activities (Norman and Kebe, 2006). Being the staple food in most low-income countries, rice is considered as poor person's food, and there exists a dichotomy between rice prices for the economic benefit of the farmers, while rice accessibility remains important for food security of many. For example, in Latin America most of the poor live in urban areas and spend 15% of their income on white rice (GRiSP, 2013), making rice access important for the welfare and quality of life for the poor.

Regulating and maintenance services

Soil submergence in paddy rice ecosystems alters the reduction-oxidation conditions, impacting nutrient cycling and the flora and fauna supported. Paddy rice typically receives fertilizer and pesticides, whose fate depend on sunlight, temperature, soil pH and redox conditions, and oxygen concentration in the floodwater. High soil and floodwater pH and temperature trigger ammonia volatilization upon application of nitrogen fertilizer. Similarly, different residual compounds may be produced from applied pesticides. Such routine management activities in rice systems influence the chemical species released to the atmosphere and water bodies. Hence the regulating service becomes critical because rice ecosystems significantly contribute in balancing water, energy, and nutrient flow in agricultural landscapes (Burkhard et al., 2015).

Water regulation and maintenance

Despite having a large water footprint, estimated to use about 24%–30% of the world's freshwater resources (Bouman, 2007), paddy rice soils play an important role in water regulation. Irrigation competes with household and industrial consumption of water resources, directly affecting regional human welfare, especially in the face of water shortages. However, paddy soils have a large water storage capacity and can control floods, especially during the monsoon season when precipitation is high. Rice fields can reduce soil erosion and through water percolation contribute to groundwater recharge. Paddy soils retain soil that is eroded from uplands, which contain nutrients. These nutrients are taken up by rice plants, thereby improving the quality of water that is drained from the rice fields into river basins. Much of the water used to irrigate rice is not lost from the landscape, as it moves through the soil to lower-lying rice fields or to waterways.

Runoff water from upland agricultural areas typically contains nitrates and phosphates, which can potentially cause eutrophication when it reaches water sources. Rice fields run along the topography allowing for water regulation, reuse, filtration, and purification from upstream to the lower plains. This provides suitable growing conditions for other plant species between rice fields, promoting landscape diversity. The moist soil conditions created by rice ecosystems with the landscape contribute to microbial biomass build up, which drives nutrient recycling within the system.

Carbon regulation and maintenance

Soil submergence favors anaerobic decomposition of organic matter, which is slower than aerobic decomposition, resulting in soil organic carbon build-up. Shirato and Yokozawa (2005) effectively modeled soil carbon in rice paddies by simply decreasing the decomposition rates in Roth-C, a model that was developed and parameterized for arable topsoil carbon dynamics for upland soils (Coleman and Jenkinson, 1996). Alberto et al. (2015) showed cumulative effect of continuous residue incorporation in a lowland rice soil, likely due to slower organic matter decomposition. In an analysis of four long-term experiments in the Philippines with two or three irrigated rice crops per year, Pampolino et al. (2008) showed positive changes in soil organic carbon even with no fertilization, although greater concentrations were observed with full fertilization. In that study, soil organic carbon slightly increased even where straw was removed, suggesting that continuous soil submergence contributes to soil organic carbon build up. This is in contrast to systems where rice is rotated with an upland crop, e.g., Majumder et al. (2008) observed a decline in soil organic carbon when no residues were added under rice-wheat cropping system in India.

Although rice paddy ecosystems are known carbon source, rather than sink, soil carbon apparently has longer residence in these systems as these long-term experiments demonstrated. This becomes an advantage in maintaining water and nutrients within the rootzone, depending on how irrigation is managed during the cropping season.

Nutrient regulation and maintenance

The plow pan in paddy soils reduces water percolation and flooding the soil and prevents nutrient leaching to deeper soil layers. Reduced soil conditions under flooding drive biogeochemical nutrient cycling in paddy soils and determine the dominating nutrient species.

Unlike upland systems, low but sustainable rice yields can be obtained without the use of nitrogen fertilizers, e.g., long-term experiments at International Rice Research Institute (IRRI) have shown rice yields of 3 t ha^{-1} $season^{-1}$ without nitrogen fertilization (Dobermann et al., 2000). This has been attributed to nitrogen inputs from biological nitrogen fixation (Grant et al., 1986; Roger and Ladha, 1992). Earlier studies estimated that the contribution of biological nitrogen fixation ranged between 15 and 50 kg N ha^{-1} $season^{-1}$ using nitrogen balance studies (App et al., 1980). This figure was revised to be approximately 40 kg N ha^{-1} $season^{-1}$, being a contribution from biological nitrogen fixation, irrigation water, and atmospheric deposition (Roger and Ladha, 1992). Despite this contribution, farmers in Asia often apply excessive amounts of nitrogen fertilizer especially after the introduction of high yielding varieties during the Green Revolution. This has resulted in environmental pollution and contributes to greenhouse gas emissions, while increasing the cost of rice farming. Site-specific nutrient management was developed to allow farmers to optimally utilize indigenous nutrients and dynamically apply nutrients to fill the gap with fertilizers to achieve a given target yield (Dobermann et al., 2002). This has resulted in increased nutrient use efficiency and achieves greater rice yields.

In a global analysis of three major cereal crops, maize, rice, and wheat Ladha et al. (2016) calculated a net positive nitrogen balance in rice, whereas it was negative in maize and wheat (Table 10.1). This was largely contributed by greater inputs from biological nitrogen fixation and lower nitrogen in harvested grain in rice systems compared with maize and wheat. However, nitrogen fertilizer losses were higher under rice than maize and wheat, most likely via ammonia volatilization in rice systems. Buresh et al. (2008) indicated nitrogen losses up to 60% of applied nitrogen fertilizers due to ammonia volatilization in paddy rice systems.

Table 10.1 Global N budget (Tg, 1961–2010) in maize, rice, and wheat production systems (Ladha et al., 2016).

	Maize	Rice	Wheat	Total
Inputs				
Fertilizer	516.9	507.5	569	1593.5
Manure	187.5	184.1	206.4	578.1
Crop residue	92.5	65.3	87.1	244.9
Biological fixation	102.7	200	175.7	478.4
Deposition	54.3	60	92.8	207.1
Seed	2.2	1.6	16.3	20
Total	956.2	1018.5	1147.3	3122
Outputs				
Crop harvest	506.4	428.7	616	1551.1
Loss from fertilizer	229.7	323.7	294.1	847.5
Loss from manure	123.8	103.1	134.2	361
Loss from residue	83.3	58.8	78.4	220.4
Loss from biological fixation	20.5	40	35.1	95.7
Loss from deposition	24.1	38.2	48	110.4
Loss from seed	0.4	0.3	3.3	4
Total	988.2	992.8	1209.1	3190.1
Change in soil N	−32	25.7	−61.8	−68.1

While crop diversification through rotations and intercropping seems to be beneficial in improving nutrient cycling in upland ecosystems, in paddy soils the introduction of an aerobic phase will trigger nitrification, forming nitrates. Nitrates being mobile can easily be leached into deeper soil layers and subsequently enter water sources. Alternatively, accumulation of nitrates during the dry phase will likely be lost through denitrification upon resaturation of the soil for paddy rice. Buresh et al. (1989) observed an accumulation of up to 91 kg NO_3^-—N ha^{-1} at the end of the season under mungbean, whereas there was none at the end of the season with rice.

While the most widespread straw management practice is burning, this causes air pollution, affecting human health, particularly causing respiratory ailments, and has been banned in most Asian countries. However, ash is rich in potassium, thus recycling the nutrient in the soil. While retention of rice straw has little effect on nitrogen and phosphorus cycling in rice ecosystems (Thuy et al., 2008), it is important for potassium since more than 70% of the potassium taken up by rice is in straw (Dobermann and Fairhurst, 2000). Soil available silicon was low in Vietnam where crop residues are removed from the fields, while in the Philippines there is high availability because of crop residue retention (Settele et al., 2018).

Biodiversity maintenance and pest regulation

Being semiaquatic, paddy fields are ideal for diverse biological organisms by offering shelter, food, breeding, and nesting habitat. The biological diversity is influenced by moisture and nutrient gradients within the rice fields and the growth stage of rice and the cropping season during which rice was grown. While agronomic practices in rice systems vary widely, paddy fields are known to support an array of life, both fauna and flora. Invertebrates, including arthropods, crustaceans, microcrustaceans, molluscs, and nematodes, inhabiting the vegetation and water dominate the fauna in paddy soil habitats (Bambaradeniya and Amerasinghe, 2004; Norton et al., 2010). A survey done in Kurunegala District in Sri Lanka found a total of 495 invertebrate species, mostly arthropods, 103 vertebrate species, and 82 species of macrophytes in rice fields (Edirisinghe and Bambaradeniya, 2010). In the Philippines, spider diversity was greater in irrigated lowland followed by rainfed lowland while rainfed upland had the lowest with a Shannon diversity index of 2.72, 2.22, and 1.98, respectively (Burkhard et al., 2015). In India a survey observed 1130 spiders belonging to 92 species, 47 genera, and 16 families in irrigated rice systems at different elevation in two cropping seasons (Sebastian et al., 2005). Species richness and diversity was higher in lowlands than high and midland areas. Rice straw provides substrate to promote flourishing of invertebrates that decompose the straw (Schmidt et al., 2015), which in turn enhances nutrient cycling in paddy soils.

As artificial wetlands, paddy ecosystems (including the irrigation ditches, bunds, and field edges) host a variety of vertebrate fauna, including fish (Katano et al., 2003), birds (Edirisinghe and Bambaradeniya, 2010), and amphibians (Machado and Maltchik, 2010), reptiles (Halwart, 2006), and rodents (Stuart et al., 2007) for at least a part of their life cycles. For example, paddy rice fields serve as substitutes for wetlands for foraging cattle egrets during the breeding season when these feed on invertebrates, fish, and amphibians (Richardson and Taylor, 2003). Katano et al. (2003) reported 19 species of fish from a survey of irrigation ditches in central Japan, with higher species richness and diversity in ditches with easy access to rice fields. The fish species collected in the study mainly feed on aquatic insects such as chironomids larvae and ephemeropteran nymphs, both of which were often reported in abundance from rice fields (e.g., Settle et al., 1996; Bambaradeniya and Amerasinghe, 2004). Machado and Maltchik (2010) surveyed rice fields in Southern Brazil and found that frog species found in paddy fields represented $\sim$75% of the species observed in the study region. They hypothesized that these frog species utilized paddy rice fields for parts of their life cycles, either as a reproductive habitat or as a movement corridor between natural wetlands.

Paddy fields are typically kept flooded throughout the cropping season, which favors growth of most microorganisms, algae, and aquatic weeds but suppresses common weeds, especially the C4-type (Roger and Bhuiyan, 1990). The rich fauna and flora of rice ecosystems contribute to ecosystem functioning and stability, providing a plethora of ecosystem services such as pest regulation, organic matter decomposition/nutrient recycling, and food provisioning via wildlife harvesting.

The high diversity of arthropods found in irrigated rice ecosystems allow for a complex food web with redundant linkages forming the basis for pest regulation (Schoenly et al., 1996). Natural enemies of arthropod pests in rice fields include arthropod predators and parasitoids, insectivorous birds, amphibians, and bats (Puig-Montserrat et al., 2015; Teng et al., 2016; Zou et al., 2017). Predators of snail pests include rodents, turtles, fish, open-billed storks, and snail kites (Stuart et al., 2007; Horgan, 2017), and natural predators of rodent pests include snakes and raptors (Labuschagne et al., 2016).

Overuse of pesticides stress the natural enemies' populations, while inducing development of pesticide resistance among the pest populations. Food chains in fields receiving early insecticide applications are shorter than unsprayed fields, indicating a disturbance of natural enemy communities (Schoenly et al., 1996). This forms the scientific basis for the recommendation of no early insecticide spray (within the first 40 days of planting). Indeed, Heong (2009) suggested that pest resurgence and secondary pest outbreak in rice ecosystems usually occur because of the breakdown of pest regulation ecosystem services offered by the natural enemies. Ecological engineering, e.g., flower or vegetable strips next to rice fields, has also been used for biological pest control by providing food and shelter for natural pest enemies (Westphal et al., 2015).

Temperature regulation and maintenance

Wetlands provide an important service of regulating air and soil temperature. In tropical regions, air in wetland areas is often several degrees cooler than in nearby nonwetland areas because of the heat absorbed in evaporative cooling processes (Sun et al., 2012). Although there has been significant research toward managing wetlands to regulate urban temperature (Zhang et al., 2016; Wong et al., 2017; Şimşek and Ödül, 2018), there is little information about how rice paddy management may contribute to this regulating service. It is possible that as water efficiency drives rice ecosystems toward less frequent standing water in the field, this will diminish their capacity to regulate temperature.

Cultural services

Rice is life—its cultivation and consumption has shaped cultures, inspired art, song, and ritual, driven social habits and landscapes, and is an economic and a political commodity. Rice is consumed in more than 100 countries throughout the world, with a unique word translation in over 80 languages. In many cultures, in fact, there is a different term for its different varieties, stages of plant growth, ways of cooking, times when eaten, and forms of consumption.

Cuisine

Rice, which is native to Asia and Africa, has become an integral component of many cultures across the globe and is celebrated through different festivities for various cultures. Its primacy as a staple is underscored in the use of the word "rice" to mean the complete meal ("*gohan*" in Japanese) and even the entire act of consumption itself ("*kanin*" in Tagalog). Rice is considered divine in many Asian cultures. Its essentialness to life is exemplified by the Indonesians who refer to "Mother Rice" (*Me Posop*), also echoed by the Aetas of the Philippines who consider rice as "breastmilk" from the heavens ("*gatas ng langit*"). More than a staple, rice is consumed in myriad ways—from snack to salad, as a side dish, for dessert, and for libation.

Different rice varieties are typically grouped into short-grained, medium-grained, and long-grained. Their use depends on cultural preferences and for different cuisine. Short-grain rice varieties, which may have a somewhat sticky texture, are suitable for dishes like rice pudding and sushi because of the ability to hold shape when molded. Medium-grained rice varieties are less glutinous but still sticky and are preferred in making dishes like risotto and paellas. Long-grained rice, such as Basmati and Jasmine, are typically 4—5 times longer than wide, stay fluffy, and separate and tend to hold the flavors in which they are cooked, hence they are commonly used for pilaffs and salads.

Rice grain can be used to make different food and nonfood products depending on its biochemical characteristics, e.g., amylose content (Juliano and Hicks, 1996). Rice with low amylose content is used in baby food while those with a high content are used in making rice noodles. Rice bran oil, which is extracted from the brown outer layer of rice, has a mild flavor and a high heating temperature and is used for deep and shallow frying in Asia. Rice bran oil lowers human plasma cholesterol and triglycerides (Cicero and Gaddi, 2001). Rice bread is gluten free and is important in diets of people with gluten intolerance, e.g., the Vietnamese baguette.

Rice is used for making alcoholic and nonalcoholic beverages. Rice beer is popular throughout West Africa, and in Nigeria a special beer is made out of rice and honey (National Research Council, 1996). Budweiser beer, brewed in the United States, is fermented from rice and barley. Sake, commonly thought of as a wine, is the quintessential Japanese alcoholic beverage made from fermented rice. Rice wine is common in Asian countries where it is embedded in traditional festivities, e.g., Tapuy, a clear alcoholic rice wine is integral for important occasions such as weddings, rice harvesting ceremonies, and other cultural events in Banaue, the Philippines.

Ritual

Rice cultivation has become embedded in different cultural traditions and milieu, and in many societies it forms part of the cultural identity (Romero-Frias, 2014). It has become part of human heritage for most rice farming communities where it forms part of historical values, which are nonutilitarian. Owing to its role in providing calories, rice is often given a feminine symbol as rice Goddess, which is more than deity of mainstream religion, representing the nurturing of a mother (Romero-Frias, 2014). Rice throwing at weddings in many cultures originally symbolized fertility and the blessing of many children; today it symbolizes prosperity and abundance. Rice plays an important role for cultural rituals, e.g., the Mende people in Sierra Leone use rice soaked in palm oil as a major part for their rituals acknowledging their ancestors (Forde, 1999). In Guinea and Northern Sierra Leone the Mandingo and Susu people use rice flour and honey to make a sweet bread featuring as centerpieces of ceremonial rituals. A modern ritual is the use of polished rice to absorb moisture from salt in a shaker or from electronic gadgets that have been accidently wetted.

Esthetic

Cultural ecosystem services derived from rice systems depend on the type of rice landscapes. For example, Tekken et al. (2017)interviewed farmers in Vietnam and the Philippines to assess their perceptions on cultural identity, landscape esthetics, and knowledge systems associate with their rice landscapes. In that study they observed that farmers in the rural mountains of the two countries using traditional rice farming practices on terraces placed more value on the cultural services provided by rice farming than their counterparts in low-lying areas where intensive rice monocropping was practiced. Some rice farming areas are a source of tourism owing to the magnificent beauty endeared to the ecosystem, e.g., the Ifugao rice terraces in the Philippines (Fig. 10.3). These rice terraces have become part of the UNESCO world heritage sites (Settele et al., 2015). Similarly, the Tanada rice terraces in Japan are motor-shaped and are a popular tourist destination because of the beautiful landscape throughout the four seasons of the year.

FIGURE 10.3

Ifugao rice terraces in the Philippines

Photo courtesy of IRRI.

Disservices of lowland rice ecosystems

Air quality and greenhouse gas emission

The most easily observable air quality problem related to rice cropping is the smoke caused by straw burning, which has caused seasonal air quality problems shortly after rice harvest, affecting millions of people in major cities, notably New Delhi, Beijing, and Manila. There are difficult-to-enforce policies against straw burning in many rice-producing countries. Straw burning is an inexpensive way to remove the bulky biomass after harvest, particularly in intensive systems where there is a small window between crops. Burning straw releases particulate matter, associated with respiratory diseases, but it releases most carbon as CO_2 (Romero-Frias, 2014), a less potent greenhouse gas than methane.

Methane with a global warming potential 25 times greater than CO_2 is produced from anaerobic decomposition of organic matter decomposition under submergence, and this is exacerbated by straw incorporation (Sander et al., 2014). Lowland rice soils contribute about 11% of the global methane emissions (IPCC, 2013), with China and India contributing 21.8% and 19.1% of the global methane emissions (FAO, 2016). When lowland rice systems are driven toward improved water-use efficiency, e.g., alternate wetting and drying, changes in soil redox potential during flooding and drying cycles may cause release of nitrous oxide, a more potent greenhouse gas than methane. Alternate wetting and drying has however reduced methane emission in paddy rice (Sander et al., 2017). However, it will be important to continue to assess the impact of changing water use patterns on both methane and nitrous oxide emission from rice fields.

Human health: water-borne diseases

Paddy rice systems, particularly in tropical and subtropical regions, provide an environment for proliferation of water-borne diseases such as malaria, leptospirosis, schistosomiasis, and Japanese encephalitis (Roger and Bhuiyan, 1990; Edirisinghe and Bambaradeniya, 2010) (Roger and Bhuiyan, 1990; Edirisinghe and Bambaradeniya, 2010). More than 137 mosquito species that breed in rice fields have been recorded worldwide, with many potentially transmitting malaria and Japanese encephalitis (Bambaradeniya and Amerasinghe, 2004). Rodents species attracted to rice fields are also known to harbor zoonoses such as leptospirosis, rat-borne typhus, and the plague (Meerburg et al., 2009; Brown et al., 2017).

Food safety challenges

Alongside the food provisioning services of paddy rice ecosystems, there are some potential food safety risks. These fall broadly into three categories: toxic metal(-loid)s, mycotoxins, and pesticide residues. A healthy rice consumer will not likely have significant negative health impact from occasionally eating affected rice, so these risks are currently not a major public health concern. However, all three risk categories become more important when rice is a large percentage of the diet or when a consumer is already suffering from malnutrition. The concentration of contaminants are usually higher in the rice bran layer than in the white rice, making products derived from rice bran the highest risk, followed by brown rice. There are global systems that detect rice contamination (FAO/WHO, 1997), safeguards for detecting contamination. In the toxic metal(loid) category the major risks for rice are arsenic, cadmium, and mercury, all of which may be present without any known source of pollution—some soils have naturally high background content. The trend toward more aerobic water management in paddy systems is expected to mitigate the risk of arsenic uptake and exacerbate the risk of cadmium uptake (Arao et al., 2009), so it is important to know the background levels of arsenic and cadmium before recommending large-scale transition to more aerobic production. With mycotoxins, most of the risk comes from postharvest practices (inadequate drying or storage) (Mannaa and Kim, 2017), although it is also suspected that improved crop management practices before harvest may mitigate risks. Pesticide residues have been found in polished white rice and are thought to primarily result from the use of pesticides late in the cropping season (e.g., fungicide to control panicle smut) or from pesticides used in storage.

Conclusion

Rice productivity has increased in the last 5 decades following the Green Revolution, which saw an increase in the use of improved high yielding varieties combined with the use of fertilizers and agrochemicals, under irrigation mostly in intensive mono-cropped systems. While this contributed to reducing hunger and poverty, particularly in Asia, provisioning food for the poor, this was associated with excessive use of fertilizers causing environmental pollution and greenhouse gas emissions. Nonetheless, paddy rice farming offers a myriad of ecosystem services in addition to food provisioning. These include nutrient cycling, carbon sequestration, and biodiversity and has shaped many cultural traditions though cuisine and rituals. Traditional rice farming offers an important role in water regulation across the landscape with reuse and filtration of water as it moves from upstream to lower plains. However, this can potentially be disturbed with the introduction of water saving technologies in the face of water shortage, which also mitigates methane emission but can trigger nitrous oxide production. Greenhouse gas emissions remain a challenge under rice farming, particularly methane, contributing 11% of global methane emission. There is need for management alternatives that mitigate both methane and nitrous oxide emission. Furthermore, water borne diseases affect many people in rice farming communities with implications on the need to find water management practices that balance rice production, mitigate greenhouse gas emissions, and reduce associated diseases.

References

Ahmed, N., Garnett, S.T., 2011. Integrated rice-fish farming in Bangladesh: meeting the challenges of food security. Food Security 3, 81–92.

Akpokodje, G.L.F., Erenstein, O., 2001. Nigeria's Rice Economy: State of the Art, the Nigerian Rice Economy in a Competitive World; Constraints Opportunities and Strategic

Choices. Final Report Presented to West Africa Rice Development Association (WARDA) Bouake, Cote d'Ivore, 55.

Alberto, M.C.R., Wassmann, R., Gummert, M., Buresh, R.J., Quilty, J.R., Correa Jr., T.Q., Centeno, C.A.R., Oca, G.M., 2015. Straw incorporated after mechanized harvesting of irrigated rice affects net emissions of CH_4 and CO_2 based on eddy covariance measurements. Field Crops Research 184, 162–175.

App, A., Watanabe, I., Alexander, M., Ventura, W., Daez, C., Santiago, T., De Datta, S., 1980. Nonsymbiotic nitrogen fixation associated with the rice plant in flooded soils. Soil Science 130, 283–289.

Arao, T., Kawasaki, A., Baba, K., Mori, S., Matsumoto, S., 2009. Effects of water management on cadmium and arsenic accumulation and dimethylarsinic acid concentrations in Japanese rice. Environmental Science and Technology 43, 9361–9367.

Bambaradeniya, C.N., Amerasinghe, F.P., 2004. Biodiversity Associated with the Rice Field Agroecosystem in Asian Countries: A Brief Review. IWMI.

Belder, P., Bouman, B., Cabangon, R., Guoan, L., Quilang, E., Yuanhua, L., Spiertz, J., Tuong, T., 2004. Effect of water-saving irrigation on rice yield and water use in typical lowland conditions in Asia. Agricultural Water Management 65, 193–210.

Bouis, H.E., Saltzman, A., 2017. Improving nutrition through biofortification: a review of evidence from HarvestPlus, 2003 through 2016. Global Food Security 12, 49–58.

Bouman, B., 2007. A conceptual framework for the improvement of crop water productivity at different spatial scales. Agricultural Systems 93, 43–60.

Brown, P., Douangboupha, B., Lao, P., Htwe, N., Myanmar, J., Kühn, J., Mulungu, L., My Phung, N., Singleton, G., Stuart, A., 2017. Control of Rodent Pests in Rice Cultivation.

Buresh, R., Flordelis, E., Woodhead, T., Cabangon, R., Shepherd, K., 1989. Nitrate accumulation and loss in a mungbean/lowland rice cropping system. Soil Science Society of America Journal 53, 477–482.

Buresh, R.J., Reddy, K.R., Van Kessel, C., 2008. Nitrogen transformations in submerged soils. Nitrogen in Agricultural Systems 401–436.

Burkhard, B., Müller, A., Müller, F., Grescho, V., Anh, Q., Arida, G., Bustamante, J.V.J., Van Chien, H., Heong, K., Escalada, M., 2015. Land cover-based ecosystem service assessment of irrigated rice cropping systems in southeast Asia—an explorative study. Ecosystem Services 14, 76–87.

Cicero, A., Gaddi, A., 2001. Rice bran oil and γ-oryzanol in the treatment of hyperlipoproteinaemias and other conditions. Phytotherapy Research 15, 277–289.

Coleman, K., Jenkinson, D., 1996. RothC-26.3-A Model for the turnover of carbon in soil. In: Evaluation of Soil Organic Matter Models. Springer, pp. 237–246.

Cruz-Garcia, G.S., Price, L.L., 2011. Ethnobotanical investigation of'wild'food plants used by rice farmers in Kalasin, Northeast Thailand. Journal of Ethnobiology and Ethnomedicine 7, 33.

Dai, B.-l., Guo, X.-j., Yuan, D.-h., Xu, J.-m., 2018. Comparison of different pretreatments of rice straw substrate to improve biogas production. Waste and Biomass Valorization 9, 1503–1512.

Demuyakor, B., Dogbe, W., Owusu, R., 2013. Parboiling of paddy rice, the science and perceptions of it as practiced in Northern Ghana. International Journal of Scientific and Technology Research 2, 13–18.

Dias, D., Lapa, N., Bernardo, M., Ribeiro, W., Matos, I., Fonseca, I., Pinto, F., 2018. Cr (III) removal from synthetic and industrial wastewaters by using co-gasification chars of rice waste streams. Bioresource Technology 266, 239-150.

Dobermann, A., Dawe, D., Roetter, R.P., Cassman, K.G., 2000. Reversal of rice yield decline in a long-term continuous cropping experiment. Agronomy Journal 92, 633—643.

Dobermann, A., Fairhurst, T.H., 2000. Rice: nutrient disorders & nutrient management. International Rice Research Institute. ISBN: 981-04-2742-5.

Dobermann, A., Witt, C., Dawe, D., Abdulrachman, S., Gines, H., Nagarajan, R., Satawathananont, S., Son, T., Tan, P., Wang, G., 2002. Site-specific nutrient management for intensive rice cropping systems in Asia. Field Crops Research 74, 37—66.

Edirisinghe, J.P., Bambaradeniya, C.N., 2010. Rice fields: an ecosystem rich in biodiversity. Journal of the National Science Foundation of Sri Lanka 34.

FAO, 2016. Available from: http://www.fao.org/faostat/en/ - data/QC. 2016.

Forde, C.D., James, W., 1999. African Worlds: Studies in the Cosmological Ideas and Social Values of African Peoples. International African Institute, ISBN 9783825830861.

Grant, I., Roger, P.A., Watanabe, I., 1986. Ecosystem manipulation for increasing biological N2 fixation by blue-green algae (cyanobacteria) in lowland rice fields. Biological Agriculture & Horticulture 3, 299—315.

GRiSP, G.R.S.P., 2013. Rice Almanac, fourth ed. International Rice Research Institute, Los Baños (Philippines), p. 283.

Haefele, S., Konboon, Y., Wongboon, W., Amarante, S., Maarifat, A., Pfeiffer, E., Knoblauch, C., 2011. Effects and fate of biochar from rice residues in rice-based systems. Field Crops Research 121, 430—440.

Halwart, M., 2006. Biodiversity and nutrition in rice-based aquatic ecosystems. Journal of Food Composition and Analysis 19, 747—751.

Heong, K.L., 2009. Are planthopper problems caused by a breakdown in ecosystem services. In: Planthoppers: New Threats to the Sustainability of Intensive Rice Production Systems in Asia, pp. 221—232.

Horgan, F., 2017. Integrated pest management for sustainable rice cultivation: a holistic approach. In: Sasaki, T. (Ed.), Achieving Sustainable Cultivation of Rice, vol. 2. Burleigh-Dodds, UK, pp. 309—342.

IPCC, 2013. In: Stocker, T.F., Qin, D., Plattner, G.-K., Tignor, M., Allen, S.K., Boschung, J., Nauels, A., Xia, Y., Bex, V., Midgley, P.M. (Eds.), Climate Change 2013: The Physical Science Basis. Contribution of Working Group I to the Fifth Assessment Report of the Intergovern- Mental Panel on Climate Change. Cambridge University Press, Cambridge, United Kingdom and New York, NY, USA, p. 1535.

IUSS Working Group, 2006. World reference base for soil resources 2006: a framework for international classification, correlation and communication. World Soil Resources Reports 103. ISBN: 92-5-105511-4.

Jahn, R., Blume, H., Asio, V., Spaargaren, O., Schad, P., 2006. Guidelines for Soil Description. FAO.

Juliano, B.O., Hicks, P.A., 1996. Rice functional properties and rice food products. Food Reviews International 12, 71—103.

Katano, O., Hosoya, K., Iguchi, K.i., Yamaguchi, M., Aonuma, Y., Kitano, S., 2003. Species diversity and abundance of freshwater fishes in irrigation ditches around rice fields. Environmental Biology of Fishes 66, 107—121.

Kauldhar, B.S., Yadav, S.K., 2018. Turning waste to wealth: a direct process for recovery of nano-silica and lignin from paddy straw agro-waste. Journal of Cleaner Production 194, 158—166.

Kirk, G., 2004. The Biogeochemistry of Submerged Soils. Wiley, Chichester, 282pp.

Kögel-Knabner, I., Amelung, W., Cao, Z., Fiedler, S., Frenzel, P., Jahn, R., Kalbitz, K., Kölbl, A., Schloter, M., 2010. Biogeochemistry of paddy soils. Geoderma 157, 1—14.

Koohafkan, P., Furtado, J., 2004. Traditional rice–fish systems as globally important indigenious agricultural heritage systems. International Rice Commission Newsletter 53, 66–73.

La Notte, A., D'Amato, D., Mäkinen, H., Paracchini, M.L., Liquete, C., Egoh, B., Geneletti, D., Crossman, N.D., 2017. Ecosystem services classification: a systems ecology perspective of the cascade framework. Ecological Indicators 74, 392–402.

Labuschagne, L., Swanepoel, L.H., Taylor, P.J., Belmain, S.R., Keith, M., 2016. Are avian predators effective biological control agents for rodent pest management in agricultural systems? Biological Control 101, 94–102.

Ladha, J., Tirol-Padre, A., Reddy, C., Cassman, K., Verma, S., Powlson, D., Van Kessel, C., Richter, D.d.B., Chakraborty, D., Pathak, H., 2016. Global nitrogen budgets in cereals: a 50-year assessment for maize, rice, and wheat production systems. Scientific reports 6, 19355.

Lim, J.S., Manan, Z.A., Alwi, S.R.W., Hashim, H., 2012. A review on utilisation of biomass from rice industry as a source of renewable energy. Renewable and Sustainable Energy Reviews 16, 3084–3094.

MA, 2005. Millennium Ecosystem Assessment. Ecosystems and Human Well-Being: Biodiversity Synthesis. World Resources Institute, Washington, DC.

Machado, I.F., Maltchik, L., 2010. Can management practices in rice fields contribute to amphibian conservation in southern Brazilian wetlands? Aquatic Conservation: Marine and Freshwater Ecosystems 20, 39–46.

Majumder, B., Mandal, B., Bandyopadhyay, P., Gangopadhyay, A., Mani, P., Kundu, A., Mazumdar, D., 2008. Organic amendments influence soil organic carbon pools and rice–wheat productivity. Soil Science Society of America Journal 72, 775–785.

Mannaa, M., Kim, K.D., 2017. Influence of temperature and water activity on deleterious fungi and mycotoxin production during grain storage. Mycobiology 45, 240–254.

Meerburg, B.G., Singleton, G.R., Kijlstra, A., 2009. Rodent-borne diseases and their risks for public health. Critical Reviews in Microbiology 35, 221–270.

Mekonnen, M.M., Hoekstra, A.Y., 2011. The green, blue and grey water footprint of crops and derived crop products. Hydrology and Earth System Sciences 15, 1577–1600.

Munder, S., 2013. Improving Thermal Conversion Properties of Rice Straw by Briquetting (Master thesis). University of Hohenheim.

Nair, C.M., Salin, K.R., Joseph, J., Aneesh, B., Geethalakshmi, V., New, M.B., 2014. Organic rice–prawn farming yields 20% higher revenues. Agronomy for Sustainable Development 34, 569–581.

National Research Council, 1996. Lost Crops of Africa, vol. 1. Grains. National academy press, Washington DC, USA.

Nguyen, V.H., Nguyen, T.N., Tran, V.T., Nguyen, V.X., Phan, H.H., 2012. Rice husk uses in the Mekong Delta of Vietnam. International Workshop on the Innovative Uses of Rice Straw and Rice Husk. Ho Chi Minh City, Vietnam December 11–13, 2012.

Nguyen, V., Topno, S., Balingbing, C., Nguyen, V., Röder, M., Quilty, J., Jamieson, C., Thornley, P., Gummert, M., 2016. Generating a positive energy balance from using rice straw for anaerobic digestion. Energy Report 2, 117–122.

Norman, J., Kebe, B., 2006. African smallholder farmers: rice production and sustainable livelihoods. International Rice Commission Newsletter 55, 33–44.

Norton, G.W., Heong, K., Johnson, D., Savary, S., 2010. Rice pest management: issues and opportunities. Rice in the global economy: Strategic research and policy issues for food security 297–332.

Pampolino, M.F., Laureles, E.V., Gines, H.C., Buresh, R.J., 2008. Soil carbon and nitrogen changes in long-term continuous lowland rice cropping. Soil Science Society of America Journal 72, 798–807.

Puig-Montserrat, X., Torre, I., López-Baucells, A., Guerrieri, E., Monti, M.M., Ràfols-García, R., Ferrer, X., Gisbert, D., Flaquer, C., 2015. Pest control service provided by bats in Mediterranean rice paddies: linking agroecosystems structure to ecological functions. Mammalian Biology-Zeitschrift für Säugetierkunde 80, 237–245.

FAO/WHO, 1997. Risk management and food safety, Report of a Joint FAO/WHO Consultation, Rome, Italy, 27 to 31 January 1997. http://www.fao.org/docrep/W4982E/w4982e00.html.

Richardson, A.J., Taylor, I.R., 2003. Are rice fields in Southeastern Australia an adequate substitute for natural wetlands as foraging areas for egrets? Waterbeds 26, 353–363.

Roger, P.-A., Bhuiyan, S., 1990. Ricefield ecosystem management and its impact on disease vectors. International Journal of Water Resources Development 6, 2–18.

Roger, P.-A., Ladha, J., 1992. Biological N_2 fixation in wetland rice fields: estimation and contribution to nitrogen balance. Biological Nitrogen Fixation for Sustainable Agriculture 41–55. Springer.

Romero-Frias, X., 2014. On the role of food habits in the context of the identity and cultural heritage of South and southeast Asia. In: Cultural Heritage and Identity International Symposium 2013. Sichuan University, Chengdu, China, 12pp.

Sander, B.O., Wassmann, R., Palao, L.K., Nelson, A., 2017. Climate-based suitability assessment for alternate wetting and drying water management in the Philippines: a novel approach for mapping methane mitigation potential in rice production. Carbon Management 8, 331–342.

Sander, B.O., Samson, M., Buresh, R.J., 2014. Methane and nitrous oxide emissions from flooded rice fields as affected by water and straw management between rice crops. Geoderma 235, 355–362.

Settele, J., Heong, K.L., Kühn, I., Klotz, S., Spangenberg, J.H., Arida, G., Beaurepaire, A., Beck, S., Bergmeier, E., Burkhard, B., 2018. Rice ecosystem services in South-east Asia. Paddy and Water Environment, pp. 211–224.

Settle, W.H., Ariawan, H., Astuti, E.T., Cahyana, W., Hakim, A.L., Hindayana, D., Lestari, A.S., Pajarningsih, Sartanto, 1996. Managing tropical rice pests through conservation of generalist natural enemies and alternative prey. Ecology 11, 1975–1988.

Schmidt, A., Auge, H., Brandl, R., Heong, K.L., Hotes, S., Settele, J., Villareal, S., Schädler, M., 2015. Small-scale variability in the contribution of invertebrates to litter decomposition in tropical rice fields. Basic and Applied Ecology 16, 674–680.

Schoenly, K., Cohen, J.E., Heong, K., Litsinger, J.A., Aquino, G., Barrion, A.T., Arida, G., 1996. Food web dynamics of irrigated rice fields at five elevations in Luzon, Philippines. Bulletin of Entomological Research 86, 451–466.

Sebastian, P.A., Mathew, M.J., Pathummal Beevi, S., Joseph, J., Biju, C.R., 2005. The spider fauna of the irrigated rice ecosystem in central Kerala, India across different elevational ranges. Journal of Arachnology 33, 247–255.

Settele, J., Spangenberg, J.H., Heong, K.L., Burkhard, B., Bustamante, J.V., Cabbigat, J., Van Chien, H., Escalada, M., Grescho, V., Harpke, A., 2015. Agricultural landscapes and ecosystem services in South-East Asia—the LEGATO-project. Basic and Applied Ecology 8, 661–664.

Shirato, Y., Yokozawa, M., 2005. Applying the Rothamsted Carbon Model for long-term experiments on Japanese paddy soils and modifying it by simple tuning of the decomposition rate. Soil Science and Plant Nutrition 51, 405–415.

Şimşek, Ç.K., Ödül, H., 2018. Investigation of the effects of wetlands on micro-climate. Applied Geography 97, 48–60.

Singleton, G.R., Belmain, S., Brown, P.R., Aplin, K., Htwe, N.M., 2010. Impacts of rodent outbreaks on food security in Asia. Wildlife Research 37, 355–359.

Stuart, A.M., Prescott, C.V., Singleton, G.R., Joshi, R.C., Sebastian, L.S., 2007. The rodent species of the Ifugao rice terraces, Philippines–target or non-target species for management? International Journal of Pest Management 53, 139–146.

Sun, R., Chen, A., Chen, L., Lü, Y., 2012. Cooling effects of wetlands in an urban region: the case of Beijing. Ecological Indicators 20, 57–64.

Sweeney, M., McCouch, S., 2007. The complex history of the domestication of rice. Annals of Botany 100, 951–957.

Tekken, V., Spangenberg, J.H., Burkhard, B., Escalada, M., Stoll-Kleemann, S., Truong, D.T., Settele, J., 2017. "Things are different now": Farmer perceptions of cultural ecosystem services of traditional rice landscapes in Vietnam and the Philippines. Ecosystem Services 25, 153–166.

Teng, Q., Hu, X.-F., Luo, F., Cheng, C., Ge, X., Yang, M., Liu, L., 2016. Influences of introducing frogs in the paddy fields on soil properties and rice growth. Journal of Soils and Sediments 16, 51–61.

Thuy, N.H., Shan, Y., Wang, K., Cai, Z., Buresh, R.J., 2008. Nitrogen supply in rice-based cropping systems as affected by crop residue management. Soil Science Society of America Journal 72, 514–523.

Westphal, C., Vidal, S., Horgan, F.G., Gurr, G.M., Escalada, M., Van Chien, H., Tscharntke, T., Heong, K.L., Settele, J., 2015. Promoting multiple ecosystem services with flower strips and participatory approaches in rice production landscapes. Basic and Applied Ecology 16, 681–689.

Wong, C.P., Jiang, B., Bohn, T.J., Lee, K.N., Lettenmaier, D.P., Ma, D., Ouyang, Z., 2017. Lake and wetland ecosystem services measuring water storage and local climate regulation. Water Resources Research 53, 3197–3223.

Xie, J., Hu, L., Tang, J., Wu, X., Li, N., Yuan, Y., Yang, H., Zhang, J., Luo, S., Chen, X., 2011. Ecological mechanisms underlying the sustainability of the agricultural heritage rice–fish coculture system. Proceedings of the National Academy of Sciences 108, E1381–E1387.

Yoon, C.G., 2009. Wise use of paddy rice fields to partially compensate for the loss of natural wetlands. Paddy and Water Environment 7, 357.

Zhang, W., Zhu, Y., Jiang, J., 2016. Effect of the urbanization of wetlands on microclimate: a case study of Xixi Wetland, Hangzhou, China. Sustainability 8, 885.

Zou, Y., De Kraker, J., Bianchi, F.J., Van Telgen, M.D., Xiao, H., Van Der Werf, W., 2017. Video monitoring of brown planthopper predation in rice shows flaws of sentinel methods. Scientific Reports 7, 42210.

CHAPTER

11 The role of ecosystem services in offsetting effects of climate change in sustainable food systems in the Zambezi Basin, Southern Africa

Chipo Plaxedes Mubaya, PhD [1,3], **Mzime Regina Ndebele-Murisa, PhD** [1,2]

[1]*Chinhoyi University of Technology, International Collaborations Office, Department of Freshwater and Fishery Science, Chinhoyi, Zimbabwe;* [2]*START International, Washington, DC, United States;* [3]*International Collaborations, Chinhoyi Uinversity of Technology, Chinhoyi, Zimbabwe*

Introduction

There are issues of larger resource management challenges that threaten to accelerate degradation and cut off vital supply of ecosystem services that directly support economic sectors and human health and well-being in the Zambezi Basin. This is representative of the situation faced by communities throughout many parts of sub-Saharan Africa (SSA) (Nicholls et al., 2013). Moreover, despite aridity, agriculture is currently the main livelihood and food production activity in a number of areas in rural southern Africa as well as Zimbabwe, including Omay, which is a case study employed in this chapter. However, food production (including agricultural) is low in the Zambezi Basin, with inadequate levels of input and technology use (Magadza, 2010). Markets for both cash and food crops (especially fruits and vegetables) are underdeveloped and sometimes under pressure because of local surpluses and competition from the commercial farming sector (Mubaya, 2006; Chikweche and Fletcher, 2009). Thus small-scale food production systems, such as agriculture and aquaculture, face serious problems and fail to reach their potential for providing a regular and adequate supply of food and reliable income to rural families. We argue that improvements appear feasible both in terms of technology development, alternative land use, community organization, and marketing, including employment of alternative adaptation measures such as fisheries and livelihood diversification, among others.

The Zambezi Basin, where our case study for this chapter is located, is vulnerable to episodes of climate occurrences such as drought, floods, and rainfall

The Role of Ecosystem Services in Sustainable Food Systems. https://doi.org/10.1016/B978-0-12-816436-5.00011-1

variability. This vulnerability of the ecosystem impacts negatively on the livelihoods of the smallholder communities living within the basin. Despite this gloomy picture on rainfall availability, agriculture remains the predominant option for the communities in the basin (Mubaya 2006; Magadza, 2010; Mubaya and Ndebele-Murisa, 2017). Moreover, local markets increasingly fluctuate under increased demand for agricultural products, bringing to the fore the fundamental problem of considerable economic and environmental risks. Volatility of the macroeconomic and general policy environments in the country worsen the situation and exacerbate agriculture and agrarian-based livelihood failure. The subsequent increasing demand for food requires a shift in the landscape activities and more efficient and environmentally friendly food production (Moyo and Chambati, 2013; Nicholls et al., 2013). The highlighted multiple stressor context, which also encompasses population growth, has triggered disproportionate pressure on the agriculture sector for greater quantities of food (protein), animal feed, biofuels, and fibers. This growing demand for food security (McNeely and Scherr, 2001; Campos, 2012) and improved livelihood can best be met through an innovative systemic approach to environment and technology nexus. Furthermore, with climate change added to the picture, there is not only increased pressure on agriculture but also trade-offs among different land uses, such as competition for arable land, water, fisheries, minerals, and other natural resources. Reduced water levels, interference with natural flow regimes, and flow channel morphology through construction of large reservoirs for hydropower production and the accompanying artificial flood releases also affect the resident and migratory wildlife of the region, leading to potential human-wildlife conflicts in the basin. Those changes have great significance for security, food production, and energy dynamics.

Nicholls et al. (2013) note that there is still limited understanding of how to implement interventions which tackle complex relationships through policies and actions. As such, framing rural development in the environment and technology nexus would perhaps provide an opportunity to address smallholder challenges through community-based natural resources and ecosystem management such as tourism, fishing, biotechnology, hydrology, animal husbandry, and agriculture to increase income and food security. In this regard, literature suggests that an integrated framework to a landscape approach to deal with these demands and trade-offs in an integrated way to create a balance between social, environmental and economic concerns is needed (McNeely and Scherr, 2001; Campos, 2012; Nicholls, et al., 2013). We believe that current events in the Zambezi Basin provide an ideal opportunity to design and implement a strategy which considers this proposed approach in full account and can be learnt and built upon for the future realization of truly holistic interventions.

The climate change problem

Climate change is one of the most serious threats to the sustainability and livelihoods of rural communities in SSA (Connolly-Bouten, 2015; Byran et al., 2013; Phiiri et al., 2016). This is mainly because of adverse effects on the environment, economic activities, and human health (Mark et al., 2008). Generally, many sub-Saharan regions are characterized by high rainfall variability, frequent droughts, dry spells, and floods (Ajani et al., 2013). In addition, most of the rainfall received in semiarid regions is lost as runoff, and very little water is harvested for plant growth or future use (Nyamadzawo et al., 2012). This does not only limit the availability of water but also causes serious erosion and nutrient loss (Nyamadzawo et al., 2012). People in the communal lands of southern Africa are more vulnerable to climate change because of their lack of adaptive capacity, and most of them rely on agriculture for food production and the sustenance of their lives (Kotir, 2011; FAO, 2012; Mugandani et al., 2012; Kutter and Westby, 2014; Conolley-Boutin, 2015). In addition, the high levels of poverty, inadequate infrastructure, and structural challenges at policy levels hamper national efforts toward adapting to a changing climate (Juana et al., 2013). As a result, these countries are the most vulnerable to unpredictable weather conditions such as, droughts and floods.

Interestingly, food systems contribute 19%–29% of global anthropogenic greenhouse gas (GHG) emissions, releasing 9800–16,900 megatonnes of carbon dioxide equivalent (CO_2e) in 2008. Agricultural production, including indirect emissions associated with land-cover change, contributes 80%–86% of total food system emissions, with significant regional variation (Vermeulen et al., 2012). Therefore there is a nexus between ecosystem services, climate change, and variability and food production and/or ecosystems. In addition to the economic crisis in Zimbabwe, rural livelihoods that mostly depend on rain-fed food production are under threat because of increasingly changing climatic patterns, yet there is little significant effort to create new rural income systems. Studies have shown that livelihood income strategies that are based on rain-fed production are negatively affected by increasingly changing climatic patterns, particularly in arid to semiarid areas where drier periods have become more prevalent (Muchena, 1991; Downing, 1992; Magadza, 1994; Matarira et al., 1995; Hulme, 1996; Chagutah, 2006; Twomlow et al., 2008; Mubaya and Ndebele-Murisa, 2017). These trends are a cause for concern, especially in this region where more than 60% of the population urban income opportunities are on the decline, and the economy's reliance on agriculture and its by-products is high.

High climate variability means that food production and yields are inconsistent from year to year, rendering rain-dependent production alone unreliable as the main means of food production and economic sustenance (Hulme, 1996). Furthermore poor households in arid to semiarid areas of SSA are vulnerable to climate change impacts because of their geographic exposure, low incomes, and greater reliance on agriculture as well as limited capacity to seek alternative livelihoods (Atiera and Koohafkan, 2008). Livelihood diversification innovations introduced postindependence in most of the countries such as nonfarm incomes derived from activities such as bee-keeping, handicraft, and small-scale mining that substantiate food purchases have mostly collapsed during the past decades because of a number of reasons, among them lack of access to markets and of course from the negative impacts of change and variability on production (Chikweche and Fletcher, 2009; Murisa, 2013; Ndebele-Murisa and Mubaya, 2019).

Climate change impacts on ecosystems

At the most basic levels of biodiversity, climate change can decrease genetic diversity of populations because of directional selection and rapid migration, which could in turn affect ecosystem functioning and resilience (Botkin et al.,2007). Beyond this the various effects on populations are likely to modify the "web of interactions" at the community level (Gilman et al., 2010; Walther, 2010). The response of some species to climate change may constitute an indirect impact on the species that depend on them. A study of 9650 interspecific systems, including pollinators and parasites, suggested that around 6300 species could disappear around the globe following the extinction of their associated species (Koh et al., 2004). In addition, for many species the primary impact of climate change may be mediated through effects on synchrony with species' food and habitat requirements. Climate change has led to phenological shifts in flowering plants and insect pollinators, causing mismatches between plant and pollinator populations that lead to the extinctions of both the plant and the pollinator with expected consequences on the structure of plant-pollinator networks (Kiers et al., 2010; Rafferty and Ives, 2010). Other modifications of interspecific relationships (with competitors, prey/predators, host/parasites, or mutualists) also modify community structure and ecosystem functions (Lafferty, 2009; Walther, 2010; Yang and Rudolf, 2010). At a higher level of biodiversity, climate can induce changes in vegetation communities that are predicted to be large enough to affect biome integrity. The Millenium Ecosystem Assessment (2005) forecasted shifts for 5%–20% of the earth's terrestrial ecosystems, cool conifer forests, tundra, scrubland, savannahs, and boreal forest (Sala et al., 2005). Of concern are "tipping points" where ecosystem thresholds can lead to irreversible shifts in biomes (Leadley et al., 2010).

The main objective of this chapter is to showcase how ecosystem services can be used to offset the effects of climate change and variability by building sustainable food systems. We use a case study of Omay Communal lands in Zimbabwe; an arid and increasingly warming area that represents a significant portion of SSA that sits on the arid belt. We reflect on impacts of climate change and variability on food systems while proffering solutions to these impacts that can help in the adaptation, coping, and mitigation against climate change and variability in this community; therefore offering ways in which sustainable food systems can be developed. Despite the threats posed by climate change and variability, the negative impacts are still reversible, and strategies and approaches exist that can reduce negative environmental externalities of food production systems. We argue that ecosystem goods and services can be used and harnessed in such a way that mitigates and builds resilience against climate change and variability for food ecosystems, despite the alarming consequences indicated for biodiversity and worst-case scenarios of extinction in the future (Barnosky et al., 2011).

Conceptualization of ecosystem services and food systems

"An ecosystem can be defined as a relatively self-contained system that includes plants, animals (including humans), microorganisms, and nonliving components of the environment as well as the interactions between them" SPC (2010). According to FAO (2018), "ecosystem services make human life possible by, for example, providing nutritious food and clean water, regulating disease and climate, supporting the pollination of crops and soil formation, and providing recreational, cultural, and spiritual benefits". Despite an estimated value of $125 trillion, these assets are not adequately accounted for in political and economic policy, which means there is insufficient investment in their protection and management. Biodiversity includes diversity within and among species and ecosystems. Changes in biodiversity can influence the supply of ecosystem services.

In this chapter we use ecosystem(s) and biodiversity interchangeably with the understanding that the mostly used definition of biodiversity based on the United Nations Convention on Biological Diversity is "the variability among living organisms from all sources including, inter alia, terrestrial, marine, and other aquatic ecosystems and the ecological complexes of which they are part: this includes diversity within species, between species, and of ecosystems" (UN 1992, Article 2, p.3). Biodiversity includes all organisms, species, and populations; the differences in their genetic building; the complex nature of the relationships between communities and their surrounding ecosystems; and how all these interact with the environment (Gaston and Spicer, 1998; Pearce and Moran, 1994; SADC, IUCN and SARDC, 2000, p. 51). Biodiversity is most often understood in terms of the number of species and their variety and can be considered at different scales that range from variation at gene level all the way up to ecosystems (Tuomisto, 2010). Central to definitions of biodiversity are its issues of variety, richness and evenness, and endemism, that is, the uniqueness of species in that they can only be found in a certain area as the key characteristics of diversity (DeLong, 1996). Therefore our understanding of ecosystems encompasses all forms of biodiversity that include agricultural and natural ecosystems that produce ecosystem goods and services and includes food production.

We define food systems as any productive system that produces food that can be derived through agricultural or natural production such as edible fruits, root, and leaves harvested from common pool resources such as natural and planted forests, orchards, fisheries, as well honey and other such activities, which can generate income that can be used to purchase food. Forests represent an important repository of food and other resources that can play a key role in contributing toward food security, especially if integrated into complex systems that are managed for multiple benefits. In some instances, such food products can be used in barter trade to procure other food items which is common practice in the rural areas of SSA (Portes, 2010). Ecosystem services are known to provide numerous benefits to society, either directly or as inputs into the production of other goods and services among which the supporting and provisioning of food stand out (see Table 11.1). Ecosystems goods and services provide play an intermediary role in mediating loss and damage from climate change and variability while boosting food and nutrition security.

The proper functioning of the entire ecosystem is important for food production. By bringing the food supply in balance with other ecosystem services, the overall

Table 11.1 Benefits of ecosystem services and goods.

Service	Definition and examples
Supporting	Include foundational processes necessary for production of other ecosystem services, including soil formation, nutrient cycling, and photosynthesis (primary production); providing plants and animals with living spaces; allowing for diversity of species; and maintaining genetic diversity, food webs of plants, and animals
Provisioning	Material benefits, e.g., supply of food (plants, fish, animals), water, fibers, genetic resources, medicines, wood and fuels, and other harvestable goods
Regulating	Benefits obtained from the regulation of ecosystem processes, e.g., the regulation of air quality and soil fertility, control of floods or crop pollination, pest and disease control, waste decomposition, and water quality regulation/water purification, waste management, coastal protection, and resilience against variability and change, as well as natural disasters
Cultural	Nonmaterial benefits people gain from ecosystems, e.g., spiritual enrichment, intellectual development, esthetic and engineering inspiration, cultural identity and spiritual well-being, recreation, and cultural and traditional heritage values

Source: FAO (2018). Ecosystem Services and Biodiversity. http://www.fao.org/ecosystem-services-biodiversity/en/.

sustainability of the system can be enhanced. Food supply in the ecosystem service approach is a provisioning service, as are fresh water and clean air (see Table 11.2). Services provided by ecosystems, such as pollination, water quality protection, pest suppression, nutrient cycling, connect food production with the surrounding landscapes and can be managed as part of a greater food system. Landscapes and their diversity provide a host of services that reduce risks and enhance resiliency along the food system value chain, whether by promoting local diversification of production systems or reducing resource inputs along the value chain.

Ecosystem-based approaches to enhance food production systems

Emerging approaches proffering community-based sustainable management of natural resources, including food systems, are gaining recognition as viable tools for climate change adaptation and mitigation through transformational and incremental adaptation and which build the resilience of socioeconomic and ecological systems. These approaches differ from traditional top-down, often hard infrastructure-based approaches to adaptation by seeking a balanced and integrated framework to reflect local conditions and community priorities as well as considering broader development goals, with the hope of better capturing the complex interdependencies between human societies and their environment. And while engineered and technological adaptation options are still common, there is growing recognition of

Table 11.2 Examples of wild and agricultural biodiversity contributions to ecosystem service categories.

Ecosystem service (TEEB classification)	How wild biodiversity generates ecosystem services	How agrobiodiversity generates ecosystem services
• Food provisioning 	• Wild caught and farmed fish and other aquatic animals • Harvested wild plants • Other wild animals (e.g., mammals, birds, insects)	• Crop provisioning; livestock provisioning • Nutritional and dietary diversity; increased number of functional traits leads to more resistant and resilient crops
• Provisioning: fresh water 	• Native vegetation influencing local to global rainfall patterns; roles in global hydrological cycle; localized water purification	• Reduced need for pesticides because of agrobiodiversity-based pest and disease control; complex vegetation structure as filter of pollutants
• Regulating: carbon sequestration 	• Ecosystems storing and sequestering greenhouse gases. Biodiversity enabling adaptation to climate change	• Increased carbon sequestration through more continuous biomass; increased soil function and carbon sequestration; and increased use of legumes reduces need for NPK use
• Regulating: biological control 	• Pests and diseases regulated through predators and parasitoids. Birds, bats, insects, and other arthropods, frogs, and fungi, all act as natural controls	• Intercropping and interspecific crop diversity providing habitat and resources for natural enemies. Intraspecific diversity suppressing pests and diseases
• Cultural: esthetic appreciation 	• Biodiversity, ecosystems, and natural landscapes have been the source of inspiration for much of our art, culture, and increasingly for science	• Landscapes, such as globally important agricultural heritage sites

Source: TEEB (2009). The Economics of Ecosystems and Biodiversity for National and International Policy Makers. Summary: Responding to the Value of Nature. http://www.teebweb.org/publication/teeb-for-policy-makers-summary-responding-to-the-value-of-nature/.

ecosystem-based, institutional and social measures to promote integrated adaptation that including building the resilience of food systems. Emerging approaches also seek to empower local people and support bottom-up participatory decision-making and planning, within a stronger institutional context.

Several, mostly holistic and participatory approaches to landscape and food ecosystem management, which are people-centered, participatory and culturally appropriate have emerged in the past couple of decades. These approaches engage communities in producing rather than prescribing solutions while seeking to address the vagaries of climate change and variability on food systems. In being inclusive the hope is to gain the acceptance, buy in, and participation by the proponents of any proposed approaches. Among such approaches are the landscape and systems approaches, ecosystem-based adaptation (EbA), and community-based adaptation (CbA) as well as innovations that seek to reduce the effects of climate change and variability on food systems. Some schools of thought advocate for the use of community and ecosystem-based approaches which have the potential to achieve sustainable food sources and reduction of climate change impacts. Two such approaches are the CbA and EbA, which provide flexible, cost-effective, and broadly applicable alternatives for building robust food systems and reducing the impacts of climate change. Both these approaches emphasize transformative adaptation and broad-based change through social and technical innovation; the formation of new structures or systems of governance; or alterations in personal belief systems that inform climate change responses. This systemic approach encourages sustainable development trajectories that combine adaptation and mitigation co-benefits as part of the interventions and solutions to climate change and variability impacts (Chevalier, 2017).

EbA is most commonly applied in the agricultural and forestry sectors, and there are multiple references in national adaptation programs of action and nationally determined contributions of emissions (IIED, 2016) to conservation, sustainable management, and the restoration of ecosystems. On the other hand, it builds on a long history of development approaches, including efforts to incorporate natural resources governance. CbA, on the other hand, characterizes the design of most disaster risk-reduction and community-based natural resource management initiatives. Capturing communities' traditional knowledge and experience and merging this with modern technical knowledge and capacities can be a valuable way in which to respond to climate variability and change. After all, communities have been adapting to changes in climate for centuries, through migration, changes in their choice of crop varieties, and livestock decisions among other strategies. CbA is community-led, that is based on communities' priorities, needs, knowledge, and capacities (Reid et al., 2009). It typically entails projects aimed at enhancing livelihood resilience (promoting, for example, hardier seed varieties, drip irrigation, expanded access to weather forecasting services, or income diversification); strengthening the capacity of local civil society and government institutions so that they can more effectively support community adaptation efforts; and increasing social mobilization to address the underlying causes of vulnerability (Dodman and Mitlin, 2013).

Closer attention to approaches that go beyond words into actions and guide policy practices are urgently needed if climate change and variability mitigation is to be achieved. An example of an innovative approach is the healthy farm index (HFI); a farm-scale biodiversity and ecosystem services assessment tool which complements existing farm assessment tools by integrating multiple metrics and outputs suitable for applied decision-making and annual evaluation by incorporating biodiversity and ecosystem services into local agricultural land use decision-making to restore biodiversity to farm systems (Quinn et al., 2013). The HFI does not encourage farmers to maximize biodiversity but rather to restore and maintain a level of diversity beneficial to the farm and local ecosystem and that contributes to local and regional conservation efforts. This is important, given that local scale analysis and local landowners' involvement has been shown as complementary to global progresses toward sustainability (MA, 2005; Persha et al., 2011). Reed et al (2017) show the potential of achieving net livelihood gains through integrating trees on farms, providing rural farmers with additional income sources and greater resilience strategies to adapt to market or climatic shocks. However, they also identified significant gaps in the discourse that demonstrate a need for larger-scale, longer term research to better understand the contribution of forest and trees within the broader landscape and their associated impacts on livelihoods and food production systems. Maintaining diversity within agricultural systems is not a novel approach but one practiced by many smallholder farmers globally, in many ways. The nutritional and livelihood benefits of diverse production systems are one way of achieving food security. Such systems are also more resilient to climate-induced events or other shocks. Sunderland (2011) shows that forests represent an important repository of food and other resources that can play a key role in contributing toward food security, especially if integrated into complex systems that are managed for multiple benefits.

Godfray et al. (2010) showed major advances in sustainable food production and that availability can be achieved with the concerted application of current technologies (given sufficient political will) and the importance of investing in research sooner rather than later to enable the food system to cope with both known and unknown challenges in the coming decades. Munir et al. (2015) point out that investments are needed for enhancing future food security; this requiring action on several fronts, including tackling climate change, preserving land and conserving water, reducing the energy footprint in food systems, developing and adopting climate resilient varieties, modernizing irrigation infrastructure, shoring up domestic food supplies, reforming international food trade, and responding to other global challenges. Another approach to ecosystem conservation which encompasses tenets of climate resilience is The Economics of Ecosystems and Biodiversity (TEEB, 2009).

Case study of Omay Communal Lands, Zimbabwe

We collected the data that we use in this chapter from Omay Communal Lands in 2016. We used mainly the qualitative methodology to collect these data through focus group discussions (FGDs), key informant interviews and in-depth case studies (see Fig. 11.1). The study area is situated in the North Eastern part of the Zambezi Valley (Murombedzi, 1992). The communal lands fall under Nyaminyami which is made up of three communal lands. The other two are Kanyati and Gatshe Gatshe communal lands (Fig. 11.2). Omay is on the Southern shore of Lake Kariba and is adjacent to Matusadonha National Park. In addition, Omay has a total area of 2870 sq. km (Taylor, 1990). Omay Communal Lands are separated from Kanyati and Gatshe Gatshe communal lands by the Matusadonha National Park. The latter is located at the center of the district and constitutes approximately 30% of the total area. Murombedzi

FIGURE 11.1

Left—women and right men discussing ecosystem services in Negande, Omay. Photo Credit @ Authors.

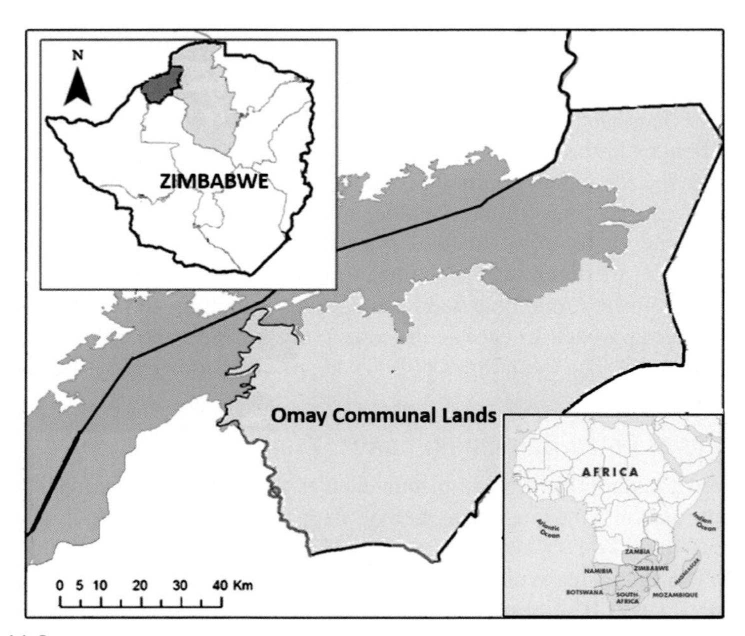

FIGURE 11.2

Location of study area with bottom right insert showing location of Zimbabwe in southern Africa and top left insert-location of study area within the Mashonaland West Province of Zimbabwe.

Map redrawn and used with permission from http://aktaylor.com/southern-africa-1/.

(1992) states that Omay is characterized by low agricultural potential and is located in agroecological zones III, IV, and V. The environment is semiarid, and there is generally a seasonal rainfall pattern that is highly variable and barely amounting to 650 mm per year. Omay has a hot climate of maximum temperatures above 40°C and rarely falling below 17°C. The soils are generally poor, resulting in consistently poor yields for the locals. Omay Communal Lands are made up of 9 wards which are under the jurisdiction of Nyaminyami Rural District Council.

The Millennium Ecosystem Assessment defines "ecosystems services" as benefits that accrue to people from ecosystems. The definition highlights benefits that are ecological, sociocultural, agricultural, recreational, and economical in nature (IUCN, 2018). The UK NEA (2011) defines "ecosystem services" as encompassing both tangible and intangible benefits that humans obtain from ecosystems. These ecosystem services are considered to be context dependent as the same feature of an ecosystem may be valued differently by different communities. Our analysis of the importance of ecosystem services that communities in Omay unravel can be classified under ecological, sociocultural, and economic (see Table 11.3).

Ecosystem services in Omay

Ecological benefits

Participants of FGDs highlighted that they benefit from ecosystems through forest products such as fruits, poles, and firewood (Table 11.3). Communities collect and sell baobab fruits, with a large proportion of the fruit being consumed locally. However, challenges such as restriction to the cutting down of trees by the Forest Commission and the unavailability of reliable markets limit full potential for these

Table 11.3 A summary of ecosystem services for communities in Omay

Resource	Services
Ecological	
Forest	Fruits, poles, firewood/wood
Wild animals	Meat, income
Land	Vegetable and other crop production
Rivers	Fish, water
Sociocultural	
Mountains	Forest products, wild animals
Infrastructure	Roads, schools, bridges
Economic	
Tourism	Income
Crop production	Income

benefits to transform the livelihoods of the concerned communities. These challenges speak to the drivers of change (in this case deforestation) and how they interact with other issues (in this case market inconsistencies) that tend to alter the way communities access benefits from their ecosystems. These participants further highlighted wild animals as some of the benefits through directly hunting for meat on their own or indirectly through hunting quotas that are set by the Nyaminyami Rural District Council (NRDC) under the Communal Areas Management Program for Indigenous Resources (CAMPFIRE) program. This picture matches the suggestions for Community Based Natural Resources Management and CbA, which are both critical for ecosystem services management (see also Ndebele-Murisa and Mubaya, 2019).

The second benefit that emerged is land, which these participants highlighted as an important component of their food system through crop production both as small-scale horticulture production and small-scale crop production in their fields. The crops grown include sorghum, maize, cowpeas, groundnuts, pearl millet, bambara nuts, pumpkins, water melons, sugar beans, and sesame, although sesame and sugar beans are grown by a few farmers. Groundnuts and sesame are also processed into butter, and sesame as well as cowpea leaves are cooked and used as both fresh and dried relish, although the groundnuts yield has significantly gone down, with the last meaningful yields realized in 2013. Reports lamented the high rate of soil erosion and along the stream banks where they mostly grow their vegetables. Rivers were also highlighted as a major source of benefits for these communities through artisanal fisheries and general water provisions for domestic and agricultural uses. One community in Negande has access to tap water which has been developed in the nearby mountains, where it comes down through gravity. The highlighted benefits indicate both wild and agrobiodiversity services as suggested by TEEB (2009) and FAO (2018).

Sociocultural benefits

Although mountains were highlighted as a very important source of benefits for the communities, there were also highlights that these "sacred" mountains have in some cases been a source of mystery and unhappiness following disappearances of community members who failed to observe cultural stipulations when visiting the mountains for one reason or another. This gives an indication that ecosystem services that communities get must conform to the everyday lives of these communities for the benefits to make a difference to their food systems. In addition, this cultural aspect of ecosystem services speaks to the highlighted importance of inclusive approaches that consider the priorities and needs of local communities as indicated by Chevalier (2017) and IIED (2016) in the EbA and CbA to ecosystem, access, use, and management.

Economic benefits

Reports emphasized tourism as an economic activity that brings significant revenue to the community. In addition to the CAMPFIRE project through the NRDC, the community highlighted that tourists visit hunting areas close to them, such as Bumi Hills, more than 60 km from Negande. It was further revealed that tourists come to Negande to see wild animals, but the collapse of Negande Dam in 2005 constructed by the District Development Fund reduced tourist activity in the area. The dam was useful as Tsetse Control Unit, and tourists used to camp at the dam. It was also used for fishing and irrigation by the local community. The discussions indicated that most of the tourists are from Canada, Namibia, and Angola and are normally accommodated at Safaris. The FGDs further revealed that revenue from hunting and tourism was allocated per ward depending on where the animals were slaughtered and the value of the slaughtered animals. Tourism opportunities in the view of these communities include house stays, and most people were keen to welcome tourists and to provide them with locally prepared food. In addition to revenue accruing to the district and communities, some households were also fortunate to receive small items as tokens of appreciation for services rendered to these tourists upon their departure. However, tourism activities under CAMPFIRE have significantly gone down since 1995 because of macroeconomic and local governance issues.

Past benefits from the CAMPFIRE project include building of schools, clinics, and compensation for destroyed crops, meat handouts from killed problem animals, including benefits from animal sales to safaris, and facilitation of free medical treatment and education. Essentially, there is an indication that these benefits transcend obvious and immediate benefits to encompass others that come as associated benefits (see Box 11.1), which are equally important for the socioeconomic sustenance of the Omay communities. There are more economic benefits from crop production (Table 11.3). For instance, sorghum and pearl millet are used to make beer and maize is processed into mealie-meal, including selling a little surplus to neighbors in times of strife. Participants highlighted that they also get services such as infrastructure (roads and bridges) that contribute to the economic well-being of the communities and households. Some households produce winnowers, baskets, and mats using locally available reeds, including cutting and selling thatch grass to Safari Operators and to people from Gokwe North. Carpentry was also identified to be another value addition activity undertaken in the area. It was revealed that "mukwa" and "mukamba" hardwood trees were used to make chairs, and on average one tree produced six chairs. However, these participants highlighted that support from government and other stakeholders would assist with income generation through promotion value addition activities in the area.

Improvement of ecosystem service benefits to the communities

To make the most of the ecosystem services than is currently the case, key informants suggested a number of solutions such as promoting water harvesting techniques and irrigation schemes, increasing community management of natural resources and their access to markets, and preserving produce. These solutions are consistent with the inclusive ecosystem service management that is suggested as key for an integrated framework. Value addition in agricultural produce emerged

Box 11.1 DIRECT AND ASSOCIATED BENEFITS FROM TOURISM AND NATURAL RESOURCES IN NEGANDE

Vegetable gardening and the drying of vegetables.
Fencing up of wild animals.
Dressing making and tailoring for sale.
Basket weaving.
Crocheting.
Cultural dance blowing the trumpet.
Thatch grass harvesting.
Brick molding.
Pottery.
Traditional dancing.
Singing for the tourists.

Table 11.4 Current and suggested value addition in agricultural produce.

Crops	Value chain/addition
Maize	*Sadza*, exchange for labor, from in June to January, beer brewing
Sorghum	*Sadza*, malt for *maheu* drink, beer brewing, and chicken feed
Groundnuts	Peanut butter, oil for those that have a small hand machine to extract oil, shelled and sold around May
Millet (grown by few farmers)	*Sadza*, malt for *maheu* drink. Feed chickens and sell. 1 gallon = 1 chicken
Cow peas	Consumed as *mutakura* and *rupiza*

as a suggestion for a way to seriously ensure food sufficiency through income generation and food provision (see Table 11.4). Participants especially highlighted value addition for agriculture as they emphasized the activity as their major livelihood source despite evidence of a multiplicity of livelihood sources, a fact that is consistent with previous studies conducted in the area and elsewhere (Mubaya, 2012). Other suggestions were around using sorghum and pearl millet for beer brewing and groundnuts in peanut butter processing. In addition, it was reported that groundnuts and bambara nuts can also be used for barter trade in exchange for groceries (food and soap), as has been the case in some situations.

Challenges limiting access to ecosystem services

The emerging challenges cited by participants in terms of full access to ecosystem services include suboptimal access to natural resources, animals being poached by other people who are not locals and who are therefore benefiting at the expense of locals, alleged lack of transparency on natural resources management by the council authorities, market inconsistencies, human wildlife conflict, low soil fertility, and prohibitive policies and lack of support for resource management. To address these challenges, participants brainstormed and suggested a suite of solutions, some of them indicated in Section 4.2. These solutions include that community-based committees should lobby and advocate for better management of resources together with council on behalf of the whole community, political leaders should in this context play a major role to ensure that the needs and priorities of the communities are considered, training in advocacy and increased community management of natural resources, including enabling policies for tourism should be jointly gazetted together with the participation of communities.

Key informants highlighted solutions that are specific to the policy context as allowing people to benefit through dialog and negotiation rather than the current one-way process which they described as top down, strengthening local management practices to access water, having cultural-based sculptures, and groupings, including cultural festivals and centers and a fishermen society/cooperation/organization that acts as a bridge between the communities and authorities (could also be used as a platform for marketing). However, it is comforting to note that while further and strengthened policies are required to ensure sustainable utilization and access to services, participants highlighted that there is also already in place a few policies and restrictions that currently and to some extent favor conservation and try to address over exploitation of services and resources, among them; preventing poisoning of fish, veldt fires, cutting down of trees, cultivation of prohibited lands, self-allocation of settlements, and the notion that everyone has equal rights to resources. More enforcement and implementation are required to ensure that the restrictions fully achieve their intended goal.

Perceptions regarding historical climate trends

Participants were asked to brainstorm and discuss to come up with trends in extreme events over the last 30 years. The periods and the related impacts were recorded, and these were presented in the form of impact diagrams. Drought according to group discussions resulted in lack of food, and people coped by resorting to consumption of wild fruits (*musiga, masarahwa, mupandakata,* etc.) and eating vegetables without any starch. Among the major impacts from the changes and variability in climate are the resultant low fruit production, decreased water for wildlife and humans, diseases, and food insecurity. Interestingly, community perceptions highlight a number of years that have had droughts and low rainfall in the past 30 years, which match the actual rainfall data (see Table 11.5).

Analysis of actual historical rainfall trends vis a vis *perceptions*

Rainfall data collection and analyses

In this section, we make use of 44-year and 24-year rainfall data from two weather stations (Kariba and Siakobvu) which are nearest to the study area. These data were sourced from the Zimbabwe Meteorological Services Department and Agricultural, Technical, and Extension Services. Kariba is a tourist town that was developed after

Table 11.5 Perceptions on extreme climatic events in the last three decades.

Period	Events	Impacts
Before 1980	1968, very cold, frost	Drying up of trees Low fruit production Fish deaths Human deaths Increased school absenteeism
1980–1990	1980, high rainfall 1982, drought 1985, late start of season 1987, drought	Flooding of rivers, food insecurity, destruction of infrastructure Low production, decreased water for wildlife High production, food secure communities, low production, food insecurity
Before 1990	1982, low rainfall 1983, low rainfall	Food insecurity Diseases increased Early marriages Increased divorces Increased school absenteeism
1991–2000	1992, low rainfall 1997, flooding 1998, low rainfall	Food insecurity Increased human wildlife conflicts Deaths Poverty absenteeism increased
2001–to date	2002, low rainfall 2004, frost 2008, low rainfall 2007, flooding 2011, frost and flooding 2015, low rainfall/high temperature	Food insecurity Reduced access to wild foods Diseases Change in onset of seasons Reduced production Increased poaching Reduced prey for wild animals Increased attacks

the creation in the late 1950s of Lake Kariba, the largest man-made reservoir, by volume, in the world. Siakobvu is the business center for Omay Communal Lands. We used rainfall data analysis to build a scientific understanding of the trends in rainfall in the study area, particularly in relation to natural resource capita; which in the context of this chapter equates to mostly ecosystems/food production systems as reported by the sampled community in Omay. The data conformed to a normal distribution (Anderson-Darling test), and therefore parametric tests were used for the analyses. Rainfall seasons in this context are defined as starting from the last third of the year (Sept–Dec) to the first third part of the following year (Jan–April). Seasonal rainfall anomalies were used to characterize rainfall seasons into wet and dry following Agnew and Chapel (1999) and Agnew (2000). Plots of series of rainfall anomalies were visually analyzed to provide limits that appropriately define different levels of wetness and dryness of the year that reflect actual observations and the procedure established five classes (see Table 11.6).

Table 11.6 Criteria for characterization of hydrological years from normalized anomalies after Agnew and Chapel (1999) and Agnew (2000).

Criteria	Level
Extreme wet year	$X_{an} > 2.0$
Wet year	$2.0 \geq X_{an} > 1.0$
Normal	$1.0 \geq X_{an} \geq -1.0$
Dry year	$-1.0 > X_{an} \geq -2.0$
Extreme dry year	$X_{an} > -2.0$

A time series (regression) analysis was employed to ascertain any trends in the rainfall over the period when rainfall data were recorded. In addition, the coefficient of variation in rainfall was calculated as a proportion (percentage) of the long-term mean. The secondary rainfall data as presented are largely useful as it reflects measurements from more than 90% of the rainfall seasons but is not uniform in terms of the years covered. This is not uncommon for remote places like Omay in Zimbabwe and indeed southern Africa, where systematic and continuous monitoring of weather and climate is often not done (Mubaya et al., 2019).

Rainfall trends

Rainfall patterns were consistent with the expectations of an arid region with a long-term annual mean total of 677 mm ± 214.2 mm for the period 1964 to 2007 for Kariba and 727 mm ± 304 mm for Siakobvu (1991–2015/16 season) and are comparable to Zimbabwe's long-term average of seasonal rainfall totals of 657 mm ± 173 mm for a 100-year period (1901–2001) though Siakobvu's 24-year average was slightly higher. Both sets of data exhibited high variability (31% and 28% respectively) ranging from 46% below to 55% above the long-term average. This is consistent with expectations as the study area lies in Natural Regions IV and V of Zimbabwe, the most arid and hottest of the regions. Despite this the majority of the years (66%) for Kariba fell within the normal range of rainfall based on the long-term average while 22 years (50%) received below average rainfall (Fig. 11.3), and of these six were drought years with 1 year (1995) exhibiting severely dry and 1 year, extremely dry conditions with no extreme dry seasons (Table 11.7). The years after 1980 were inclined toward dryness being below the long-term average when compared with previous years. Wet conditions were experienced in 7 years with four moderately wet years and three severely wet years, but none of the years was extremely wet (Table 11.7). In contrast the rainfall patterns in Siakobvu exhibited a ~2-year cycle of wet and dry years. Out of the 24 years when rainfall data were available, half (50%) of the seasons exhibited rainfall events in the normal range, a third (6 out of 24 years) were dry years; one season (1996–97) was a wet, and yet another (2006–07) was an extremely wet one (Table 11.7). In contrast, dry conditions occurred at an average frequency of 4.5 years for the Kariba data, but

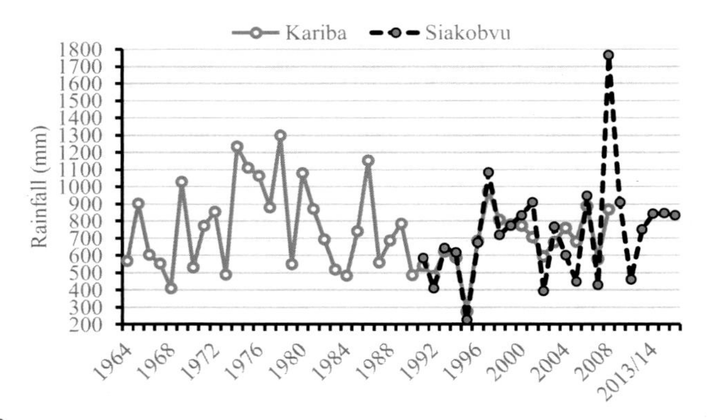

FIGURE 11.3

Rainfall trends recorded at Kariba and Siakobvu around Omay. Source: Zimbabwe Meteorological services and AGRITEX.

Table 11.7 Characterization of hydrological rainfall seasons (SPI 12 months classification) of Kariba and Siakobvu; Bold years represent similar and those in italics show different characterization.

Rainy season characterization	Rainy season (Kariba)	Rainy season (Siakobvu)
Extreme wet	–	*2007–08*
Severely wet	*1974–75, 1978*	–
Moderately wet	*1969, 1976, 1980, 1986*	*1996–97*
Normal	*1964–67, 1970–73, 1977, 1979, 1981–82, 1985, 1987–89*, **1991, 1993–94, 1996-2007**	**Jan-March 1991**, *1992–93*, **1993–94**, **Jan-May 1996, 1997–98, 1998–99, 1999–2000, 2000–01, 2002–03, 2003–04, 2005–06**, *2010–11, 2012–13, 2013–14, 2014–15, 2015-16*
Moderately dry season	*1983-84, 1990,* **1992**	*1991-92, Oct- Dec 1995*, 2001–02, 2004–05, 2006–07, 2011–12
Severely dry	*1968*	–
Extreme dry season	*1995*	–

MSD, *Meteorological Services Department;* AGRITEX, *Agricultural, Technical, and Extension Services*
Source of rainfall data: MSD and AGRITEX

the majority of the drought events were clustered between 1983 and 1995, with droughts (SPI ≤ -1) occurring in the years 1968, 1983–84, 1990, 1992, and 1995. The frequency of occurrence of drought intensified between 1990 and 1995 at an average frequency of 2 years. Extremely dry conditions (SPI ≤ -2) were recorded in 1995.

Since 1996 the Kariba area has experienced near normal rainfall conditions. The lowest annual total rainfall recorded between 1964 and 2007 was 272 mm (received in 1995), and the highest total amount recorded during the same period was 1299 mm (received in 1978). With the exception of 1969 and 1986 the majority of the wet conditions were experienced in the 1970s. For both sets of data the rainy season began in October up to April which were characterized by high temperatures, while the cool, dry season occurred from May to July, whereas August and September were hot and dry. This was expected, as rainfall trends show restrained decreases in annual rainfall (Kruger, 2006; Maure et al., 2018) as well as increased interannual rainfall variability across some parts of southern Africa since the late 1960s with more intense and widespread droughts (Fauchereau et al., 2003). Rainfall anomalies demonstrate high rainfall variability year-in-year-out across the region, as a persistent feature of the region's climate for several decades now. The alternating patterns of above-normal/below-normal rainfall periods show that prevalent rainfall cycles in southern Africa with extremes of wet and dry years are associated with floods and droughts as demonstrated in the study area. The differences between the Kariba and Siakobvu rainfall data are likely as a result of differences in the time periods when rainfall data are available and analyzed; however; the rainfall patterns exhibited during common years showed similar patterns (Fig. 11.3).

Interestingly, community perceptions pointed out a number of drought years and low rainfall as well as their impacts in the past 30 years which match the actual rainfall data (see Table 11.5). The dry years oscillate with national trends of dry seasons, but the two wet years in Siakobvu exhibited were experienced in different years to those around Zimbabwe. The rainy season of 2007–08 was an abnormally wet year for Siakobvu with the highest rainfall levels of more than double the long-term average (1 764 mm) recorded followed by a moderately wet year during the 1996–97 rainy season. The driest season was 1991–92 (225 mm) followed by 2001–02 (395 mm) then 2006–07, 2004–05 and 2011–12, which had comparable rainfall totals of 429, 449 and 460 mm respectively (Fig. 11.3). A linear regression on the seasonal rainfall totals reflected an insignificant increase in rainfall over time in Siakobvu while for Kariba a time series showed a 2-year moving average of −0.63 mm per year illustrating that rainfall declined by 27.1 mm from 1964 to 2007 at a rate of 6.3 mm per decade. What is common for both sets of data is that the rainfall totals (amount) vary widely from one rainy season to another. The rainfall patterns exhibited are expected as the study area Zimbabwe lies in the semiarid belt of southern Africa of which the region is predominantly semiarid with high rainfall variability.

In addition, we also observed much variation in the start/onset and the cessation of the rainy season as over the years, this oscillated between August to December and January to May, respectively. More recently the rainfall season has shifted from when it would begin from as early as August-Sept-Oct to Dec–Jan and from ending much later (March–May) than the previous Jan–March period. Such observations of changes and delay in the onset of the start and finish of the rainy season have been observed in other districts of Zimbabwe (Ndebele-Murisa and

Mubaya, 2015; Mubaya et al., 2016). These observations also appear to be in tandem with perceptions regarding changes in weather parameters by communities in the study area.

Toward a framework for ecosystem-based sustainable food systems under a changing climate in Southern Africa

This chapter takes the premise that attention needs to be focused on adaptation to climate change in the context of ecosystem services by paying attention to communities at the local level and how their food security can be enhanced. In broad terms, adaptation strategies to climate change and threatened food security, especially for the most vulnerable poor in SSA, need to be urgently developed. These strategies need to review land use plans, food security programs, fisheries, and forestry policies to protect the poor from the adverse effects of climate change on food production. Furthermore, there is need for a comprehensive process of identifying climate change resilient livelihood strategies that can be implemented successfully at local levels. Income diversification is the key to attaining household food security under a changing climate, and underpinning this process is the need for strategies that ensure a balance between livelihood sustainability and conservation of food biodiversity that is maintaining the health and integrity of ecosystems. Food insecurity and climate change can be addressed together by diversifying livelihoods as well as transforming food production systems through adopting practices that are "climate-smart". These are practices such as intensive small-scale fish farming that sustainably increasing productivity, resilience (adaptation), reduce GHGs (mitigation), and enhance achievement of national food security and development goals (FAO, 2008).

However, for optimum adaptation as highlighted in the preceding paragraph, we base our argument on the various ecosystem-based approaches that we review in this chapter, and the challenges and opportunities that are described for Omay Communal Lands to concur with recent assertions that flexible and modern approaches that consider local level needs and priorities are key. We submit that no single approach can be followed in identifying problems and designing solutions to these problems in ecosystem services management. Rather, we suggest an eclectic approach that considers that for instance EbA and CbA work together in conjunction to address challenges to ecosystem services and benefits for local communities such as Omay.

Three facts come into play in this context that we are building an understanding for in the chapter. First, it is important to identify homegrown and unique characteristics of ecosystem services, which generally tend to be context specific, rather than relying on other documented contexts for suggestions to policy and strategy development and implementation. In-depth understanding of these characteristics is important especially given the wide ranging and associated benefits that transcend the superficial benefits. This is crucial to building an understanding of the various opportunities that interventions can ride on in order to ensure sustainable and resilient food systems. Second, two or more approaches are important for analysis of the context given that two approaches such as the community-based adaptation (CbA) and ecosystem-based adaptation (EbA), among others, provide flexible, cost-effective, and broadly applicable alternatives for building robust food systems and inclusive alternatives to ecosystem management as a replacement to top-down approaches that tend to thwart community priorities. While EbA is mostly applied in the agricultural and forestry sectors, CbA characterises the design of most disaster risk-reduction and community-based natural resource management initiatives, making a combination of the two effective in addressing challenges in all the ecosystem-based systems. Both approaches already have a history of community-based designs that consider communities' traditional knowledge and experience, merged technical knowledge and capacities, both of which are crucial for responding to climate variability and change. We highlight in the Omay case how participants emphasise the importance of their voice in ecosystem management and highlight that they have responded to variability over a long period of time, which can be important in assisting the design of an adaptation framework suited for the locals. Third, participants views indicate that a framework that considers community led committees and organisations is key in ensuring that both existing and new policies for ecosystem management foster sustainability of services and resources. This is possible if the communities recognise and support their own committees that represent community views and aspirations, otherwise there may be challenges of communities failing to abide by restrictions and laws that govern these resources.

References

Agnew, C.T., 2000. Using the SPI to Identify Drought. Drought Network News (1994-2001).

Agnew, C.T., Chappell, A., 1999. Drought in the sahel. Geojournal 48, 299–311.

Kutter, A., Dwight, L., 2014. Managing rural landscapes in the context of a changing climate. Westby Development in Practice 24 (4), 544–558. https://doi.org/10.1080/09614524.2014.907241.

Ajani, E.N., Mgbenka1, R.N., Okeke, M.N., 2013. Use of Indigenous Knowledge as a Strategy for Climate Change Adaptation among Farmers in sub-Saharan Africa: Implications for Policy Asian. Journal of Agricultural Extension, Economics & Sociology 2 (1), 23–40, 2013; Article no. AJAEES.2013.003.

Atiera, M.A., Koohafkan, P., 2008. Enduring Farms; Climate Change, Smallholders and Traditional Farming Communities. Third Worlds Network, Penang, Malaysia, pp. 1–63.

Barnosky, A.D., Matzke, N., Tomiya, S., Wogan, G.O.U., Swartz, B., Quental, T.B., et al., 2011. Has the Earths sixth mass extinction already arrived? Nature 471, 51–57.

Bryan, E., Ringler, C., Okoba, B., Roncoli, C., Silvestri, S., Herrero, M., 2013. Adapting agriculture to climate change in Kenya: Household strategies and determinants. Journal of Environmental Economics and Management 114, 26–35.

Campos, J.J. (2012) Sustainability in our region. p. 16. In: Sustainability Report 2011. Inter-American Development Bank, Washington, USA. www.iadb.org/en/topics/sustainability/2011-sustainabilityreport,1520.html (accessed 10 May 2013)

Chagutah, T., 2006. Environmental Reporting in the Zimbabwean Press: A Case of the Standard and the Sunday Mail. University of Zimbabwe, Unpublished MA dissertation.
Chikweche, T., Fletcher, R., 2009. Understanding factors that influence purchases in subsistence markets. Journal of Business Research 63, 643–650.
Chevalier, M., 2017. A new paradigm to interpret southern African palaeorecords: the Tropical-Temperate Interactions. https://chevaliermanuel.wixsite.com/webpage/single-post/2017/10/04/Paper-SWP-climate#!.
Connolly-boutin, L., 2015. Climate change, food security, and livelihoods in sub-Saharan Africa. Regional Environmental Change 16 (2), 385–399.
DeLong Jr., D.C., 1996. Defining biodiversity. Wildlife Society Bulletin 24 (4), 738–749.
Dodman, D., Mitlin, D., 2013. Challenges for community-based adaptation: discovering the potential for transformation. Journal of International Development 25 (5), 640–659, 2013.
Downing, T.E., 1992. Climate Change and Vulnerable Places: Global Food Security and Country Studies in Zimbabwe, Kenya, Senegal and Chile. Research Report No. 1. Environmental Change Unit, University of Oxford, Oxford, 1992.
FAO, 2012. Climate-smart agriculture sourcebook. Food and Agriculture Organization, Rome.
Fauchereau, N., Trzaska, S., Rouault, M., Richard, Y., 2003. Rainfall variability and changes in Southern Africa during the 20th Century in the global warming context. Natural Hazards 29, 139–154. https://doi.org/10.1023/A:1023630924100.
FAO, 2018. Ecosystem Services and Biodiversity. http://www.fao.org/ecosystem-services-biodiversity/en/.
Food and Agricultural Organisation (FAO), 2008. Climate Change Adaptation and Mitigation in the Food and Agriculture Sector. High Level Conference on World Food Security - Background Paper HLC/08/BAK/1. ftp://ftp.fao.org/docrep/fao/meeting/013/ai782e.pdf.
Gaston, K . 1. & Spicer, 1. I. (1998). Biodiversity, an introduction. Blackwell, Oxford.
Gilman, E., Gearhart, J., Price, B., Eckert, S., others, 2010. Mitigating sea turtle by-catch in coastal passive net fisheries. Fish Fish 11, 57–88.
Godfray, C.J.R., 2010. Beddington and I R Crute 2010 Food Security: The Challenge of Feeding 9 Billion People Science, 812, p. 327. https://doi.org/10.1126/science.1185383.
Hulme, M., 1996. Climate Change and Southern Africa, an Exploration of Some Potential Impacts and Implications in the SADC Region. WWF International, Avenue du Mont Blanc, Gland, Switzerland.
IUCN 2018 Protected planet report 2018 : tracking progress towards global targets for protected areas Cambridge : UNEP-WCMC; Gland : IUCN; Washington, D.C. : NGS, 2018 ISBN: 978-92-807-3721-9
Kruger, A.C., 2006. Observed trends in daily precipitation indices in South Africa: 1910–2004. International Journal of Climatology 26 (15), 2275–2285.
IIED, July 2016. Ecosystem-based Adaptation: A Win–Win Formula for Sustainability in a Warming World? Briefing. http://pubs.iied.org/pdfs/17364IIED.pdf.
Juana, J.S., Kahaka, Z., Okurut, F.N., 2013. Farmers' perceptions and adaptations to climate change in sub-Sahara Africa: A synthesis of empirical studies and implications for public policy in African agriculture. Journal of Agricultural Science 5 (4), 121–121.
Kotir, J.H., 2011. Climate change and variability in sub-Saharan Africa: a review of current and future trends and impacts on agriculture and food security. Environ Dev Sustain 13 (3), 587–605. https://doi.org/10.1007/s10668-010-9278-0.

Koh, L.P., Sodhi, N.S., Brook, B.W., 2004. Species Coextinctions and the Biodiversity Crisis Biotropica 36, 272 (2004).

Lafferty, K.D., 2009. The ecology of climate change and infectious diseases. Ecology 90, 888–900.

Leadley, P., Pereira, H.M., Alkemade, R., Fernandez-Manjarres, J.F., Proenca, V., Scharlemann, J.P.W. et al. 2010. Biodiversity scenarios: projections of 21st century change in biodiversity and associated ecosystem services. In: Secretariat of the Convention on Biological Diversity (ed. Diversity SotCoB). Published by the Secretariat of the Convention on Biological Diversity, Montreal, p. 1–132. Technical Series no. 50.

Magadza, C.H.D., 1994. Climate change, some multiple impacts in southern Africa. In: Downing, T.E. (Ed.), Climate Change and Food Security, NATO ASI Series. Series 1, Global Environmental Change, vol. 37, p. 662.

Magadza, C.H.D., 2010. Indicators of above normal rates of climate change in the middle Zambezi Valley, Southern Africa. Lakes and Reservoirs: Research and Management 15, 167–192.

Matarira, C.H., Makadho, J.M., Mwamuka, F.C., 1995. Zimbabwe: Climate Change Impacts on Maize Production and Adaptive Measures for the Agricultural Sector. Interim Report on Climate Change Country Studies.

Mark, W.R., Mandy, E., Gary, Y., Lan, B., Saleemul, H., Rowena, V.S., 2008. Climate change and agriculture: Threats and opportunities. Federal Ministry for Economic Cooperation and Development, Germany.

Maure, G.A., Pinto, I., Ndebele-Murisa, M.R., Muthige, M., Lennard, C., Nikulin, G., Dosio, A., Meque, A.O., 2018. The Southern African climate under 1.5° and 2°C of global warming as simulated by CORDEX models. Environmental Research Letters 13 (6). https://doi:10.1088/1748-9326/aab190.

McNeely, J.A., Scherr, S., 2001. Common Ground, Common Future: How Ecoagriculture Can Help Feed the World and Save Wild Biodiversity. IUCN, Gland. Report No. 5/01.

Millennium Ecosystem Assessment (MA), 2005. Ecosystems and Human Well-Being: Biodiversity Synthesis. World Resources Institute, Washington, DC. http://www.millenniumassessment.org/documents/document.354.aspx.pdf. Viewed 26 April 2013.

Moyo and Chambati 2013. Land and Agrarian Reform in Zimbabwe : Beyond White-Settler Capitalism. Edited by Sam Moyo & Walter Chambati. Dakar, CODESRIA & AIAS, 2013, 372 p., ISBN 978-2-86978-553-3

Mubaya, C.P., 2006. Understanding Policy and Legislative Policies for Natural Resource Management: An Assessment of Governance, Uses and Livelihood Options Surrounding Selected High Value Resources by Small Scale Communities in Africa (The Case of Omay Communal Lands, Zimbabwe) CIRAD-CASS Report.

Mubaya, C.P., Njuki, J., Mutsvangwa, E.P., Mugabe, FT., Nanja, D, 2012. Climate variability and change or multiple stressors? Farmer perceptions regarding threats to livelihoods in Zimbabwe and Zambia. Journal of Environmental Economics and Management 102, 9–17.

Mubaya, C.P., Ndebele-Murisa, M.R., 2017. Beyond agriculture: a landscape approach towards adaptation under a changing climate in Omay Communal Lands, Zimbabwe. In: Zinyengere, N., Theodory, T.F., Gebreyes, M., Ifejika-Speranza, F. (Eds.), Beyond Agricultural Impacts: Multiple Perceptions on Adaptation to Climate Change and Agriculture in Africa. Academic Press, Elsevier, UK Pgs, pp. 101–123.

Mubaya, C.P., Mutopo, P., Murisa-Ndebele, M.R., Ngepah, N., 2016. Community Experiences. Weather and Climate Information Dissemination in Farming Systems in Zimbabwe. Chinhoyi University of Technology Printing Press, Chinhoyi, Zimbabwe.

Muchena, P., 1991. Implications of Climate Change for Maize Yields in Zimbabwe. Plant Protection Research Institute. Department of Research and Specialist Services, Harare. Zimbabwe, pp. 1–9.

Mugandani, R., Makarau, Wuta M., Chipindu, B., 2012. Reclassification of Agro-ecological regions of Zimbabwe in conformity with climate variability and change African. Crop Science Journal 20 (s2), 361–369. ISSN 1021-9730/2012.

Munyaradzi Chenje. State of the Environment in the Zambezi Basin 2000. Maseru, Lusaka and Harare: SADC, IUCN, ZRA, and SARDC, 2000. xxvi + 334 pp. $39.95 (paper), ISBN 978-1-77910-009-2.

Munir, T.M., Perkins, M., Kaing, E., Strack, M., 2015. Carbon dioxide flux and net primary production of a boreal treed bog: Responses to warming and water-table-lowering simulations of climate change Biogeosciences 12, 1091–1111, 2015.

Murisa, T., 2013. Prospects for smallholder agriculture in southern Africa. In: Hendricks, F., Ntsebeza, L., Helliker, K. (Eds.), Promised Land. Undoing a Century of Dispossession in South Africa of Land. Jacana Media, South Africa, pp. 185–213.

Murombedzi, J., 1992. Decentralization or recentralisation?. In: Implementing CAMPFIRE in the Omay Communal Lands of Nyami Nyami District. CASS Occasional Paper. University of Zimbabwe, Harare.

Ndebele-Murisa, M.R., Mubaya, C.P., 2015. The climate change scare: livelihoods and adaptation options for smallholder farmers in Zimbabwe. In: Murisa, T., Chikweche, T. (Eds.), Beyond the Crises: Zimbabwe's Transformation and Prospects for Development. TrustAfrica and Weaver Press, Harare, Zimbabwe, pp. 154–197. ISBN: 978-1-77922.

Ndebele-Murisa, M.R., Mubaya, C.P., 2019. Climate change: a threat to community-based interventions on socio-ecological systems in southern Africa. In: Svyatets, E., Chatterjee, M. (Eds.), Environment Issues: Science, Policy, and Diplomacy. Cognella Publishing, San Diego, California, USA in press.

Nicholls, C.I., Rios, L., Altieri, M.A., 2013. Agroecologia y resiliencia socioecologica: adaptandose al cambio climatico. Red IberoAmericana de Agroecologia para el Desarrollo de Sistemas Agricolas Resilientes al Cambio Climatico (REDAGRES). Medellin, p 207.

Nyamadzawo, G., Nyamugafata, P., Wuta, M., Nyamangara, J., Chikowo, R., 2012. Inltration and runo- losses under fallowing and conservation agriculture practices on contrasting soils, Zimbabwe. Water SA 38 (2), 233–240.

Pearce, D., Moran, D., 1994. Earthscan, London (Review). International Environmental Affairs 8 (2), 178–181.

Persha, L., Agrawal, A., Chhartre, A., 2011. Social and ecological synergy: Local rulemaking, forest livelihoods, and biodiversity conservation. Science 331 (6024), 1606–1608.

Portes, P.R., 2010. Food Prices and Rural Poverty. World Bank Group. http://siteresources.worldbank.org/INTRANETTRADE/Resources/Pubs/Food_Prices_Rural_Poverty.pdf.

Quinn, J.E., Brandle, J.R., Johnson, R.J., 2013. A farm-scale biodiversity and ecosystem services assessment tool: The healthy farm index. Papers in Natural Resources 535. http://digitalcommons.unl.edu/natrespapers/535.

(Pdf), 2015 (Pdf) Sustainable use of game population in a zimbabwean communal area : production of cheap edible meat for local communities. Available from: https://www.researchgate.net/publication/273771835_sustainable_use_of_game_population_in_a_zimbabwean_

communal_area_production_of_cheap_edible_meat_for_local_communities [accessed Sep 11 2019].
Phiri, G.K., Egenu, A., 2016. Climate change and agriculture nexus in sub Saharan Africa. The agonizing reality for smallholder farmers. Health Care Science Journal Impact Factor.
Rafferty, N., Ives, R., 2010. Effects of experimental shifts in flowering phenology on plant—pollinator interactions Ecology Letters (2011) 14, 69—74.
Reid, H., et al., 2009. Community-based Adaptation to Climate Change: An Overview, vol. 60. Participatory Learning and Action, p. 13.
Reed, J., Van Vianen, J., Foli, S., Clendenning, J., Yang, K., Macdonald, M., Petrokosfsky, G., Padoch, C., Sunderland, T., 2017. Trees for life: The ecosystem service contribution of trees to food production and livelihoods in the tropics. Forest Policy and Economics n. pag. Web.
Sala, O.E., van Vuuren, D., Pereira, H.M., Lodge, D., Alder, J., Cumming, G. et al. 2005. Chap 10: biodiversity across Scenarios. In: Millenium Ecosystem Assesment (ed. Island press, NY). Millenium Ecosystem Assesment, Volume 2: Scenarios assessment, pp. 375—408.
Vermeulen, S.J., Campbell, B.M., Ingram, J.S.I., 2012. Climate change and food systems, 37, pp. 195—222 (Volume publication date November 2012) First published online as a Review in Advance on. (Accessed 30 July 2012).
SPC, 2010. Importance of agricultural biodiversity in the Pacific. https://lrd.spc.int/our-work/forest-and-agriculture-diversification/15/importance-of-agricultural-biodiversity-in-the-pacific.
Sunderland, T.C.H., 2011. Food security: why is biodiversity important? International Forestry Review 13 (3) online.
Taylor, R.D., 1990. Socioeconomic aspects of meat production from impala harvested in a Zimbabwean communal land. 2nd International Game Ranching Symposium. Edmonton, Canada, 182—193. (15).
Tuomisto, H., 2010. A Consistent Terminology for Quantifying Species Diversity Yes, It Does Exist. Oecologia 164, 853—860.
TEEB, 2009. The Economics of Ecosystems and Biodiversity for National and International Policy Makers. Summary: Responding to the Value of Nature. http://www.teebweb.org/publication/teeb-for-policy-makers-summary-responding-to-the-value-of-nature/.
Toby Kiers, E., Palmer, T.M., Ives, A.R., Bruno, J.F., Bronstein, J.L., 2010. Mutualisms in a changing world: an evolutionary perspective Ecology Letters. https://doi.org/10.1111/j.1461-0248.2010.01538.x.
Twomlow, S., Shiferaw, B., Cooper, P., Keatinge, J.D.H., 2008. Integrating Genetics and natural resource management for technology targeting and greater impact of agricultural research in the semi-arid tropics. Experimental Agriculture 44, 235—256.
UK, N.E.A., 2011. UK National Ecosystem Assessment: Technical Report [United Nations Environmental Programme— World Conservation Monitoring Centre (UNEP-WCMC)]. Cambridge.
United Nations (UN), 1992. Convention on Biological Diversity convention_on_biological_-diversity_-_1992.pdf.
Walther, G.R., 2010. Community and ecosystem responses to recent climate change. Philosophical Transactions of The Royal Society B Biological Sciences 365 (1549), 2019—2024. https://doi.org/10.1098/rstb.2010.0021.
Yang, L.H., Rudolf, V.H.W., 2010. Phenology, ontogeny and the effects of climate change on the timing of species interactions. Ecology Letters 13, 1—10.

Further reading

Altieri, A., 1999. The ecological role of biodiversity in agroecosystems. Agriculture, Ecosystems & Environment 74, 19–31.

CBD, 2009. Biodiversity, Development and Poverty Alleviation: Recognizing the Role of Biodiversity for Human Well-Being. Secretariat of the Convention on Biological Diversity, Montreal.

Davis, C.L., 2011. Climate Risk and Vulnerability: A Handbook for Southern Africa. Council for Scientific and Industrial Research, Pretoria, South Africa, p. 92.

Hempel, E., 2006. Regional trends in fisheries and aquaculture. In: ICEIDA/UNU - FTP Workshop on Fisheries and Aquaculture in Southern Africa Development and Management. 21–24 August 2006, Windhoek, Namibia.

MA (Millennium Ecosystem Assessment), 2005. Ecosystems and Human Well-Being: Synthesis. Island Press, Washington D.C.

McNeill, C., Shei, P., 2002. A Framework for Action on Biodiversity and Ecosystem Management. WEHAB Working Group, WSSD, Johannesburg.

Musona, M., 2011. An Exploration of the Causes of Social Unrest in Omay Communal Lands of Nyaminyami District of Zimbabwe: A Human Needs Perspective. Department of Academic Administration, Examination section, Summerstrand Nelson Mandela Metropolitan University, Port Elizabeth, South Africa.

UNDP Human Development Report (HDR), 1999. Development with a Human Face. Published for the United Nations. Development Programme. (UNDP). Oxford University Press, New York; Oxford.

CHAPTER

Accounting for the invisible value of trees on farms through valuation of ecosystem services

12

Brian Chiputwa, BSc, MSc, PhD [1], Hanna J. Ihli[2], Priscilla Wainaina[3], Anja Gassner[4]

[1]Livelihoods and Gender Specialist, Research Methods Group, World Agroforestry (ICRAF), Nairobi, Kenya; [2]Decision Analyst, Systems Theme, World Agroforestry (ICRAF), Nairobi, Kenya; [3]Post-doctoral Fellow, Governance Theme, World Agroforestry (ICRAF), Nairobi, Kenya; [4]Senior Livelihood Specialist & Head of Research Methods, World Agroforestry (ICRAF), Philippines

Introduction

The demand for food around the world is increasing rapidly. This increased demand is driven by complex dynamics including a steeply rising global population, rapid globalization and urbanization, as well as steadily improving living standards. These factors all have profound impacts on consumer preferences and human diets, and ultimately on the global food system (Charles et al., 2014; Chiputwa et al., 2015). Based on current trends, Tilman et al. (2011) estimates that, overall, 1 billion hectares of land will be cleared globally for agricultural production by the year 2050. This will significantly raise greenhouse gas emissions and chemical fertilizer use, further enhancing agriculture's negative impact on the environment. For food systems to become more sustainable, agricultural production must be able to meet the global demand for sufficient, safe, and nutritious food, while exerting minimum damage to essential ecosystem services (ES) (Allen and Prosperi, 2016).

Agroforestry, defined as the deliberate integration of trees and shrubs with crops and livestock on the same land management unit, is widespread across the humid tropics and is increasingly recognized for its contributions to sustainability in temperate areas as well (Moreno et al., 2018; Chazdon et al., 2018). However, across the tropics, such integrated systems have been challenged by the expansion of intensified land use systems that offer simplification in terms of management, labor, and risk.

Agroforestry is a land use system that can contribute to sustainable food systems by proposing significant economic, environmental, and social contributions and benefits. At the farm level, agroforestry systems contribute directly to household well-being through improved food security, nutrition, and income generation, by providing edible products such as fruits and other tree products that can be used for domestic purposes, or by selling tree products including fuelwood, minor timber,

The Role of Ecosystem Services in Sustainable Food Systems. https://doi.org/10.1016/B978-0-12-816436-5.00012-3

medicinal plants, and fodder in the markets (Mercer, 2004; Noordwijk et al., 2016). Agroforestry systems also provide important ES that have supporting and regulating functions through the provision of carbon storage, increased biodiversity, reduced erosion, and direct support for better performance of seasonal crops (Vira et al., 2015). Beyond the individual plot and farm scales, agroforestry systems can play a significant role in the provision of environmental services and support food production at landscape scale as well (Duguma, 2013; Smith and Mbow, 2014). Agroforestry systems have a high carbon sequestration potential and therefore contribute directly to both global climate mitigation and food security compared to other land use systems that do not include trees (Mbow et al., 2014). Despite agroforestry being recognized as a land use system that contributes to food and nutrition security, investments into the practice still remain insufficient. Part of this underinvestment can be related to the general assumption that agroforestry as a "traditional system" has been practiced by farmers around the world anyway, and that local knowledge must hence be sufficient. However, intensified agroforestry systems that optimize the synergies between the different components by carefully selecting species as well as spatial and temporal arrangements need substantial investments. Technical assistance in the form of well-trained extension services, especially when new systems and species are introduced, is often missing (Meijer et al., 2015). Farmers need readily available information on the right species combinations and training on appropriate management practices to ensure that synergies between the agroforestry components are optimized and competition minimized. High-quality germplasm for the tree components are often not available, especially for indigenous species (Cornelius and Miccolis, 2018), leading both to the selection of readily available exotic species, as well as lower productivity and lower farm gate prices (Do and Mulia, 2018). To ensure that farmers have access to the right species with the right quality, substantial investments in seed collection and nurseries are needed. The markets for most tree products are not well developed, particularly the value chains for wood and timber products from single trees grown on private lands. Underdeveloped markets contribute to farmers depending on middlemen who dictate economically unattractive farm gate prices.

While the potential contribution of agroforestry has received growing attention, there is lack of evidence-based research that demonstrate the true value of trees. The most recognized contribution of agroforestry systems are mostly from amenities that provide direct benefits such as income from marketable tree products and household food security, nutrition and health such as edible fruits and medicinal tree products (Vira et al., 2015). However, there is little recognition of the less direct and non-marketed ES that contribute to conservation of soil and water, nutrient recycling, greater biodiversity, erosion control, improved water quality and carbon sequestration. Providing decision-makers with improved value estimates that account for the full value of trees on farms is an important step in ensuring better and more targeted policies and incentives that promote biodiversity conservation for more sustainable food systems. In addition, this provides guidance for donors and private companies that want to consider investing in programs and interventions that contribute to both livelihoods and the environment (Godfray et al., 2010). The aim of the chapter is to provide an overview of the state-of-the-art methodologies to quantify and value both marketed and nonmarketed goods and ES from agroforestry. We hope that researchers will find this chapter useful in guiding them on the appropriate methods to use for measuring, quantifying, and valuing the contribution that agroforestry systems make towards sustainable food production. More empirical studies on quantifying the value of ES will provide the evidence base to encourage decision-makers, governments, donors, and multinational companies to invest in agroforestry systems as effective strategies to support food security. The chapter is divided into six sections.

Measuring, quantifying, and valuing ecosystem services

ES are defined as the benefits that humans realize from their interaction with the natural environment. They can be classified into four broad categories: (1) provisioning services such as production of food and water; (2) supporting services such as nutrient cycles, primary production of oxygen, and soil formation; (3) regulating services such as the control of floods and diseases; and (4) cultural services such as spiritual and recreational benefits (MEA, 2005). With the growing global demand for food in recent years, there is increased attention on the relationship between intensive agricultural production, which is commonly invoked to meet this demand, and potential further loss in ES. The need to quantify and assign market and nonmarket values to ES has become an important step when taking all externalities from farm gate to consumption into account. ES are also considered in terms of the impacts that, strictly speaking, are not externalities but constitute effects of social concern such as climate change. The consideration of ES can help decision-makers to take corrective measures for such environmental problems (Atela et al., 2017; TEEB, 2018). For example, the carbon sequestration potential of different agroforestry systems can be expressed in financial terms, based on a measurement and valuation of carbon stocks both above and below the soil. Similarly, by valuing the ecological benefits of ecosystems services from agroforestry, such as nutrient cycling, water conservation, improved soil nutrients, increased species diversity, and so on, we able to assess, to a high degree, the profitability of agroforestry systems. Most ecosystems services from agroforestry may not be invisible to society since there are no functional markets that exist for these services.

Market valuation of ES illuminate the economic and social contribution of agroforestry systems and provide important information that can be used to inform policy. In addition, if ecosystem values in agroforestry are better understood and integrated into formal decision-making processes, agroforestry might become a more economically attractive option for farmers, landowners, and governments (Namirembe et al., 2015).

Similarly, sustainable harvesting of tree products is often promoted among communities living within forested areas by demonstrating and making the values of tree species products more apparent by quantifying their economic, social and ecological importance (Atela et al., 2017). However, such valuation needs to take into account that quantified ES and associated values have varying appeal and importance to different people, institutions, and ecologies—and that the ES valuation approach needs to be tailored to fit the context.

With use of proper ES valuation techniques, it is also possible to apportion the contribution that a particular actor (e.g., farmers, private companies) or sector (e.g., mining, industry, transport) makes towards greenhouse gas emissions (Atela et al., 2017). Quantification and valuation also inform the rights to use or invest in natural capital among various stakeholders, including farmers, governments, and others. For instance, if the economic, environmental, and social contribution of ES from agroforestry systems are clearly measured and articulated, it becomes justifiable for policy makers in the government to put in place policies that are biased toward protecting biodiversity.

The importance of agroforestry systems

As previously mentioned, agroforestry systems provide both direct and indirect benefits for farmers in terms of food security with effects at different scales from plot to farm to landscape and globally. Agroforestry is important in rural livelihoods as it provides a range of ES with additional benefits such as keeping farmers more food secure through more diversified food and cash crop outputs (fruit tree products, other nontimber forest products, food crops) and improving their resilience to environmental or socioeconomic shocks through on-farm livelihood diversification and enhancement of regulating ES for yield stability (Namirembe et al., 2015; Vira et al., 2015). Agroforestry systems have been shown to result in higher yields within maize-based systems. Fertilizer trees such as *Gliricidia sepium* and *Faidherbia albida*, for instance, resulted in higher maize yields in Malawi (see Coulibaly et al., 2017). However, other studies reported lower crop yield within agroforestry systems compared to unshaded systems, especially within coffee and cocoa agroforestry systems. For example, a study in Ethiopia by Mitiku et al. (2018) showed that coffee yields from the agroforestry systems were lower than the yields from unshaded coffee, which was mainly attributable to the low use of agrochemicals and fertilizers within the agroforestry systems, competition for nutrients, as well as low density of coffee trees within the agroforestry system. On the other hand, other direct benefits attributable to agroforestry systems include timber, wood fuel, honey, charcoal, wild fruits, medicine, among others, in Africa and globally (e.g., Namirembe et al., 2015; Gockowski et al., 2013; Reichhuber and Requate, 2012). Beyond that, many indirect benefits are attributable to agroforestry, including carbon sequestration, biological pest and disease control, pollination, higher species diversity, higher vegetative diversity, soil fertility and nutrient cycling, water treatment and regulation, and so on (Namirembe et al., 2015; Temesgen et al., 2018).

Several studies were conducted that aimed at valuing ecosystems within agroforestry systems in Africa. Some studies focused on the valuation of specific ES drawn from agroforestry; for example, Bedimo et al. (2008) looked at the valuation of biological disease control (coffee berry disease) within coffee agroforestry systems in Cameroon. However, most studies focused on valuing different ES within agroforestry systems, while others went further to conducting cost–benefit analysis and computing net present values (NPVs) for the various agroforestry systems (e.g., Namirembe et al., 2015; Duguma, 2013; Gockowski et al., 2013; Reichhuber and Requate, 2012).

Total economic values of agroforestry systems

In its simplest form, the total economic value (TEV) of ES distinguishes between use and nonuse values. Use values can be associated with private or quasi-private goods, for which market prices usually exist, while nonuse values from ecosystems are those values that do not involve the direct or indirect use of the ES in question (Pascaul et al., 2010). Fig. 12.1 presents the classification of the use and nonuse values of ES. The use values are further categorized into:

a) Direct use values: These relate to the benefits obtained from the direct use of an ecosystem. For agroforestry systems, these include direct products which have market values such as timber, poles, charcoal, gum arabica, medicine, as well as other nontimber forest products (NTFPs) such as wild fruits, honey, fodder, crop harvests, recreation value, and so on.

b) Indirect use values: These are usually associated with regulating services. In the context of agroforestry systems, these include services such as carbon sequestration, water treatment and regulation, soil erosion control, pollination, and so on.

c) Option values: These include valuing ES for the option of future use such as medicinal purposes. An example is the *Prunus africana* tree specie which can be integrated in shaded coffee systems and its leaves, roots, and stem have wide applications in African traditional medicines.

Nonuse values, on the hand, are categorized into:

a) Bequest value: This captures the value arising from the satisfaction of knowing that future generations will access nature's benefits. This value is concerned with intergenerational equity.

b) Altruist value: This value concerns intragenerational equity, i.e., the satisfaction of knowing that other people can also access nature's benefits.

c) Existence value: This value is derived from the satisfaction of knowing a certain species exists. For example, indigenous trees, endangered species, and medicinal plants.

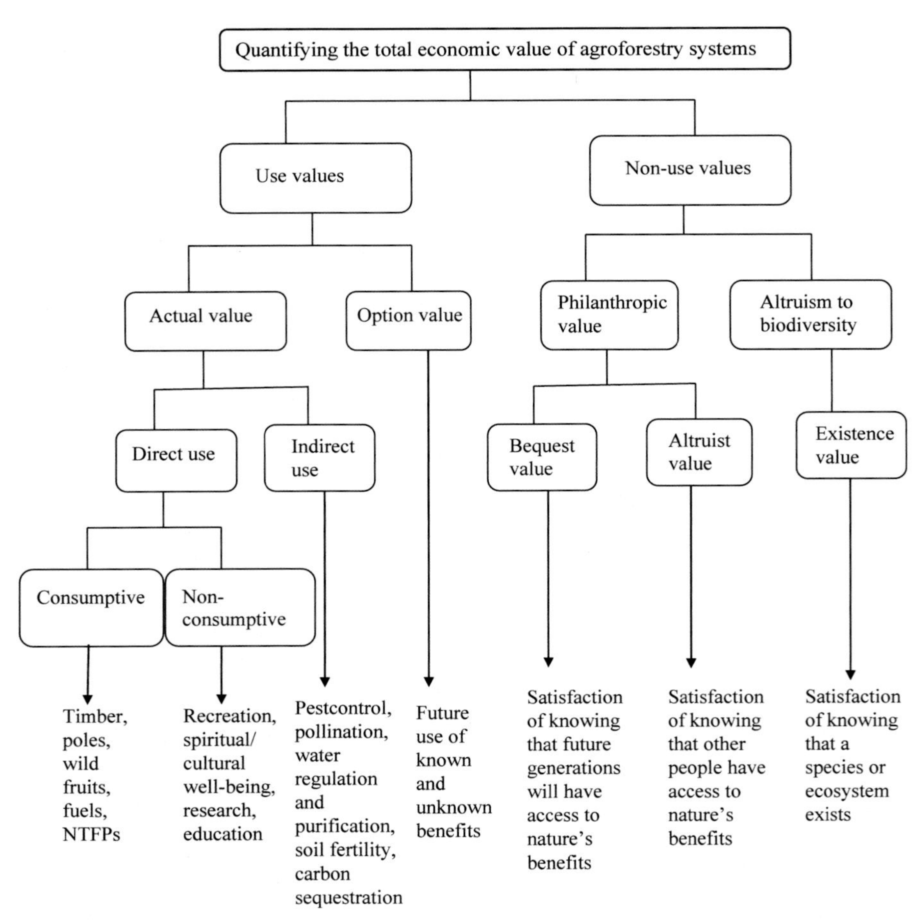

FIGURE 12.1

Classification of economic values of agroforestry systems.

Adapted from Pascual, U., Muradian, R., Brander, L., Gómez-baggethun, E., Martín-lópez, B., Verma, M., et al. (2010). Chapter 5 the Economics of Valuing Ecosystem Services and Biodiversity.

Table 12.1 presents the link between use and nonuse values, and the appropriate technique that can be used for valuation of marketed and nonmarketed goods and services from agroforestry systems. To value direct use products such as wood products and nonwood products, market analysis is most suitable since there is already a functioning market for these products. Some of the nonconsumptive, direct market values, including recreational or cultural values, may be valued with the help of other approaches such as the hedonic pricing or the contingent valuation method (CVM). For indirect use values, such as nutrient cycling, air pollution, microclimate regulation, approaches like replacement/restoration cost, avoidance cost, or

Table 12.1 Examples of links between value category, functions, and valuation tools for agroforestry systems.

USE	USE values			NONUSE values	
	Direct use value	**Indirect use value**	**Option value**	**Bequest value**	**Existence value**
Functions	Wood products (timber, fuel) Nonwood products (food, medicine, genetic materials) Human habitat Educational, recreational and cultural uses.	Nutrient cycling Air pollution reduction Carbon storage Microclimate regulation	Possible future uses of goods and services (use values) by actual stakeholders	Possible future uses of goods and services (use values) by future generations.	Biodiversity Culture and heritage Benefits to stakeholders of only knowing the existence of goods and services without using them.
Valuation tool proposed					
Valuation tools	Market analysis Travel cost method Contingent valuation method (CVM) Hedonic pricing	Production function approach Replacement/restoration cost Avoidance cost	CVM	CVM	CVM

Source: Lette and Boo (2012).

production function can be used. For the nonuse values, stated preference methods including the CVM can work, since the respondents can state their willingness to pay (WTP) for those services based on a hypothetical market.

Economic valuation methods using the TEV approach

Within the TEV framework, monetary values, if available, are derived from information about individual behavior provided by market transactions that are directly related to a given ES. In the absence of such information, price information must be obtained through non-market techniques associated indirectly with the good to be valued. If both direct and indirect price information on ES are absent, hypothetical markets may be created in order to elicit values. These situations correspond to a common categorization of the available techniques used to value ES:

(a) Direct market valuation approaches
(b) Cost-based approaches
(c) Revealed preference approaches
(d) Stated preferences approaches
(e) Benefit transfer approaches
(f) Valuation under risk and uncertainty

Tables 12.2–12.4 provide lists of empirical studies that use market, nonmarket, and experimental techniques in valuing agroforestry systems, respectively.

Direct market valuation approaches

For those ES that involve market transactions and can be quantified and valued in monetary terms, while others do not involve such market transactions and need different methods of valuation. Where data from actual markets are available, direct market valuation approaches are preferred. These are typically applied when valuing provisioning services but are also frequently used to value cultural services.

Market methods

In cases where the ES is traded in through markets, then the economic valuation can be directly obtained from what consumers are willing to pay for the associated good or commodity. Provisioning services such as food, fuel, timber, wood and honey associated with tree species can easily be valued based on their going prices in the market. For example, in valuing the provisioning services that emanate from agroforestry systems of coffee farms in Ethiopia and cocoa farms in Ghana, Namirembe et al. (2015) used the prevailing prices of all associated commodities in the market. These included timber, fruit tree products, cocoa or coffee, and food crops. The value for all the provisioning services from agroforestry systems for coffee in Ethiopia was estimated at $1100 per ha for semi-forest and $2450 per ha for garden coffee.Agroforestry provisioning services in cocoa sytems in Ghana, over a 20 year

Table 12.2 Summary of market valuation methods.

Valuation tools/ methods applied	Full reference of the study	Specific ES valued	Type of AF systems	Challenges encountered in the valuation process	Excerpt from the publication on the valuation process	Comment
Market methods	Namirembe et al. (2015)	Supporting services: carbon stocks in agroforestry systems	Cocoa agroforestry systems in Ghana; Coffee agroforestry systems in Ethiopia	Cannot be used to value those Ecosystem services that have no market value	For carbon valuation, the existing market price may not adequately provide the true value of carbon. Hence it was important to incorporate the social price of carbon since it captured all the externalities associated with carbon.	Market methods provide reliable estimates when the market price of a commodity exists.
Production methods	Bedimo et al. (2008), Classen et al. (2014)	Regulating services: biological disease control Regulating services: Biological pest control and pollination services	Coffee agroforestry systems in Cameroon Coffee agroforestry systems in Tanzania	When sufficient data on key inputs are missing, it might yield biased estimates.	Yield loss from coffee berry diseases/pests was lower in agroforestry systems. Thus, the value of avoided loss was interpreted as the value of biological disease/ pest control associated with agroforestry systems.	This method requires good understanding of production technology and collection of sufficient data to allow for comparisons to be made. Longitudinal data collected over several periods may be needed.

Table 12.3 Summary of nonmarket valuation methods.

Valuation tools/ methods applied	Full reference of the study	Specific ES valued	Type of AF systems	Challenges encountered in the valuation process	Excerpt from the publication on the valuation process	Comment
Choice experiments	Mercer and Snook (2004)	Hypothetical	Hypothetical agroforestry systems	Systematic and thorough qualitative investigation of the attributes that are relevant in the decision context, and of the attribute levels and appropriate framing, is crucial in the process of designing a choice experiment	Analysis of how farmers value different attributes of agroforestry systems and which combinations of attributes (technical assistance, labor input, products produced, source of tree seedlings, and the impact on future forest environments) are most likely to be adopted	Choice experiments provide information to assist in the design of new agroforestry systems and projects that would be more attractive to farmers
Benefit transfer	Temesgen et al. (2018)	Supporting services: nutrient cycling, soil erosion control, water regulation and treatment, pollination, soil formation Cultural services: aesthetic utility	Agroforestry systems in Ethiopia	Availability, reliability, and distribution of data on the various ecosystem services across the various ecosystems are a challenge.	Applicability of the method depends exclusively on whether two sites are tolerably similar, a requirement that often challenges the applicability of the method and is seriously debated in the literature on ecological economics	Benefit transfer translates the monetary value determined for one place and time to make inferences about the economic value of ES in another place and/or at time. It is useful when budget and time are a constraint

Table 12.4 Summary of valuation methods that consider risk and uncertainty.

Valuation tools/ methods applied	Full reference of the study	Specific ES valued	Type of AF systems	Challenges encountered in the valuation process	Excerpt from the publication on the valuation process	Comment
Mean-variance analysis	Ramirez and Sosa (2000)	Supporting: Shade	Shaded coffee	- Lack of time-series data on shaded coffee system yields imposes limitations on the financial risk analysis - Reliability of risk assessment depends on how closely the estimated profit cumulative distribution function, which is obtained from the simulated price and yield probability distributions functions, resembles the true underlying cumulative distribution function for profit	Financial risk and return analysis of diversified coffee production systems in Costa Rica. All systems for shaded coffee production are profitable, on average, when both fixed and variable costs are considered.	Analysis of agricultural systems' risk requires estimates of the probability distribution functions that reflect the basic statistical behavior of the key risk-producing variables

Continued

Table 12.4 Summary of valuation methods that consider risk and uncertainty.—*cont'd*

Valuation tools/methods applied	Full reference of the study	Specific ES valued	Type of AF systems	Challenges encountered in the valuation process	Excerpt from the publication on the valuation process	Comment
Stochastic dominance (SD)	Castro et al. (2013)	Supporting: agrobiodiversity	Shaded coffee and maize	- Assumptions underlying SD might not be reasonable for farmers, who are known to be risk-averse - Consideration of mutually exclusive land use options lead to excessive compensations, which appear to be not realistic	Analysis of implications of uncertainty on conservation payments (CPs) to preserve shade coffee production in Ecuador. CPs derived by SD were at least twice the amount calculated by mean-variance analysis.	Mean-variance analysis provides more cost-effective CPs than SD
Real options	Frey et al. (2013)	Supporting: Carbon sequestration	Hardwood forestry systems and alley cropping agroforestry systems	- Model uses historical data and may not accurately reflect future risk scenarios - Model assumes fixed agricultural returns and timber prices, which could have an impact on the results	A real options analysis examined the impact of flexibility in decision making under agriculture, forestry, and agroforestry and demonstrated that adoption of forestry or agroforestry systems is less feasible than would be predicted by deterministic models	Option value plays an important role in land use decisions

cycle, was US$ 1400 per hectare. One of the main ecological benefits from agroforestry systems is carbon sequestration. To value the amount of above and below-ground carbon storage in agroforestry systems, different studies have used either the market price or the social price of carbon or both. For example, Namirembe et al. (2015) valued the above-ground carbon stock in coffee agroforestry system in Ethiopia at approximately US$ 3370 per hectare when using the market price of carbon, and US$ 21,000 per hectare when using the social price of carbon.

Production function approaches

The production function approaches are applicable in cases which ES may contribute to the increased production of goods or commodities traded in markets. ES in this case are considered as an input that go into the production process of a market good and hence their value can be inferred by taking into account changes in production attributed to the ES. This approach can be applied when valuing ES from agroforestry systems and how they improve soil conservation, increase nitrogen fixation or increase pollinators and the resulting effect on crop yields (Bright, 2004). Bedimo et al. (2008), applied this approach to value incidences of diseases reduction (biological control service) as an ES arising from agroforestry systems in Cameroon. Specifically, the authors observed lower incidences of coffee berry disease in coffee-agroforestry systems compared to full sun coffee systems. The value of reduction in yield damage was captured as the value of biological disease control attributable to agroforestry systems. The challenge with this approach is that it is data-intensive and requires observations over a period of time and across a wide geographical space.

Dose response function

The dose response function approach can be applied in valuing the effect that a change in environmental quality and the resulting effect on the desired output, e.g., agricultural productivity or health. For example, the function allows to estimate the impact of air quality (global warming) affect agricultural production of different crops. Once the physical dose and response relationship is established, monetary values can be derived by multiplying the change in output (or the change in a physical indicator of damage) with the market or shadow price or value of the change in crop output (TEEB, 2010a). This approach can also be used to value negative health impacts resulting from the loss of clean air due to deforestation, which can be converted to value or days/hours lost from work based on prevailing wage rates.

Cost-based approaches

Cost-based approaches are often used to value regulating services such as water regulation, erosion control, air regulation, and so on (TEEB, 2010a). They include:

Replacement/restoration cost

This method considers the value of ES in terms of the cost of replacing them using alternative or artificial means (Pascual et al., 2010; TEEB, 2010a). Replacement cost/restoration cost techniques approximate the benefits of environmental quality by estimating the costs that would be incurred by replacing/restoring ES using alternative means. It can be applied only if the replacement is possible i.e. there should be alternative and cost-effective ways to produce equivalent amounts and quality of the ES in question. This approach can be used to estimate the value of pollination, one of the ecological services derived from agroforestry systems. For example, Allsopp et al. (2008) estimated the value of both wild and managed pollination services in South Africa. The replacement approach was artificial pollination, which requires that pollen have to be collected from appropriate cross-pollinating cultivars, and then applied either by hand or mechanical means (e.g., pollen dusting). The cost of pollen dusting was estimated to be the value of insect pollination.

Avoidance cost

The principle behind this method is that the value of ES can be determined by the cost that is incurred to avoid or replace loss of the ES. It includes all the expenses incurred in order to avoid an undesired outcome. This approach may be applied in estimating the health costs associated with consumption of low quality water due to loss watershed trees that protect water qaulity. This is done by capturing the total cost incurred through the uptake of preventive measures, in this case, the costs associated with precautionary measures taken to reduce direct exposure toavoid the negative health effects of drinking contaminated water e.g., costs of purifying water.

Residual imputation approaches

The residual imputation approach is most commonly used to judge the productivity of a resource that is not easily measured in direct terms (e.g., impact of management practice, good quality land, use of a particular farming technology, value of irrigation water, etc.) TEEB (2018). The residual value represents the maximum amount the producer is willing to pay for a resource for which they are already paying for like land, water for irrigation, and well-drained soils) and still cover all other factors or input costs.

Revealed preference approaches

In cases where direct market data are not easily available or markets do not exist, the revealed preference or stated preference methods are used. Revealed preference approaches include

Travel cost

Valuations of site-based amenities are implied by the costs people incur to enjoy them. The travel cost method, first used by Hotelling (1947), estimates the value of recreational sites, which may be public or quasi-public goods (e.g., recreational value of agricultural landscapes). The model uses actual expenditures and other costs (including the value of time) incurred by individuals when visiting a specific recreational site to estimate the value of the benefits obtained from that site. One of the assumptions of the travel cost method is that there is a clear perceived relationship between the environmental attribute in question and visitors' travel patterns. This, however, may not be true. In many cases, visitors know the quality of a site only after they visit; it can therefore be difficult to value changes in recreational or environmental quality. In addition, the method is rather data intensive and can be complicated if a tourist visits multiple sites on a single trip. While the method can be used to obtain use values, it does not give reliable results if the sites or travel zones are very close to each other, or if there is not enough variation in the explanatory variables. The method is also very sensitive to the type of statistical analysis chosen and to how the opportunity cost of time is assessed.

Hedonic methods

Land productivity depends on various attributes, including agronomic variables, the neighborhood it is located in, as well as environmental and policy variables. Land prices, in turn, indicate the value that consumers or producers are willing to pay for these attributes. Two different pieces of land may look identical but their characteristics and environmental attributes (e.g., soil quality, biodiversity) may be different, and thus the price that a potential buyer is willing to pay may be different. The price differential between the lands due to a difference in one such characteristic can be used as a measure of the marginal value of the characteristic. This is called the hedonic price method. This technique is widely used to measure various characteristics such as the implicit price for soil, the impact of soil erosion, the recreational benefit of open space, the value of clean air, and so on. The value of a service is implied in what people will be willing to pay for the service through purchases in related markets, such as housing markets (e.g., open-space amenities). A limitation of this approach is that agricultural markets are rarely as dynamic as housing markets thus limiting its use in valuing most agricultural services. The data requirements can be quite intensive as well. The method works well if markets can pick up quality differentials, which may not be the case for agricultural land, due to the nonobservability of some attributes (TEEB, 2018).

Stated preference approaches

These approaches use hypothetical/simulated markets to elicit values through the assessment of individuals' WTP to obtain a given ES, or their willingness to accept (WTA) compensation for losing access to that ES. They include

Contingent valuation method

People are directly asked about their WTP or to accept compensation for a change in an ecological service such as cleaner air. According to this approach, individual preferences are directly elicited through individual surveys using a stated preference approach by simulating hypothetical markets. The survey aims to understand the preferences of an individual by describing a scenario (i.e., describing the good, provision of the good, existing state of the environment), and how the provision would change under different management responses or hypothetical alternatives. Respondents are then asked to state their WTP to avoid loss of the service or WTA this change using different elicitation methods attached to different payment vehicles such as taxes, user fees, one-time payments. This method is also used to capture the nonuse values of ES such the value of biodiversity conservation resulting from maintaining agroforestry systems. A limitation of CVM is that it is complex, data intensive, costly to implement, and requires carefully designed surveys to gather unbiased information. In addition, respondents may understate or overstate their WTP/WTA depending on their beliefs and other factors not related to valuation (TEEB, 2018).

Choice experience/modeling

In choice experiments, rather than presenting a single scenario, respondents face a sequence of choice sets. These present different environmental attributes of varying quantity and quality, including the cost of providing the specific good, or the price the consumer or user may have to pay to obtain the good. The respondent's preferred options, in response to the change in attribute levels, are modeled to determine the individual's WTP or WTA for the changes in different levels or quality of attribute under consideration. Thus, it is possible to see how people trade one attribute or preference against the other and the resulting welfare changes can be calculated. The success of this method highly depends on the selection of the appropriate attributes and levels. Too few alternatives or too many alternatives may give incorrect estimates as the respondent may end up choosing the alternatives that are simpler, rather than the ones that are preferred. Choice experiments are a quantitative, econometric-based method for ex ante analysis of the adoption potential of new agroforestry systems. Respondents evaluate alternative land use systems and make tradeoffs among various features of these land use systems. In doing so, they select combinations of attributes (features) that they prefer to others. The analysis can be used to assess which economic and noneconomic criteria farmers use in, or which factors contribute to, how they manage their lands. It reveals how farmers value different attributes of various land use systems, how these values affect adoption and management behavior, and allow to determine the characteristics of agroforestry systems most likely to be adopted (Mercer and Snook, 2004).

There are a number of empirical applications that use choice experiments to evaluate preferences for several agroforestry attributes from the perspective of landowners and the public. Birol et al. (2006) applied a choice experiment to estimate the private benefits that farmers derive from four agrobiodiversity components found in Hungarian home gardens. These include richness of crop varieties and fruit trees, crop landraces, integrated crop and livestock production, and soil microorganism diversity. Results from the study show that there are differences in the private values that households attached to home gardens and these differences also persist across different regions. These findings contribute to a better understanding of the potential role that home gardens can play in conserving agrobiodiversity and the regions where there are valued more. Marenya et al. (2014) designed a choice experiment to elicit preferences among four major policy options that provide incentives for adopting agroforestry-based conservation practices among smallholder farmers in Malawi. These include cash payments, an ideal crop insurance contract, an index insurance contract with basis risk, cash payments, and fertilizer subsidies. The results showed that most farmers preferred cash payments to index insurance contracts, even when the insurance contracts offered substantially higher expected returns. Further, more risk-averse farmers were more likely to prefer cash payments than less risk-averse and risk-loving farmers. Shrestha and Alavalapati (2005) used a choice experiment to estimate Florida cattle ranchers' WTA incentive payments to adopt silvopasture practices. The attributes presented to farmers on the choice cards include water quality improvement, carbon sequestration, and wildlife habitat protection. The results showed that the ranchers would be willing to adopt silvopasture if an additional $0.19 per pound price premium was paid for cattle raised on silvopasture lands. Mercer and Snook (2004) also used a choice experiment to assess how farmers value different attributes of agroforestry systems in Campeche, Mexico. The attributes included labor input, technical assistance, products produced, source of tree seedlings, and the impact on future forest conservation. The results showed that farmers valued future environmental impacts strongly. Other important factors for adopting agroforestry were access to technical assistance, means of acquiring seedlings, and access to a mix of products including timber, crops, and fruit trees.

Conjoint analysis

This technique is used to determine how people value different attributes that make up an individual service. The main objective is to come up with a limited combination of attributes that drives the decision-making or choices made by respondents. An example is when respondents are presented with different ecological scenarios (e.g., choosing between wetlands scenarios with differing levels of flood protection and fishery yields) described using their characteristics or attributes and asked either to rank or choose between them.

Benefit transfer approach

The benefit transfer method is used to estimate economic values for ES that were elicited from one location or at a given time and make inferences about the economic value of ES in another location and/or at another time. In the absence of site-specific valuation information, benefit transfer is an alternative to estimating nonexisting values. It adapts existing valuation information to a new context (location or time) and it is principally useful when there are budget and time constraints with collection of primary data (Temesgen et al., 2018). For example, Temesgen et al. (2018) applied the benefit transfer method to value different ES within agroforestry systems and other land use types in Ethiopia. Among the ES valued figure provisioning services (e.g., food, water, raw material, medicinal purposes), regulating services (e.g., water regulation, water treatment, erosion control, climate regulation), supporting services (e.g., nutrient cycling, pollination, soil formation, habitat), and cultural services (recreation and cultural). Since valuation results are heavily reliant on the social, cultural, and economic contexts, it is important in cases where there are significant differences between the context from which elicited values are drawn from to the one they will be transferred to, then correcting values may be necessary.

Valuation of agroforestry ecosystem services under risk and uncertainty

Following economic theory, farmers who are price-sensitive tend to invest in agroforestry when the expected returns are higher than for all other alternatives to use their land, labor, and capital. Other determinants of farmers' adoption include household preferences, resource endowments, market incentives, biophysical factors, and risk and uncertainty (Mercer, 2004; Mercer and Pattanayak, 2003). In particular, risk and uncertainty may play an important role when considering investing in agroforestry trees on farms compared to annual cropping innovations due to the longer planning horizons required for planting and managing trees (Pannell, 2003; Pattanayak et al., 2003). Specifically, most agroforestry systems take 3 to 6 years before the first benefits are realized, compared to new annual crops that usually take a few months (Franzel and Scherr, 2002).

Variability in returns and farmers' ability to change or postpone decisions in order to adapt to changing conditions are important decision criteria in the agroforestry adoption process. Deterministic models are only able to incorporate changing conditions related to expected future changes to current conditions and can permit decisions that would allow to adapt to these circumstances. However, these models cannot predict which decisions would be optimal under risky or uncertain conditions. In deterministic models, decision-makers are assumed to have perfect foresight of future conditions. In contrast, stochastic models can be used to incorporate risky or uncertain future conditions. Stochastic models can be used to estimate the value created by having flexibility when facing uncertain future conditions. Applying both types of models is advantageous and can provide important insights about financial decisions (Frey et al., 2013). A variety of methods and examples from the agroforestry literature are discussed in the following paragraphs based on Mercer et al. (2017).

Mean-variance analysis

The mean-variance analysis is a classical approach to risk management, which is able to identify the set of "efficient" alternatives that either minimize risk for any given level of returns or maximize returns for any given level of risk (Markowitz, 1959). It is usually referred to as the expected value/variance (EV) because the return is derived from the EV of the random portfolio return, while risk is quantified through the variance of the portfolio return. EV is considered as a powerful decision-making tool. Although the underlying assumptions about utility functions and distributions of returns often conflict with economic theory, EV is appropriate under various utility functions and levels of risk aversion (Kroll et al., 1984). Lilieholm and Reeves (1991) and Babu and Rajasekaran (1991) applied EV to analyze the efficient allocation of agroforestry within the whole farm and showed that adopting agroforestry can be optimal for certain levels of risk aversion. Ramirez et al. (2001) and Ramirez and Sosa (2000) evaluated expected returns and financial risk based on EV analyses of estimated cumulative distribution functions of the NPVs. Ramirez et al. (2001) compared financial returns, stability, and risk of six cacao-laurel-plantain systems, while Ramirez and Sosa (2000) assessed the financial risk and return tradeoffs for coffee agroforestry systems in Costa Rica. Both studies demonstrated that it is important to allow for nonnormality of the variables in NPV analyses. Frey (2009) applied economic analyses of agroforestry systems on private lands in Argentina and in the United States and used EV to examine the tradeoffs between returns and risk, as well as the potential to reduce risk by adopting agroforestry systems. The results showed that the whole-farm risk reduces substantially if annual row crop production is diversified with cottonwood, pecan, and silvopasture. However, there is a tradeoff between reducing variation of returns (with agroforestry) and maximizing the EV of the returns (annual cropping).

Stochastic dominance

Stochastic dominance (SD) is a method to compare different distributions of outcomes among alternatives. It can be applied when EV is inappropriate due to less restrictive assumptions and the use of partial information. SD encompasses the entire probability distribution of outcomes and does not require normality for the utility functions. Further, it requires only minimal assumptions about preferences (Hadar and Russell, 1969). The SD results are less deterministic than EV results, but only provide a partial ranking of efficient and inefficient alternatives. Therefore, SD is typically used for initial screenings of alternatives to provide a partial ordering based on partial information (Hildebrandt and Knoke, 2011). Castro et al. (2013) applied SD to analyze the uncertainties associated with using conservation payments (CPs) to preserve shade coffee in Ecuador. They investigated the effects of land use diversification on CPs by allowing different combinations of coffee agroforestry and monoculture maize production on farms. They found that CPs were much higher when calculated with SD compared to maximizing a concave utility function and concluded that the assumptions underlying SD are inappropriate for risk-averse farmers.

Real options

The real options (RO) approach is a method that takes strategic management options of alternatives, and the flexibility to exercise or abandon the options at different points in time, into account. Management flexibility varies widely between land use alternatives. The ability to change or postpone actions can be important for the decision to adopt one land use system rather than another. Although deterministic models can incorporate changing future conditions and optimize decisions that adapt to these circumstances, they are inappropriate under risky or uncertain conditions because they assume perfect foresight. Stochastic analyses using RO techniques can estimate the value of flexibility given uncertain future conditions. RO assumes that decisions made in the current year can be put off until the future. For example, a farmer can put off timber harvest and reforestation decisions based on current conditions (Frey et al., 2013).

Frey et al. (2013) used RO to analyze how the variability of returns and the flexibility to change or postpone decisions (option value) affect the economic potential of forestry and agroforestry systems in the United States to keep private land in production while continuing to provide ES. An RO analysis examined the impact of flexibility in decision-making for agriculture, forestry, and agroforestry and demonstrated that adoption of forestry or agroforestry systems is less feasible than would be predicted by deterministic capital budgeting models. Behan et al. (2006) used RO and showed that it is optimal for Irish farmers to wait longer to reforest or afforest than suggested by standard discounted cash-flow analyses because of high establishment costs and the relative irreversibility of switching to forestry. Mithofer and Waibel (2003) used RO to analyze investment decisions for tree planting in Zimbabwe and found that indigenous fruit tree planting is affected by tree growth rates and the cost of collecting fruits from communal forests. Wolbert-Haverkamp and Musshoff (2014) applied RO analysis to examine farmers' options to integrate short rotation woody crops into traditional agricultural cropping systems in Germany. They found that the switch triggers calculated with RO analysis are higher than predicted by classical investment theory. Further, risk-averse farmers would be expected to switch earlier from rye production to short rotation woody crops than risk-neutral farmers.

Empirical example that combines different valuation techniques in agroforestry systems

In this section, we describe different methods, namely, an attribute-based choice experiment, an experiment on risk preference, and an experiment on time preference for ex ante analysis of the adoption potential of new agroforestry systems and provide a brief example of its application in eastern Uganda (Ihli et al., 2019). The objectives of this case study project were (1) to elicit coffee farmers' risk and time preferences using lottery-based experiments in the Mt. Elgon region in Uganda and (2) to investigate key attributes (or features/characteristics) of companion trees in coffee agroforestry systems that are preferred by farmers using a choice experiment. To demonstrate the effects of risk and time preferences on the adoption of companion trees, these experimental data were coupled with the results from the choice experiment, which allowed to evaluate farmers' preferences for the included companion tree attributes. A better understanding of farmers' preferences in terms of the features/characteristics of companion trees that they like and dislike is important to design context-specific agroforestry options that aim to increase adoption of trees on farms in a way that is in line with and responsive to farmers' needs and preferences.

Choice experiment

A choice experiment was used to analyze farmers' preferences for different features of companion trees in coffee-banana farming systems. In a choice experiment, respondents are presented with alternative descriptions of a good, differentiated by their attribute levels, and are asked to choose one of the alternatives (Holmes and Adamowicz, 2003). In order to identify contextually relevant attributes and their levels, key informant interviews and focus group discussions with farmers during a preliminary field visit to the study area were conducted. Based on their feedback, six attributes that they deemed important in a companion tree with two to six levels were selected (Table 12.5). The first attribute relates to the products provided by companion trees, namely fruits, timber, fuelwood, and fodder. Regulating ES provided by companion trees is the second attribute. The four levels are microclimate

Table 12.5 Overview of attributes and levels used in the choice experiment.

Attributes	Definition	Attribute levels
Tree products	Products provided by companion trees	**1.** Fruits **2.** Timber **3.** Fuelwood **4.** Fodder
Ecosystem services	Regulating services provided by companion trees (i.e., microclimate, soil fertility, pests and diseases control, and weed control)	**1.** Buffering temperature extremes and conserving soil moisture **2.** Producing mulch and controlling erosion **3.** Fewer problems of White Coffee Stem Borer and Coffee Leaf Rust **4.** Suppressing weed growth
Tree growth rate	Growth rate of companion tree species	**1.** Slow-growing **2.** Medium-growing **3.** Fast-growing
Seedling price	Cost of one tree seedling of companion tree species	**1.** 0 USh **2.** 200 USh **3.** 500 USh **4.** 1000 USh **5.** 1500 USh
Shade quality	Shade quality of companion tree species	**1.** Light, mottled shade **2.** Dense shade
Tree height	Tree height of companion tree species	**1.** Short (<5 m) **2.** Tall (>5 m)

(i.e., buffering temperature extremes and conserving soil moisture), soil fertility (i.e., producing mulch and controlling erosion), pests and diseases control (i.e., decreasing incidence of white coffee stem borer and coffee leaf rust), and weed control (i.e., suppressing weed growth). The third attribute is the growth rate of companion trees and considers three levels: slow-, medium-, and fast-growing. The fourth attribute is the seedling price, categorized in five levels: 0 USh, 200 USh, 500 USh, 1000 USh, and 1500 USh. The fifth attribute concerns the provision of quality shade for coffee in two levels: light and mottled shade, as well as dense shade. The last attribute in the choice experiment is the tree height of the companion tree, either short (<5 m) or tall (>5 m).

The six attributes and their different levels lead to a full factorial design with 960 ($5 \times 4^2 \times 3 \times 2^2$) combinations. Theoretically, each unique combination of attribute levels represents a specific companion tree species. To produce a more manageable experiment, a d-optimal design was used to generate a subset of companion tree species that covers the range of variability between all possible combinations (Hensher, Rose, and Greene, 2015). The d-optimal design takes into consideration all main effects as well as potential interactions between the attributes. In total, 32 choice sets were included in our design. The choice sets were further subdivided into four subsets containing eight choice sets each. To reduce the response burden and to avoid fatigue, respondents were randomly assigned one of these four subsets, with an even number of households allocated to each of the subsets. A choice set consisted of two alternative companion tree species (A and B) and a status quo ("none of the trees") option. The status quo option is provided because a respondent might not have a preference for either of the companion tree species listed. Moreover, illustrations were included in the choice sets to increase respondents' comprehension of the attributes and levels (Fig. 12.2). Before conducting the choice experiment, it was explained to the respondents that the drawings used hypothetical companion tree species rather than real ones. The attributes and levels used were carefully explained. Respondents were also informed that the choices they made in the experiment would not have any immediate consequence. It was clarified that the results would be used more generally to better understand farmers' preferences for particular characteristics of companion trees that may inform project design or future project implementation.

Experiment on risk preferences

A series of lottery-based experiments was used to elicit behavioral characteristics related to risk and potential losses. The experiment used in the study is based on those introduced in Tanaka et al. (2010) and Liu (2013). This experimental design, which takes the form of a Multiple Price List (MPL) design, had previously been tested among individual respondents in different developing countries (Liebenehm and Waibel, 2014; Nguyen, 2011; Ward and Singh, 2015). According to this method, respondents are confronted with an array of paired lotteries (including options A and B) and one of these two options has to be chosen, which implies that the other has to

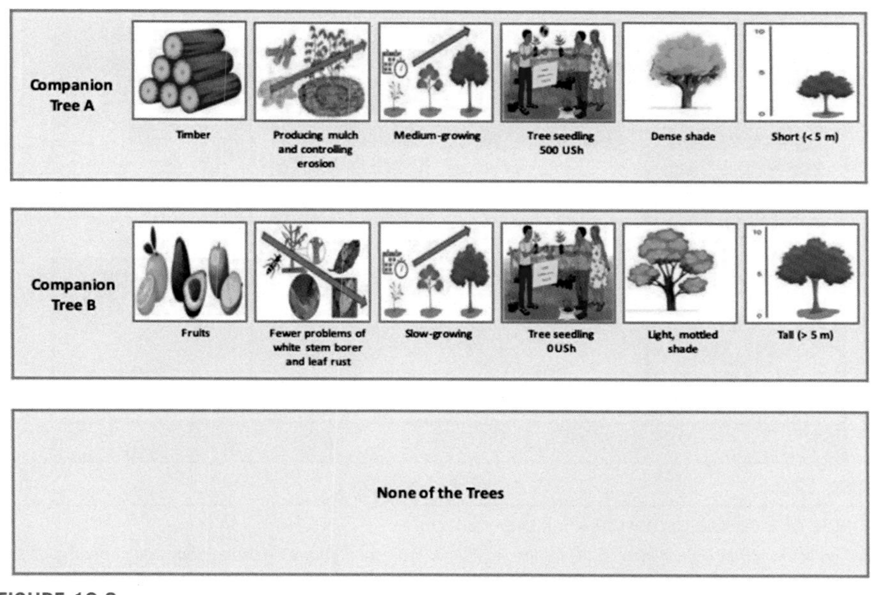

FIGURE 12.2

Example of a choice card.

Reproduced from Ihli, H.J., Winter, E., Gassner, A., 2019. Protocol of Implementing Behavioral Experiments for Trees On Farms Options. World Agroforestry Centre Working Paper.

be rejected. The switching points are used to estimate the respondents' risk preference parameters, namely, risk aversion, loss aversion, and probability weight. While the experiment maintained the general design of previous studies, a few adaptations were made to improve contextual suitability. For instance, payoffs were specifically calibrated to the context of Ugandan smallholder farmers. Furthermore, the overall experiment was framed in a way that is familiar to these farmers, rather than keeping it hypothetical. Specifically, risk preference was determined based on the respondents' choice between two types of tree species that promise different levels of income depending on the weather conditions.

The risk experiment consisted of three series of paired lotteries. In each series, the respondent has to choose between two options ("Tree species A" and "Tree species B"), where each option is a lottery (Fig. 12.3). The probabilities were explained using a fair ten-sided dice, numbered 1 to 10, with different rewards for each option. The numbers 1 to 10 represent 10 years of weather ("good rains" or "bad/no rains"). The respondent makes a choice based on single choice cards illustrating each lottery pair. For example, "Tree species A" gives 4000 USh as income from production in times of "good rains" (in 3 out of 10 years) and 1000 USh in times of "bad/no rains" (in 7 out of 10 years). Alternatively, "Tree species B" gives 15,000 USh as income from production in times of "good rains" (in 1 out of 10 years) and 500 USh in times

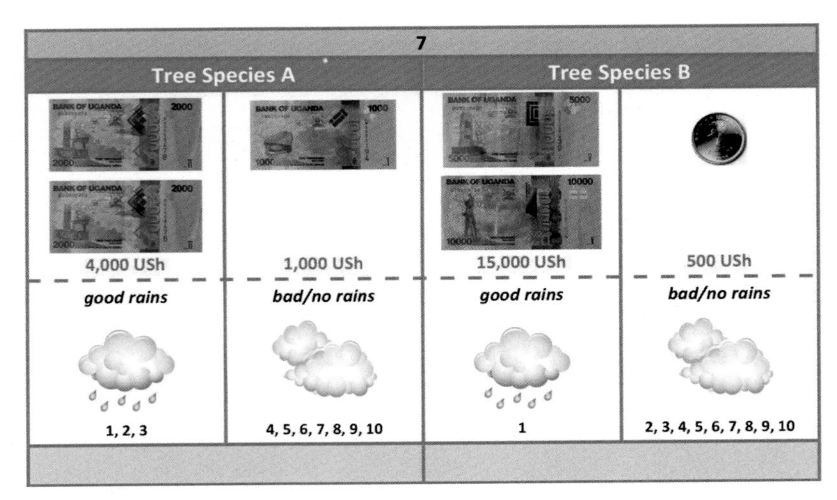

FIGURE 12.3

Example of a choice card in the risk experiment.

Reproduced from Ihli, H.J., Winter, E., Gassner, A., 2019. Protocol of Implementing Behavioral Experiments for Trees On Farms Options. World Agroforestry Centre Working Paper.

of "bad/no rains" (in 9 out of 10 years). One would note that "Tree species B" pays more in times of "good rains", but less in times of "bad/no rains". In total, there were 35 choices to make. These were grouped in three independent series, each of which contained between 7 and 14 choices Table 12.6. At the end of the experiment, one pair of lotteries was randomly selected to be played for real money to encourage participants to reveal their true preferences (Andersen, Harrison, Lau and Rutström, 2006; Holt and Laury, 2002).

Experiment on time preferences

The time experiment consisted of 15 series of five choices between a smaller reward delivered immediately (Option A) and a larger reward delivered at a later specified time (Option B) (Nguyen, 2011; Tanaka et al., 2010). In total, respondents had to make 75 choices, which are partially presented in (Table 12.7). The table shows only the first three series in which the same range of five immediate rewards (Option A) is contrasted with the same delayed reward at three different points of time in the future (Option B). The future reward varies between 3000 USh (approximately $0.8) and 30,000 USh (approximately $8), and the delay varies between 3 days and 3 months. Within each series, the respondent had to decide, whether he or she preferred Option A or Option B. Respondents made choices based on single choice cards illustrating both options (Fig. 12.4). Again, monotonic switching was enforced. At the end of the experiment, one choice card was randomly selected to be played for real money .

Table 12.6 Design of risk experiment (in Ugandan shillings).

		Option A		Option B	
		Probability		Probability	
Series 1	**Choices**	**30%**	**70%**	**10%**	**90%**
	1	4000	1000	6800	500
	2	4000	1000	7500	500
	3	4000	1000	8300	500
	4	4000	1000	9300	500
	5	4000	1000	10,600	500
	6	4000	1000	12,500	500
	7	4000	1000	15,000	500
	8	4000	1000	18,500	500
	9	4000	1000	22,000	500
	10	4000	1000	30,000	500
	11	4000	1000	40,000	500
	12	4000	1000	60,000	500
	13	4000	1000	100,000	500
	14	4000	1000	170,000	500
Series 2	**Choices**	**90%**	**10%**	**70%**	**30%**
	1	4000	3000	5400	500
	2	4000	3000	5600	500
	3	4000	3000	5800	500
	4	4000	3000	6000	500
	5	4000	3000	6200	500
	6	4000	3000	6500	500
	7	4000	3000	6800	500
	8	4000	3000	7200	500
	9	4000	3000	7700	500
	10	4000	3000	8300	500
	11	4000	3000	9000	500
	12	4000	3000	10,000	500
	13	4000	3000	11,000	500
	14	4000	3000	13,000	500
Series 3	**Choices**	**50%**	**50%**	**50%**	**50%**
	1	2500	−400	3000	−2100
	2	400	−400	3000	−2100
	3	100	−400	3000	−2100
	4	100	−400	3000	−1600
	5	100	−800	3000	−1600
	6	100	−800	3000	−1400
	7	100	−800	3000	−1100

Table 12.7 Design of time experiment (in Ugandan shillings).

Series	Choices	Option A	Option B
1	1	2000 USh today	12,000 USh in 1 week
	2	4000 USh today	12,000 USh in 1 week
	3	6000 USh today	12,000 USh in 1 week
	4	8000 USh today	12,000 USh in 1 week
	5	10,000 USh today	12,000 USh in 1 week
2	6	2000 USh today	12,000 USh in 1 month
	7	4000 USh today	12,000 USh in 1 month
	8	6000 USh today	12,000 USh in 1 month
	9	8000 USh today	12,000 USh in 1 month
	10	10,000 USh today	12,000 USh in 1 month
3	11	2000 USh today	12,000 USh in 3 months
	12	4000 USh today	12,000 USh in 3 months
	13	6000 USh today	12,000 USh in 3 months
	14	8000 USh today	12,000 USh in 3 months
	15	10,000 USh today	12,000 USh in 3 months
...	...	...	...

Conclusions and future research

The global demand for food is rising and exerting tremendous pressure on agricultural production systems. Agricultural intensification involving the use of agrochemicals like pesticides and fertilizers can increase productivity but is also widely known to contribute to environmental degradation and social harm. Agroforestry is recognized as a farming practice that can supply important ES that are beneficial to the environment and society by combining a focus on increased food production with consideration for sustainable land and other natural resource use. However, most of these ES from trees on farms are not traded in markets and hence their economic value is not taken into account when considering uptake decisions.

While the general importance of valuing agroforestry systems is recognized, empirical studies that attempt to value and compare the ES benefits of agroforestry ecosystems have remained few—and the evidence base has remained thin. Most existing studies have used conventional nonmarket valuation methods, including contingent valuation technique and the travel cost method, to value ES from agroforestry systems. However, such methods do not incorporate or consider risk and uncertainty. Yet, tree-based interventions have longer planning horizons than most other crops, risk, and uncertainty are more important than for annual crops. This implies that ES are typically undervalued, and purely market-based cost-benefit analyses are likely to attribute higher returns to enterprises that have shorter cycles, e.g., annual crop production.

Meijer, S.S., Catacutan, D., Ajayi, O.C., Sileshi, G.W., Nieuwenhuis, M., 2015. The role of knowledge, attitudes and perceptions in the uptake of agricultural and agroforestry innovations among smallholder farmers in sub-Saharan Africa. International Journal of Agricultural Sustainability 13 (1), 40–54.

Mercer, D.E., 2004. Adoption of agroforestry innovations in the tropics: a review. Agroforestry Systems 61 (1–3), 311–328.

Mercer, D.E., Pattanayak, S., 2003. Agroforestry adoption by small holders. In: Sills, E., Abt, K. (Eds.), Forests in a Market Economy. Kluwer Academic Publishers, Dordrecht, Netherlands, pp. 283–300.

Mercer, D.E., Snook, A., 2004. Analyzing ex-ante agroforestry adoption decisions with attribute-based choice experiments. In: Alavalapati, J.R.R., Mercer, D.E. (Eds.), Valuing Agroforestry Systems: Methods and Applications. Kluwer Academic Publishers, Dordrecht, Netherlands, pp. 237–256.

Mercer, D.E., Li, X., Stainback, A., Alavalapati, J., 2017. Chapter 4: valuation of agroforestry services. In: Schoeneberger, M.M., Bentrup, G., Patel-Weynand, T. (Eds.), Agroforestry: Enhancing Resiliency in U.S. Agricultural Landscapes under Changing Conditions. Gen. Tech. Report WO-96.

Mithofer, D., Waibel, H., 2003. Income and labour productivity of collection of indigenous fruit tree products in Zimbabwe. Agroforestry Systems 59 (3), 295–305.

Mitiku, F., Nyssen, J., Maertens, M., 2018. Certification of semi-forest coffee as a land-sharing strategy in Ethiopia. Ecological Economics 145, 194–204. https://doi.org/10.1016/j.ecolecon.2017.09.008.

Moreno, G., Aviron, S., Berg, S., Crous-Duran, J., Franca, A., de Jalón, S.G., Hartel, T., Palma, J.H.N., Paulo, J.A., Re, G.A., Sanna, F., Thenail, C., Viaus, V., Burgess, P.J., 2018. Agroforestry systems of high nature and cultural value in Europe: provision of commercial goods and other ecosystem services. Agroforestry Systems 92 (4), 877–891.

Namirembe, S., Mcfatridge, S., Duguma, L., Bernard, F., Minag, P., 2015. Agroforestry: An Attractive REDD+ Policy Option? World Agroforestry Centre (ICRAF), Nairobi, Kenya.

Nguyen, Q., 2011. Does Nurture Matter: Theory and Experimental Investigation on the Effect of Working Environment on Risk and Time Preferences. Journal of Risk and Uncertainty 43 (3), 245–270.

Noordwijk, M.V., Coe, R., Sinclair, F., 2016. *Central hypotheses for the third agroforestry paradigm within a common definition.* Working paper 233, Bogor, Indonesia: world agroforestry Centre (ICRAF) southeast Asia regional program. https://doi.org/10.5716/WP16079.PDF.

Pannell, D.J., 2003. Uncertainty and adoption of sustainable farming systems. In: Babcock, B.A., Fraser, R.W., Lekakis, J.N. (Eds.), Risk Management and the Environment: Agriculture in Perspective. Kluwer Academic Publishers, Dordrecht, Netherlands, pp. 67–81.

Pascual, U., Muradian, R., Brander, L., Gómez-baggethun, E., Martín-lópez, B., Verma, M., et al., 2010. Chapter 5 the Economics of Valuing Ecosystem Services and Biodiversity.

Pattanayak, S.K., Mercer, D., Sills, E., et al., 2003. Taking stock of agroforestry adoption studies. Agroforestry Systems 57 (3), 173–186.

Ramirez, O.A., Sosa, R., 2000. Assessing the financial risks of diversified coffee production systems: an alternative nonnormal CDF estimation approach. Journal of Agricultural and Resource Economics 25 (1), 267–285.

Ramirez, O.A., Somarriba, E., Ludewigs, T., Ferreira, P., 2001. Financial returns, stability and risk of cacao-plantain-timber agroforestry systems in Central America. Agroforestry Systems 51 (2), 141–154.
Reichhuber, A., Requate, T., 2012. Alternative use systems for the remaining Ethiopian cloud forest and the role of Arabica coffee - a cost-benefit analysis. Ecological Economics 75, 102–113. https://doi.org/10.1016/j.ecolecon.2012.01.006.
Shrestha, R.K., Alavalapati, J.R., 2005. Estimating ranchers' cost of agroforestry adoption. In: Alavalapati, J., Mercer, D. (Eds.), Valuing Agroforestry Systems. Kluwer Academic Publishers, Dordrecht, Netherlands, pp. 183–199.
Smith, M.S., Mbow, C., 2014. Editorial overview: sustainability challenges: agroforestry from the past into the future. Current Opinion in Environmental Sustainability 6, 134–137. https://doi.org/10.1016/j.cosust.2013.11.017.
Tanaka, T., Camerer, C., Nguyen, Q., 2010. Risk and time preferences: Linking experimental and household survey data from Vietnam. American Economic Review 100 (1), 557–571.
TEEB, 2010. The Economics of Ecosystems and Biodiversity. Initiatives. https://doi.org/10.1093/erae/jbr052.
TEEB, 2018. TEEB for Agriculture & Food: Scientific and Economic Foundations.
Temesgen, H., Wu, W., Shi, X., Yirsaw, E., Bekele, B., Kindu, M., 2018. Variation in ecosystem service values in an agroforestry dominated landscape in Ethiopia: implications for land use and conservation policy. Sustainability 10 (4), 1126. https://doi.org/10.3390/su10041126.
Tilman, D., Balzer, C., Hill, J., Befort, B.L., 2011. Global food demand and the sustainable intensification of agriculture. Proceedings of the National Academy of Sciences. https://doi.org/10.1073/pnas.1116437108.
Vira, B., Wildburger, C., Mansourian, S. (Eds.), 2015. Forests, Trees and Landscapes for Food Security and Nutrition. A Global Assessment Report, vol. 33. IUFRO World Series, Vienna, 172 p.
Ward, P.S., Singh, V., 2015. Using Field Experiments to Elicit Risk and Ambiguity Preferences: Behavioural Factors and the Adoption of New Agricultural Technologies in Rural India. Journal of Development Studies 51 (6), 707–724.
Wolbert-Haverkamp, M., Musshoff, O., 2014. Is short rotation coppice economically interesting? An application to Germany. Agroforestry Systems 88 (3), 413–426.

Further reading

Binswanger, H., 1980. Attitudes toward risk: experimental measurement in rural India. American Journal of Agricultural Economics 62 (3), 395–407.
Bruinsma, J., 2003. In: Bruinsma, J. (Ed.), World Agriculture: Towards 2015/2030: an FAO Perspective, vol. 20. Earthscan, London. https://doi.org/10.1016/S0264-8377(03)00047-4.
Coller, M., Williams, M.B., 1999. Eliciting individual discount rates. Experimental Economics 2 (2), 107–127.
Dave, C., Eckel, C., Johnson, C., Rojas, C., 2010. Eliciting risk preferences: when is simple better? Journal of Risk and Uncertainty 41 (3), 219–243.
Eckel, C.C., Grossman, P.J., 2002. Sex differences and statistical stereotyping in attitudes toward financial risk. Evolution and Human Behavior 23 (4), 281–295.

FAO. 2013. Advancing Agroforestry on the Policy Agenda: A Guide for Decision-Makers, by G. Buttoud, in Collaboration with O. Ajayi, G. Detlefsen, F. Place & E. Torquebiau. Agroforestry Working Paper No. 1. Food and Agriculture Organization of the United Nations. FAO, Rome. 37 pp.

Feder, G., Just, R.E., Zilberman, D., 1985. Adoption of agricultural innovations in developing countries: a survey. Economic Development and Cultural Change 33 (2), 255–298.

Ghaley, B.B., Vesterdal, L., Porter, J.R., 2014. Quantification and valuation of ecosystem services in diverse production systems for informed decision-making. Environmental Science & Policy. https://doi.org/10.1016/j.envsci.2013.08.004.

Harrison, G.W., Lau, M.I., Williams, M.B., 2002. Estimating individual discount rates in Denmark: a field experiment. The American Economic Review 92 (5), 1606–1617.

Luedeling, E., Kindt, R., Huth, N.I., Koenig, K., 2014. Agroforestry systems in a changing climate—challenges in projecting future performance. Current Opinion in Environmental Sustainability 6, 1–7. https://doi.org/10.1016/j.cosust.2013.07.013.

Namirembe, S., Leimona, B., van Noordwijk, M., Bernard, F., Bacwayo, K.E., 2014. Co-investment paradigms as alternatives to payments for tree-based ecosystem services in Africa. Current Opinion in Environmental Sustainability 6, 89–97. https://doi.org/10.1016/j.cosust.2013.10.016.

Neugarten, R.A., Langhammer, P.F., Osipova, E., Bagstad, K.J., Bhagabati, N., Butchart, S.H.M., Dudley, N., Elliott, V., Gerber, L.R., Gutierrez Arrellano, C., Ivanic, K.-Z., Kettunen, M., Mandle, L., Merriman, J.C., Mulligan, M., Peh, K., Raudsepp-Hearne, C., Semmens, D.J., Stolton, S., Willcock, S., 2018. Tools for Measuring, Modelling, and Valuing Ecosystem Services: Guidance for Key Biodiversity Areas, Natural World Heritage Sites, and Protected Areas. IUCN, Gland, Switzerland.

Pascual, U., Muradian, R., Brander, L., Gómez-Baggethun, E., Martín-López, B., Verma, M., Armsworth, P., Christie, M., Cornelissen, H., Eppink, F., Farley, J., Loomis, J., Pearson, L., Perrings, C., Polasky, S., McNeely, J.A., Norgaard, R., Siddiqui, R., David Simpson, R., Kerry Turner, R., Simpson, R.D., 2012. The economics of valuing ecosystem services and biodiversity. In: The Economics of Ecosystems and Biodiversity: Ecological and Economic Foundations. https://doi.org/10.4324/9781849775489.

Tengnas, B., 1994. Agroforestry Extension Manual for Kenya. International Centre for Research in Agroforestry, Nairobi.

Zomer, R., Trabucco, A., Coe, R., Place, F., van Noordwijk, M., Xu, J., 2014. Trees on Farms: An Update and Reanalysis of Agroforestry's Global Extent and Socio-Ecological Characteristics (No. 179). https://doi.org/10.5716/WP14064.PDF. Bogor, Indonesia.

CHAPTER

Challenges in maximizing benefits from ecosystem services and transforming food systems

13

Leonard Rusinamhodzi, BSc, M.Phil, PhD
Scientist - Systems Agronomist, Sustainable Intensification Program, CIMMYT, Kathmandu, Nepal

Introduction

Huge increases in demand for food because of population pressure while the size and quality of agricultural land is decreasing require new ways of thinking to produce more food to meet the dietary needs of a dynamic population. Sustainable intensification has been suggested as the best strategy where crop yields can be increased using fewer inputs, without land expansion and with minimal adverse environmental impact (Dore et al., 2011). In addition, these cropping systems should withstand the devastating effects of climatic variability and change. This approach is currently the basis upon which the prospects of feeding more than nine billion people in ca. 2050, while improving the environment are premised (Tilman et al., 2011). Sustainability is complex and takes many forms, including environmental, social, and economic aspects, though environmental sustainability is thought to be the basis on which the other two pillars of sustainability can be realized. For the large proportion of the population that rely on agriculture an increase in this production is needed to give economic surplus required for households to make new investments which can lead to a positive spiral of development. The strategy is to improve water and plant nutrient use efficiencies by using the best technical means (http://www.yieldgap.org/).

The Food and Agriculture Organization defines sustainable food systems as an internet of things encompassing the entire production, aggregation, processing, distribution, consumption, and disposal of food products that originate from agriculture, forestry, or fisheries and parts of the broader economic, societal, and natural environments in which they are embedded. Emphasis is placed on the need to guarantee economic, social, and environmental sustainability to ensure that the food security and nutrition of future generations are not compromised. The relationship between ecosystem services (ESs) and sustainable food systems (SFSs) is two-way. SFSs defined in preceding chapters, but simply meaning to ensure sustainability before, within, and after production until consumption, can have a role in sustaining and amplifying ESs. On the other hand, ESs play a

The Role of Ecosystem Services in Sustainable Food Systems. https://doi.org/10.1016/B978-0-12-816436-5.00013-5

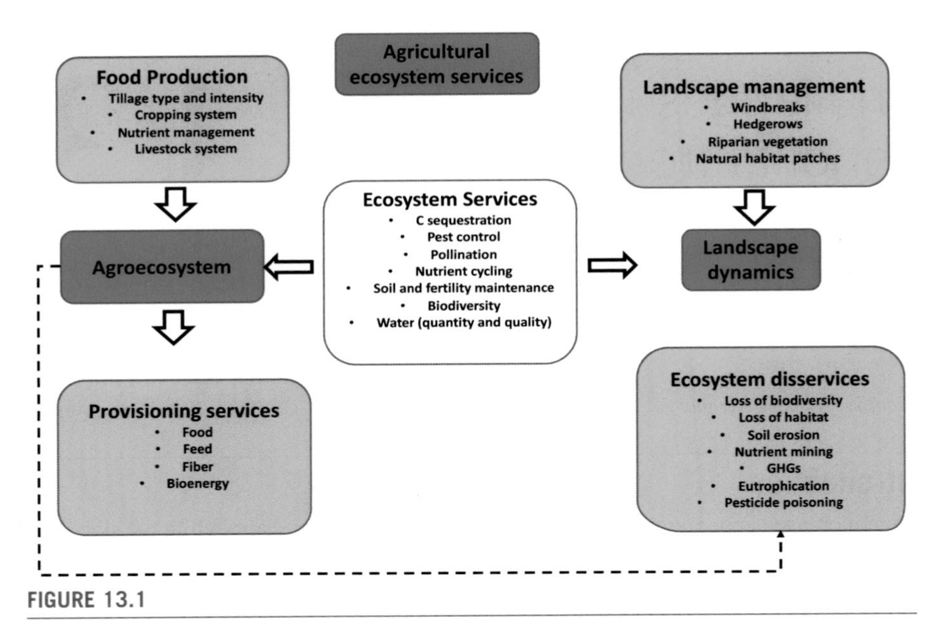

FIGURE 13.1

Impacts of food production and landscape management on the flow of ecosystem services and disservices to and from agroecosystems. *GHGs*, Greenhouse gases.

Adapted from Power, A.G., 2010. Ecosystem services and agriculture: tradeoffs and synergies. Philosophical Transactions of the Royal Society of London B Biological Sciences 365, 2959–2971.

reciprocating role by facilitating high productivity for crops and animals. The realization of these benefits is not a stroll in the park, and it is generally agreed that appropriate agricultural practices are the cornerstone of the realization of ES and their maintenance (Fig. 13.1). Poor agricultural practices can lead to severe ecosystem disservices with implications on human, animal, and environmental health.

Giller et al. (2009) in their assessment of the potential of conservation agriculture in the drylands of sub-Saharan Africa (SSA) provide a comprehensive analogy of the factors that potentially prevent widespread practice of CA, a case study chapter in the book. These factors or barriers are not peculiar to SSA alone but manifest in many forms across the world especially as long as agriculture is rainfed and managed by poorly resourced farmers.

The challenges to ES take many forms and different scales. At farm or plot level, multiple nutrient deficiencies hamper crop responses to better management resulting in prolonged nutrient mining—leading to soil degradation. At landscape level the inability to enact by-laws that are enforceable mean that some practices are unlikely to be widely adopted. A good example is the practice of CA, where community level decisions are needed to solve the conflicts of communal grazing and crop residue uses. The availability of ecosystem services (ES) at higher scales and accrual to others creates trade-offs between primary food production and other community interests. On the other hand, climate variability and change exacerbate the trade-offs because of significant effects on both crop and animal productivity. Rainfall variability characterized by frequent droughts currently causes the major effects on crop productivity resulting in acute food and feed shortages (Challinor et al., 2007).

At national level, policies are fragmented and largely difficult to enforce—partly due to limited congruency on the drivers of some of the technologies being driven e.g., in Africa. For example, it is not known with absolute certainty which components of CA drive yield increases and whether such factors are exclusive to CA. Crop rotation has been practiced for ages but is often considered the cornerstone of CA, yet it is also present in CT situations. Such dilemma has had a negative effect on development agencies on what form of messages they have to push and in what context. Depending on the focus of a particular institution and the political economy, blanket recommendations have characterized how extension messages are pushed, yet it has been proven that agroecology, management decisions, and resource ownership interact strongly to determine the fate of these technologies.

The Green Revolution in the 1960s was based on the use of high yielding varieties, large quantities of fertilizer, pesticides, herbicides, and intensive management to meet the food demands of an increasing population (Douglas et al., 2002). The side-effects of intensification soon followed marked by reduction in biodiversity, increased incidence of pests and diseases because of monocropping, reductions in soil quality, and pollution (Conway and Barbie, 1988; Matson et al., 1997). Concerns about the long-term sustainability and environmental consequences of such high input use and management required new approaches to sustainable agriculture (Conway and Barbie, 1988; Matson et al., 1997; Tscharntke et al., 2005). The paradigm of ecological intensification of agriculture means manipulating nature's functions to design sustainable production systems that use less inputs and leads to positive biophysical and socioeconomic outcomes (Cassman, 1999; Doré et al., 2011; Hochman et al., 2013; Tittonell and Giller, 2013). It entails efficient use of pesticides, chemical fertilizers, water and fossil fuels, and development of locally adapted varieties. Thus the major pillar of ecological intensification is increasing resource use efficiency, i.e., resource capture efficiency × resource conversion efficiency (Trenbath, 1986; de Wit, 1992).

Sustainable intensification in developed and developing countries will take different pathways in the short term. In the developed countries reductions in fertilizer inputs are needed because of environmental concerns (e.g., Tamminga, 2003). In Africa fertilizer inputs need to be increased (ca. 50 kg ha based on Abuja declaration) to sustain large crop productivity while minimizing negative effects on the environment. In less favorable environments such as those in southern Africa, feasible ecological intensification options include integration of crop and livestock production, increased crop diversification, and agroforestry systems that promote nutrient and soil conservation (e.g., Cassman and Harwood, 1995; Mafongoya et al., 2006). Tittonell and Giller (2013) also suggested that manipulating planting dates, crop spacing, cultivar selection, and weeding intensities may be important to achieve large crop productivity. Recently, Ryschawy et al. (2012) identified mixed crop-livestock systems as the most suitable pathway to environmental and economic sustainable agriculture although there is no guarantee that inputs are always used efficiently and that outputs are always positive on the environment.

Knowledge about ecosystems services

In many regions of the world where food production occurs, most producers are not aware of ES—their scope, extent, and importance (McKenzie et al., 2014). Farmers are actively involved in soil fertility, erosion control, and pest and disease control and are aware of the direct effect on productivity. The benefits that occur at higher scales do not receive the same attention although they may be beneficial to the farm. This is proven by a recent publication which suggested that landscape management and control of pests such as stemborer and fall armyworm overrides local farm efforts because of easy migration of these pests (Kebede et al., 2019).

An example of a situation where farmers are largely not aware of the ES is the carbon credit scheme where farmers are rewarded for returning carbon to the soil. While many participate in these schemes possibly because of the financial incentives, it is hard to sustain these arrangements outside funded projects. The immediate question often asked is who pays and how to sustain such arrangement, including the trade-offs between C input and food production at the farm level (Rusinamhodzi et al., 2016b). The allocation of crop residues for livestock feed meets two out of three critical objectives; it ensures feed during the dry season and improves the quantity and quality of manure to restore soil fertility but does not ensure permanent soil cover (Rusinamhodzi et al., 2016a) which is needed for effective carbon sequestration.

In low-input systems where food shortages are acute the sustenance of ecosystem services (ES) is subservient to food production leading to more pressure on the environment. In extensification systems, land expansion is the only way to increase productions albeit with finite land sizes. Such systems lead to destruction of natural vegetation often with burning the cleared forests. The result is loss of C from the system and works against harnessing ES. Thus in low-input systems, production of food competes with ecosystem services (ES), and it is unlikely that these systems can enhance ES as land quality decreases with age because of nutrient mining. Intensification is the opposite, an increase in the productivity of existing land through increased inputs of external resources in the production of food and cash crops and livestock (Table 13.1).

Food production especially in low input systems is characterized by accelerated soil quality decline despite differences in initial soil quality and crop management (Fig. 13.2), and it is always unlikely to achieve the C content of the original virgin land even if restorative efforts are done.

Table 13.1 Contrasting characteristics between extensification and intensification of agricultural production systems.

Characteristics	Extensive systems	Intensive systems
Fallow length	Long	Short
Productivity	Low	High
Efficiency	High	Variable but lower
Population density	Low	High
Technology	Simple	Often complex
Fertilizing of soil	None or little	High
Land tenure	Communal ownership	Individual/family
Economic systems	Usually subsistence	Usually market
Socio-political complexity	Generally less	Generally greater

Adapted from Boserup, E., 1965. The Conditions of Agricultural Growth: The Economics of Agrarian Change under Population Pressure, Aldine, Chicago.

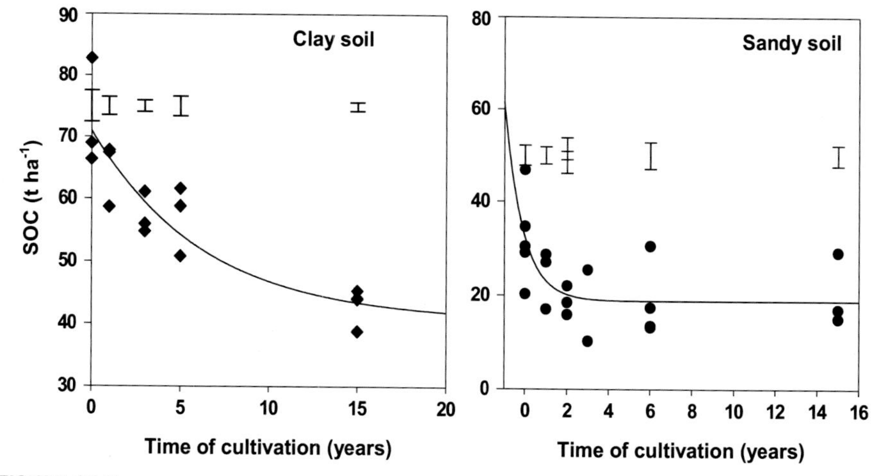

FIGURE 13.2

Changes in soil organic carbon (SOC) concentration with time to show the decrease in land quality with increasing duration of cultivation/intervention of intensification. Plots based on unpublished chronosequence work carried out in central Mozambique in 2010.

Quantification of ecosystems services

It was clear in Chapter 12, that the identification and quantification of ecosystem services (ES) is a major challenge. A hybrid approach for valuing ecosystem services (ES) is considered appropriate and should incorporate: (1) assessment of monetary values for those values that directly or indirectly relate to monetary inputs/outputs and (2) ranking and mapping others that cannot be readily monetized as suggested by Sangha et al. (2017). In many developed countries, food production maybe intrinsically linked to other cultural norms and traditions which may not allow assignment of a monetary value. Some ecosystem goods and services cannot be valued because they are not quantifiable or because available methods are not appropriate, adaptable, or reliable.

In literature the major challenge of assigning a value on ecosystem services (ESs) lies in providing an explicit description and adequate assessment of links between the structures and functions of natural systems and the benefits (i.e., goods and services) derived by humanity (NRC, 2004). The critical challenge is how to link the ecological production function with the economic valuation function with absolute certainty and reproducibility. Where an ecosystem's services and goods can be identified and measured, it is often possible to assign values to them by employing existing economic (primarily nonmarket) valuation methods.

Transforming food systems to maximize ecosystem services

Modern cropping systems should satisfy social, economic, and ecological or environmental goals of food production. The traditionally used cropping systems are monoculture, intercropping, crop rotations, fallows systems/shifting cultivation, strip cropping, multiple cropping, contour strip cropping, and cover crops. The cropping systems characteristics are described briefly in Table 13.2.

Transformation of current cropping systems to modern in southern Africa will be underpinned by crop diversity and intensity within a single field to achieve the multiple goals highlighted earlier. Cropping intensity is defined as the number of crops harvested per year. Increasing cropping intensity will be a game changer as recent assessment suggest that southern Africa has an intensity of only 0.45 or 45% (Siebert et al., 2010). However, high crop intensities are not sustainable under rain-fed conditions, as southern Africa has a long dry season of up to 7 months. Investments in irrigation development to support high cropping intensities are key to achieve significant transformation in smallholder agriculture.

Available evidence suggests that integration of multipurpose legumes in the predominant cereal cropping systems of southern Africa is the most appropriate pathway. Inclusion of legumes as intercrops requires rearrangement of the planting patterns through substitutive or additive designs to maintain the productivity of the main crop (Giller, 2001). Moreover, crop diversification through a legume-based approach is considered to be more beneficial than cash crop-based diversification pathways (Maggio et al., 2018). Competition can also be reduced by staggering the planting dates of the companion crops in the intercropped system (Francis et al., 1982). Staggered planting is also used for reducing risk of total crop failure when expected rainfall is uncertain and within-season fluctuations are common (Cooper et al., 2008). When intercropped with cereals, larger quantities of better-quality organic matter inputs are produced leading to greater productivity benefits compared with continuous maize monocrops (Rochester, 2011). Multipurpose grain legumes such as pigeon pea (*Cajanus cajan* (L.) Millsp.) have shown potential to be

Table 13.2 Characteristics of some of the most common historical and current cropping systems.

Cropping system	Characteristics
Monocropping	The growing of a single crop in pure stands on the same piece of land for successive cropping seasons. In literature it can also be referred to as sole cropping or monoculture. Productivity has been enhanced mostly though increased fertilizer use, chemicals, and elite crop hybrids.
Intercropping	The growing of two or more crops together or one after another on the same land in the same cropping season. Intercropping can be in distinct rows, strips, or haphazard. In Africa as a part of traditional farming system, intercropping is practiced in small farms but often with very limited resources.
Crop rotations	Crop rotation is the growing of different species of crops in succession on the same piece of land, i.e., the opposite of monocropping. Crop rotation can be used to control pests and diseases, as well as improving the soil fertility status.
Fallow systems	A fallow system involves leaving a piece of land without crops for a couple of seasons for self-regeneration of soil fertility while cropping is moved to another piece of land—shifting cultivation. Owing to rapid population pressure, this option is increasingly becoming irrelevant.
Cover cropping	A form of intercropping in which a secondary crop is planted to cover the area between the rows of the main crop. Cover crops are primarily intended to stop soil erosion but can also be used for pests and diseases, soil fertility restoration, and enhancement of biodiversity.

included in cereal-legume rotations in the tropics (Giller et al., 2009; Rusinamhodzi et al., 2012).

The reality is that crop mixtures managed by smallholder farmers are poorly understood by scientists. These mixtures may appear haphazard to those not familiar with the system but have evolved over many years of practice, satisfying food, and nutritional needs of the households. Yet, fertilizer recommendations and other management needs are often developed for monocrops and are invariably focused on maize in East and Southern Africa, leaving the orphan crops (neglected and underused crops) to rely on residual nutrients in the soil. More nutrients are removed in intercrops compared with sole crops leading to more rapid depletion of natural soil fertility (Mason et al., 1987) creating the need for better nutrient management in such systems. Even with the inclusion of legumes in the intercrops the legume may end up depending on applied fertilizer or N released by the soil because its nitrogen fixation capacity is inhibited by other multiple nutrient deficiencies (Rusinamhodzi et al., 2006). Thus the dynamics of nutrient uptake and utilization in complex traditional crop mixtures managed by smallholder farmers are poorly understood. Developing appropriate nutrient management for intercrops is challenging because of many factors that need to be considered, i.e., the residual nutrients, contribution of companion legume crops, addition of residues by component crops, temporal differences in nutrient requirement, spatial differences in foraging, and differential response of crops to nutrients. The scope of complementarity in these complex mixtures is not known; information on the interaction between the root systems of the component crops (root and cereal crop roots), the soil layers in which they forage, nutrient dynamics, and depletion of soil moisture is scant.

Another simple but transformative intervention for cropping system improvement is achieving optimum plant density. Generally, smallholder farmers in southern Africa plant below optimum plant densities with a significant negative impact on productivity (Fig. 13.3). Advances in crop breeding allows the deployment of crop varieties tolerant to high densities and responsive to improvements in fertilizer and water management practices to maximize productivity (Shi et al., 2016). The easy entry points for transformation are the provision of knowledge so that farmers are aware of the optimum plant population in their locality.

Cropping systems managed by farmers involve trade-offs, i.e., balancing two or more opposing outcomes. The primary trade-off associated with intercropping is the competition with the main crop, i.e., the need to preserve yield of main crop while ensuring a reasonable yield from the companion crop(s). However, a focus on yield in low-input, low-output systems that dominate SSA may not capture the trade-offs perceived by farmers or the broader ecological benefits. Successful intercropping even with multiple crops requires that farmers design efficient

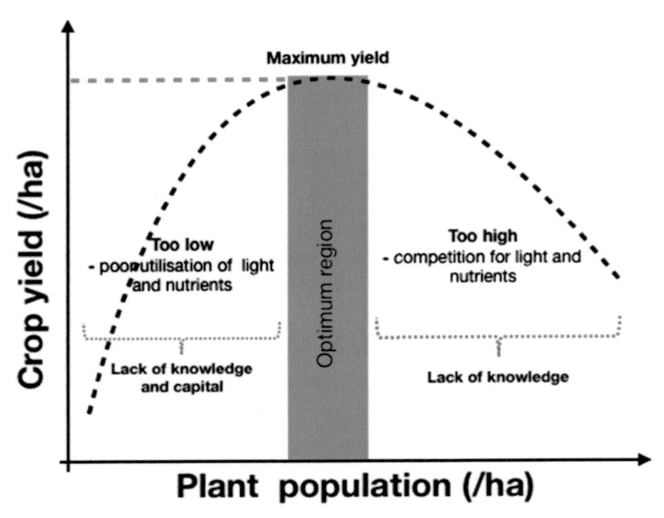

FIGURE 13.3

Hypothetical relationship between increasing plant density and crop yield. Initial relationship can be linear and reaches maximum around optimum density range before it plummets due to competition. In southern Africa limited capital and knowledge can singly or in combination affect the ability of farmers to achieve optimum plant population.

systems, in which complementary effects of intercropping on net returns exceed competitive effects. Therefore assessment and resolution of trade-offs at higher scales than the farm are needed to identify the complete benefits such as efficient use of labor or erosion control at landscape level. Future work and resources need to be channeled toward understanding the issues highlighted above in selected major dominant cropping systems and design cereal-legume systems that work for most farmers.

Agricultural policies that deliberately support crop diversification are needed, with well-defined targeted type of diversification and policy measures (Maggio et al., 2018). The region is dominated by maize production; thus diversification is mostly targeted at integration of multipurpose grain legumes. Market development for grain legumes to play a role in providing additional income will be key to stimulate their widespread production. Sustained availability and accessibility of quality fertilizers including micronutrients in adequate quantity and in time will require enhanced performances and functions of governments and the private sector along the sustainable food production value chain. To achieve a competitive market growth the production system should provide a surplus at the farm level. Investments are needed in irrigation development in combination with water-saving measures to support high cropping intensities. There is need to streamline administrative and managerial procedures to improve efficiency with a deliberate focus on supporting farmer-owned and farmer-operated irrigation.

In most countries the seed systems of legume are fragmented and dominated by informal systems in which quantities rarely meet demand, and quality is very low. At present as much as 90% of smallholder farmers obtain seed from the informal system (McGuire and Sperling, 2016). Such scenarios require urgent attention as good quality seed is one of the key bases for improved productivity. The extension support systems in most of SSA are weak and will need to be strengthened given the knowledge intensity for successful intercropping especially at the local level. Recent developments in agriculture and elsewhere require that extension transfer new food crop technology to farmers and help the rural poor by promoting agricultural diversification, as well as access to markets (Eicher, 2007).

Conclusions

Food production depends on ecosystem services (ESs) but can also amplify and sustain ecosystem services (ESs). Water and land ecosystems have interrelated functions that provide a wide range of important goods and services. In recent years such services can be assigned a value—which may be used to show the importance of these goods and services. The state of these natural ecosystems determines the quality of goods and services they can supply. However, rapid population growth and the need to produce more food often results in negative impact on the environment because of excessive extraction and use of water and introduction of chemicals which often alters negatively the biodiversity in a given location.

Owing to limited knowledge, the importance of ecosystem functions and services can be taken for granted and overlooked in environmental decision-making. Thus expansion of human activities often leads to degradation of ESs, and informed decisions are needed to resolve these trade-offs. However, the true and complete economic values of these ecosystem goods and services to society have to be known, so that they can be compared with the economic values of activities that may compromise them, and improvements to one ecosystem can be compared with those in another (NRC, 2004). More work is needed on designing context specific tools to identify and measure ESs.

Opportunities to improve ecosystem services (ESs) in food production systems is underpinned by crop diversification. Small and dwindling farms do not allow fallowing and may limit crop rotation. Successful cropping system design and improvement will depend on a full understanding of constraints and opportunities in a particular domain, co-design, and development. Cereal-legume intercropping, use of short season varieties, increasing cropping intensity along with irrigation development are suggested strategies for cropping system improvement under rainfed conditions. Partnerships between users, development scientists, government agencies, and private partners especially market agencies will be key.

References

Boserup, E., 1965. The Conditions of Agricultural Growth: The Economics of Agrarian Change Under Population Pressure. Aldine, Chicago, 1965.

Cassman, K.G., 1999. Ecological intensification of cereal production systems: yield potential, soil quality, and precision agriculture. Proceedings of the National Academy of Sciences of the United States of America 96, 5952–5959.

Cassman, K.G., Harwood, R.R., 1995. The nature of agricultural systems: food security and environmental balance. Food Policy 20, 439–454.

Challinor, A., Wheeler, T., Garforth, C., Craufurd, P., Kassam, A., 2007. Assessing the vulnerability of food crop systems in Africa to climate change. Climatic Change 83, 381–399.

Conway, G.R., Barbie, E.B., 1988. After the Green Revolution: sustainable and equitable agricultural development. Futures 20, 651–670.

Cooper, P.J.M., Dimes, J., Rao, K.P.C., Shapiro, B., Shiferaw, B., Twomlow, S., 2008. Coping better with current climatic variability in the rain-fed farming systems of sub-Saharan Africa: An essential first step in adapting to future climate change? Agriculture. Ecosystems & Environment 126, 24–35.

de Wit, C.T., 1992. Resource use efficiency in agriculture. Agricultural Systems 40, 125–151.

Doré, T., Makowski, D., Malézieux, E., Munier-Jolain, N., Tchamitchian, M., Tittonell, P., 2011. Facing up to the paradigm of ecological intensification in agronomy: revisiting methods, concepts and knowledge. European Journal of Agronomy 34, 197–210.

Douglas, G., Stephen, P., Robert, E., 2002. The Green Revolution: An End of Century Perspective. Department of Economics, Williams College.

Eicher, C.K., 2007. Agricultural extension in Africa and Asia; literature review. Cornell University, New York.

Francis, C.A., Prager, M., Tejada, G., 1982. Effects of relative planting dates in bean (Phaseolus vulgaris L.) and maize (Zea mays L.) intercropping patterns. Field Crops Research 5, 45–54.

Giller, K.E., 2001. Nitrogen Fixation In Tropical Cropping Systems. CABI Publishing, New York.

Giller, K.E., Witter, E., Corbeels, M., Tittonell, P., 2009. Conservation agriculture and smallholder farming in Africa: the heretics' view. Field Crops Research 114, 23–34.

Kebede, Y., Bianchi, F.J.J.A., Baudron, F., Tittonell, P., 2019. Landscape composition overrides field level management effects on maize stemborer control in Ethiopia. Agriculture, Ecosystems & Environment 279, 65–73.

Mafongoya, P.L., Bationo, A., Kihara, J., WaswaB, S., 2006. Appropriate technologies to replenish soil fertility in Southern Africa. Nutrient Cycling in Agroecosystems 76, 137–171.

Maggio, G., Sitko, N., Ignaciuk, A., 2018. Cropping system diversification in Eastern and Southern Africa: Identifying policy options to enhance productivity and build resilience. FAO Agricultural Development Economics Working Paper 18-05. FAO, Rome. Licence: CC BY-NC-SA 3.0 IGO.

Mason, W.K., Pritchard, K.E., 1987. Intercropping in a temperate environment for irrigated fodder production. Field Crops Research 16, 243–253.

Matson, P.A., Parton, W.J., Power, A.G., Swift, M.J., 1997. Agricultural intensification and ecosystem properties. Science 277, 504–509.

McGuire, S., Sperling, L., 2016. Seed systems smallholder farmers use. Food Security 8, 179–195.

McKenzie, E., Posner, S., Tillmann, P., Bernhardt, J.R., Howard, K., Rosenthal, A., 2014. Understanding the use of ecosystem service knowledge in decision making: lessons from international experiences of spatial planning. Environment and Planning C: Government and Policy 32, 320–340.

NRC, 2004. Valuing Ecosystem Services: Toward Better Environmental Decision-Making. National Academies Press, 500 Fifth Street, NW, Washington, DC, 20001.

Power, A.G., 2010. Ecosystem services and agriculture: tradeoffs and synergies. Philosophical Transactions of the Royal Society of London B Biological Sciences 365, 2959–2971.

Rochester, I.J., 2011. Sequestering carbon in minimum-tilled clay soils used for irrigated cotton and grain production. Soil & Tillage Research 112, 1–7.

Rusinamhodzi, L., Murwira, H.K., Nyamangara, J., 2006. Cotton-cowpea intercropping and its N2 fixation capacity improves yield of a subsequent maize crop under Zimbabwean rain-fed conditions. Plant Soil 287, 327–336.

Rusinamhodzi, L., Corbeels, M., Nyamangara, J., Giller, K.E., 2012. Maize-grain legume intercropping is an attractive option for ecological intensification that reduces climatic risk for smallholder farmers in central Mozambique. Field Crops Research 136, 12–22.

Rusinamhodzi, L., Corbeels, M., Giller, K.E., 2016a. Diversity in crop residue management across an intensification gradient in southern Africa: system dynamics and crop productivity. Field Crops Research 185, 79–88.

Rusinamhodzi, L., Dahlin, S., Corbeels, M., 2016b. Living within their means: reallocation of farm resources can help smallholder farmers improve crop yields and soil fertility. Agriculture, Ecosystems & Environment 216, 125–136.

Ryschawy, J., Choisis, N., Choisis, J.P., Joannon, A., Gibon, A., 2012. Mixed crop-livestock systems: an economic and environmental-friendly way of farming? Animal FirstView 1–9.

Sangha, K.K., Russell-Smith, J., Morrison, S.C., Costanza, R., Edwards, A., 2017. Challenges for valuing ecosystem services from an Indigenous estate in northern Australia. Ecosystem Services 25, 167–178.
Shi, D., Li, Y., Zhang, J., Liu, P., Zhao, B., Dong, S., 2016. Increased plant density and reduced N rate lead to more grain yield and higher resource utilization in summer maize. Journal of Integrative Agriculture 15, 2515–2528. https://doi.org/10.1016/S2095-3119(16)61355-2.
Siebert, S., Portmann, F.T., Döll, P., 2010. Global Patterns of Cropland Use Intensity. Remote Sensing 2, 1625–1643.
Tamminga, S., 2003. Pollution due to nutrient losses and its control in European animal production. Livestock Production Science 84, 101–111.
Tilman, D., Balzer, C., Hill, J., Befort, B.L., 2011. Global food demand and the sustainable intensification of agriculture. Proceedings of the National Academy of Sciences 108, 20260–20264.
Tittonell, P., Giller, K.E., 2013. When yield gaps are poverty traps: the paradigm of ecological intensification in African smallholder agriculture. Field Crops Research 143, 76–90.
Trenbath, B.R., 1986. Resource use in intercrops. In: Francis, C.A. (Ed.), Multiple Cropping Systems. Macmillan, New York, pp. 57–81.
Tscharntke, T., Klein, A.M., Kruess, A., Steffan-Dewenter, I., Thies, C., 2005. Landscape perspectives on agricultural intensification and biodiversity – ecosystem service management. Ecology Letters 8, 857–874.

Index

'*Note:* Page numbers followed by "f" indicate figures and "t" indicate tables and "b" indicate boxes.'

Printed in the United States
By Bookmasters